MOLECULAR BIOLOGY
INTELLIGENCE
UNIT

# Retroviruses and Primate Genome Evolution

Eugene D. Sverdlov

Institute of Molecular Genetics
of Russian Academy of Sciences
Moscow, Russia

CRC Press
Taylor & Francis Group
Boca Raton London New York

CRC Press is an imprint of the
Taylor & Francis Group, an **informa** business

CRC Press
Taylor & Francis Group
6000 Broken Sound Parkway NW, Suite 300
Boca Raton, FL 33487-2742

# CONTENTS

# EDITOR

## Eugene D. Sverdlov
Institute of Molecular Genetics of Russian Academy of Sciences
Moscow, Russia
e-mail:sverd@humgen.siobc.rus.ru
*Chapters 1, 2, 4, 5*

# CONTRIBUTORS

Jonas Blomberg
Section of Virology
Department of Medical Science
Uppsala University
Uppsala, Sweden
email: Jonas.Blomberg@medsci.uu.se
*Chapter 11*

Natalia E. Broude
Center for Advanced Biotechnology
   and Department of Biomedical
   Engineering
Boston University
Boston, Massachusetts, U.S.A.
e-mail: nebroude@bu.edu
*Chapter 2*

Gerhard Hunsmann
Department of Virology
   and Immunology
Deutsches Primatenzentrum
Goettingen, Germany
e-mail: ghunsma@gwdg.de
*Chapter 3*

Patric Jern
Section of Virology
Department of Medical Science
Uppsala University
Uppsala, Sweden
e-mail: patric.jern@medsci.uu.se
*Chapter 11*

Aris Katzourakis
Department of Biological Sciences
Imperial College
Ascot, Berkshire, U.K.
e-mail:a.katzourakis@imperial.ac.uk
*Chapter 10*

Yuri B. Lebedev
Laboratory of Structure and Functions
   of Human Genes
Shemyakin-Ovchinnikov Institute
   of Bioorganic Chemistry
Russian Academy of Science
Moscow, Russia
e-mail: yuri@humgen siobc.ras.ru
*Chapter 8*

Christine Leib-Mösch
GSF-National Research Center
   for Environment and Health
Institute of Molecular Virology
Oberschleissheim, Germany
e-mail: leib@gsf.de
*Chapter 7*

Dixie L. Mager
Terry Fox Laboratory
B.C. Cancer Agency
Department of Medical Genetics
University of British Columbia
Vancouver, British Columbia, Canada
e-mail: dmager@bccrc.ca
*Chapter 6*

Patrik Medstrand
Department of Cell and Molecular
    Biology
Lund University
Lund, Sweden
e-mail: patrik.medstrand@medkem.lu.se
Chapter 6

Eugene V. Nadezhdin
Laboratory of Structure and Functions
    of Human Genes
Institute of Bioorganic Chemistry
Russian Academy of Sciences
Moscow, Russia
e-mails: neugene@humgen.siobc.ras.ru
eugene_nadezhdin@mail.ru
Chapter 4

Christian Roos
Working Group Primate Genetics
Deutsches Primatenzentrum
Goettingen, Germany
e-mail: croos@dpz.gwdg.de
Chapter 3

Ulrike Schön
Alopex GmbH
Kulmbach, Germany
e-mail: uschoen@alopexgmbh.de
Chapter 7

Wolfgang Seifarth
Medical Clinic III
Faculty of Clinical Medicine Mannheim
University of Heidelberg
Mannheim, Germany
e-mail: seifarth@rumms.uni-
    mannheim.de
Chapter 7

Michael Tristem
Department of Biological
Imperial College
Ascot, Berkshire, U.K.
e-mail: a.katzourakis@ic.ac.uk
Chapter 10

Dmitrijs Ushameckis
Section of Virology
Department of Medical Science
Uppsala University
Uppsala, Sweden
e-mail: Jonas.Blomberg@medsci.uu.se
Chapter 11

Louie N. van de Lagemaat
Terry Fox Laboratory
B.C. Cancer Agency
Vancouver, British Columbia, Canada
e-mail: lvandela@bccrc.ca
Chapter 6

Tatiana V. Vinogradova
Laboratory of Structure and Functions
    of Human Genes
Shemyakin-Ovchinnikov Institute
    of Bioorganic Chemistry
Russian Academy of Science
Moscow, Russia
e-mail: tv@humgen.siobc.ras.ru
Chapter 9

Hans Zischler
Institute for Anthropology
University of Mainz
Mainz, Germany
e-mail: zischler@mail.uni-mainz.de
Chapter 3

# PREFACE

Three famous questions: "Who we are?" "Where are we from?" and "Where are we going?" together with a more general one "What is life?" were asked by people of every culture. When it became clear that certain genes are responsible for certain phenotypic traits, one more question was added to these four: "Which genes make us human?". One could hope to find the answers by studying primates, their communities and differences from humans, as well as their evolutionary relations, succession of appearance and accumulation of differences after the divergence of the human lineage from lineages of the extant human close relatives, great apes (chimpanzees, gorillas, and orangutans). Humans and our closest living relatives, the apes, including both "great apes" and "lesser apes" (gibbons and siamangs) form the *Hominoidea* superfamily. These *Hominoidea* are remarkably similar and at the same time dramatically different. They are different not only in their appearance but also in such characteristics as behavior and resistance to various diseases, including cancer and AIDS. Many lines of evidence indicate that all of them originated from a common ancestor about 17 Mya (million years ago), and that the last common ancestor of human and great apes, i.e., of human and chimpanzee, extincted about 5 Mya. It is a great challenge to reconstruct its genetic architecture and then to understand the ways of its transformation into two closely related, but different architectures of human and chimpanzee. What events caused their divergence in evolution? What genes and regulatory systems were involved in branching hominids off from their closest relatives, chimpanzees and bonobo and then in their proceeding to the *Homo* genus crowned with extant *Homo sapiens*, that is humans with their brain size of at least 600 cubic centimeters, extended period of childhood growth and development, possession of language and many other human specific traits? And what processes step by step shaped the modern human, *Homo sapiens sapiens* during 5 Myr (million years) of its progress after the divergence from chimpanzee?

It is widely believed that the evolutionary history of a species is reflected in its genome sequence, and therefore the most straightforward way to study primates is sequence comparisons of various primate genomes. The sequencing of the human genome has already contributed a great deal to such an analysis, and as soon as the sequencing of the chimpanzee genome is finished, we will have enormous information to work on. Multiple differences of various kinds that can be envisioned between the two genomes will inevitably puzzle researchers trying to find the genetic reasons for human speciation and rapid phenotypic evolution. How can one single out a relatively small number of the differences that have been actually or most possibly involved in speciation from the multitude of just random neutral mutations accumulated during millions of years?

One way is to try to identify the regions positively selected in evolution. In the case of coding stretches of the genome, an enhanced rate of nonsynonymous substitutions compared to synonymous ones is a widely accepted indication of positive selection. However, the situation with regulatory regions or regions which encode noncoding RNAs is much more tangled. Their conservation indicates that these regions were important for a sufficiently long period(s) of evolution, but generally speaking could be of no importance in other periods. Clearly, this criterion may not be applicable to the regulatory units that appeared after the human-chimp lineages divergence. But precisely these units might be the acquisitions that played a major role in shaping our human phenotype.

Therefore, it seems inevitable to resort to rather traditional hypothesis-driven approaches, when the research starts from the hypotheses aimed at explaining why and how the most significant human features, such as language or cognitive capacities, could emerge. In this approach, only particular loci will be taken for interspecies comparison. This "last line of research" will undoubtedly be stimulated by new information on genome-wide comparisons.

Obviously, the chances to reconstruct the succession of the genetic events which had occurred during millions of years of evolution are negligible. But what we can hope to gain as a result of such a comparative research is a deeper understanding of the mechanisms governing the modern genome and the role of particular elements in the networks responsible for the functional integrity of the genome. We will also certainly be able to reveal differences in spatial-temporal networks of the events determining the development of different species and thus to form a basis for the second order hypotheses related to the genetic basis of differences in the phenotypes of extant species. The achievement of this goal will be a great step towards the understanding of what life is in general, and what its peculiarities are regarding our presently prospering, but still endangered, species as well as how these peculiarities could evolve.

What kind of differences might promote speciation? Since a classical work of King and Wilson, who in 1975 undertook a thorough comparison of the molecular data available on chimpanzee (*Pan troglodytes*) and human (*Homo sapiens*), it is widely accepted that, as put by the authors: "...a relatively small number of genetic changes in systems controlling the expression of genes may account for major organizational differences between human and chimpanzees". It is now known as the regulatory hypothesis. Later, it became a major constituent of the Evo-Devo concept suggesting that evolution depends on heritable changes in the development and, according to Duboule and Wilkins (1998), "...the primary source of developmental differences... will prove to be not unique gene products but rather the way that comparable, or the same, gene functions are differentially deployed in their development.... Many so called heterochronic shifts altering developmental programs and morphologies involve no more than alteration in the times and cellular site at which particular regulatory molecules are expressed rather than alteration in those molecules themselves". In metazoan evolution, these processes have been brought under intercellular control regarding the time, place, and conditions of functioning. It seems logical to propose that such developmental functional shifts could be caused by changes in gene regulation, which in turn could result from addition of a new regulatory module(s) to the pre-existing gene regulatory system.

The title of this volume, *Retroviruses and Primate Genome Evolution,* reflects its goal to conceive the role of the obligate inhabitants of all vertebrate genomes—endogenous retroviruses, especially those emerged in genomes rather recently, during primate evolution. Although a special focus in the volume is put on human endogenous retroviruses (HERVs), some attention is also paid to other retroelements (REs), like LINEs and Alu to give a more comprehensive view of the evolutionary potential of these perpetually mobile entities now occupying almost a half of the human genome.

Keeping in mind that REs are jumping carriers of the regulatory *cis*-elements adapted for RNA-polymerase II or III transcription regulation, it is quite reasonable to put them on the list of highly probable candidates for evolutionarily significant changes, capable of affecting the regulation of the genes in the vicinity of which they were inserted. Such changes could quite possibly occur in the developmental regulatory machinery thus causing the above mentioned developmental shifts.

Indeed, the data obtained for different species clearly demonstrate that REs insertions can change not only the structure of genes, and hence their products, but also their regulation. Moreover, transposable elements can have their own genes and thus enrich the genome with new genetic information, like genes of reverse transcriptase or viral resistance. Although the newly inserted elements are known to mostly cause deleterious effects including hereditary diseases, the host cells sometimes exploit the ability of REs to generate variations for their own benefit. Among other REs, HERVs are considered to be the most sophisticated. ERV-related sequences are believed to represent footprints of ancient germ-cell retroviral infections which now occupy up to 8% of the human genome. They have excitingly diverse tools of affecting the human genome functioning originated from exogenous retrovirus systems of successful life cycle. They can change the host genome function through expression of retroviral genes, human genome loci rearrangements due to retropositions of HERVs, or by the ability of their long terminal repeats (LTRs) to regulate nearby genes. A multitude of solitary LTRs comprise a variety of transcription regulatory elements, such as promoters, enhancers, hormone-responsive elements, and polyadenylation signals. This feature makes LTRs potentially able to strongly affect the expression patterns of neighboring genes. It can be imagined that the appearance of such invaders in the genome can change some functions relevant to development and thus provide new traits for subsequent natural selection. They can therefore be considered prime suspects for being a major class of causative agents in speciation.

Individual chapters in this book are devoted to specific areas of research into human genome evolution and possible involvement of REs in the processes related to evolution and includes REs own evolution which was prime interdependent with the host genome evolution.

The book opens with four chapters giving a general insight into human genome structure and function analysis and ideas on the genome evolution. Chapter 1, "A glance at evolution through the genomic window" (by E. Sverdlov), describes the status of whole genome sequence comparisons which, for the first time, opened a possibility to analyze evolutionary changes at a whole genome level. The intra- and interspecies comparisons of the sequenced genomes demonstrated that the genome complexities did not directly correlate with the number of genes and suggested the importance of combinatorial interactions in the cells and organisms as a major player in the complexity of live systems. The whole genome comparisons allowed one to elucidate the role of gene duplications, gross genome rearrangements, transposable elements and other genomic changes in divergence of genomes, thus forming a solid basis for understanding genetic mechanisms of evolution. The whole genome analyses developed in parallel and interdependently with the development of new concepts of evolution, such as evolutionary developmental biology (Evo-Devo), aimed at explaining how developmental processes and mechanisms become modified in evolution, and how these modifications produce changes in animal morphology. This review considers new data and trends and supports the idea that transposable elements play a role of a major pacemaker in evolution being a "depot" of evolvability factors.

Chapter 2 entitled "Complex Genome Comparisons: Problems and Approaches" by N. Broude and myself provides a brief outline of the experimental approaches to genome-wide interindividual, interpopulation, and interspecies comparisons. Such comparisons form a unique background for deciphering spatial and temporal genomic regulatory networks and their changes during evolution. They are also indispensable for understanding genetic and environmental contributions to complex diseases afflicting modern society. The chapter describes also a range of modern approaches to genome-wide complex genome comparisons with their advantages and disadvantages.

Chapter 3 by H. Zischler and his colleagues is devoted to primate evolution. Information on evolutionary events and relations of different primate extant species is indispensable for understanding the role of particular genomic elements in evolution. The authors review the modern status of investigations on primate phylogeny with all its problems and contradictions. An emphasis is made on the divergence of non-human primates, relevant interpretations of the fossil record and molecular evidence, including retropositional evidence. Whereas a congruent view is emerging concerning phylogenetic relationships among primate taxa at a higher taxonomic level, e.g., primate infraorders, there is still considerable debate on primate origins or very recent splits in primate evolution. Obtaining more clarity about primate origins is to a large degree hampered by the sparseness of the critical fossil record. If both molecular and fossil evidence is available for a certain splitting, many interpretations based on these two completely different approaches seem to be remarkably compatible. Attention is also paid to problems of modern human evolution.

Chapter 4 "How different is the human genome from genomes of the great apes?" by E. Nadezdin and E. Sverdlov gives an account on sequence and chromosomal organization differences between highly related genomes of humans and the African great apes that were accumulated during Hominoid evolution. Some of them certainly form a genetic basis for recently evolved, specifically human traits. Human genome sequencing revealed its characteristic features, and the ongoing sequencing of the chimpanzee genome continuously widens the possibilities of large-scale systematic comparison of the two genomes. Now it is more and more apparent that most probably hundreds and thousands of genes were involved in the divergence of even the most closely related species of human and chimpanzee. The divergence might be caused by changes in gene regulation and by modifications of protein biochemical functions, gene duplications, losses and acquisitions. A great challenge will be to single out functionally significant differences from the mess of all changes accumulated during evolution.

These four general chapters are followed by reviews devoted to various aspects of evolution, interactions with the host genome, and involvement of REs in various human diseases. This series opens with a very brief introduction, Chapter 5, "Endogenous Retroviruses and other retroinsiders", which I wrote to give general information about retroviruses, their endogenous counterparts, and other retroelements in the human genome. I hope it will make the reading of the following more specialized reviews easier.

The next two chapters, by Dixie Mager et al (Chapter 6, "Genomic Distributions of Human Retroelements") and by Christine Leib-Mosch et al (Chapter 7, "Influence of human endogenous retroviruses on cellular gene expression") focus on distribution and function of REs in the human genome. Chapter 6 reviews the studies performed in the last 20 years on chromosomal arrangements of human retroelements including endogenous retroviruses. Biological mechanisms or evolutionary forces that might influence their modern distribution patterns are also discussed. Chapter 7 discusses a variety of effects of newly inserted REs on adjacent genes. These effects include not only impairment of gene function, but also enhancement of transcription, changes in tissue specificity of gene expression, and creation of new gene products with modified functions, e.g., via alternative splicing. The conclusion is that retrotransposable elements may have served as catalysts of genomic evolution and possibly played a role in primate speciation and adaptation.

The reviews by Yu. Lebedev, "Genome-wide search for human specific retroelements" (Chapter 8), and Tatyana Vinogradova, "Approaches to genome wide analysis of human gene expression: application to analysis of expression of human endogenous retroviruses in normal and cancerous tissues" (Chapter 9), provide an insight into experimental techniques used for revealing species specific REs and analysis of their functional status.

Chapter 10 by A. Katzourakis and M. Tristem, "Phylogeny of human endogenous and exogenous retroviruses", is somewhat different in its style from other reviews in this book. And although it is rather a research article than a review, this chapter successfully demonstrates the state-of-the art for attempts to reconstruct the correct phylogeny of endogenous retroviruses with all their problems and difficulties. Quite a number of assumptions made to smooth evident contradictions of grouping based on just the level of sequence identity make the resulting tree appreciably dependent on the researcher's intuition and prejudice. Similarly, the results obtained with even more sophisticated tools of modern computer-based phylogeny analysis are by no means final or indisputable. Unfortunately, our past seems to be almost as cloudy as our future. With all their assumptions, Tristem and his colleagues found 31 HERV families in the human genome. Currently, it is probably the most extensive survey of HERVs diversity. I think that sequencing of other primate genomes will reveal even more HERV families.

Finally, the title of the last chapter by J. Blomberg et al, "Evolutionary aspects of human endogenous retroviral sequences (HERVs) and diseases", precisely reflects its content. Also discussed is the impact of retroviruses on a variety of human diseases.

Taken together, these partially overlapping chapters hopefully provide a balanced and accurate overview of our current knowledge of the complex interplay of the human genome with its mobile inhabitants, retrotransposons.

To conclude the Preface, I would like to stress that hardly a particular genomic constituent or even a numerous group of constituents like REs has caused such a pronounced phenotypic difference between human and chimpanzee. Undoubtedly, hundreds or even thousands of changes occurred during 5 million years after the two species diverged, and that eventually made us humans. However, the chain of events leading to human might well be initiated by some very first genomic perturbations, and this initiation could be caused by retroviral integration(s) and/or by other RE retropositions into a critical site of the ancestor genome.

Exact dating of RE integrations and comparative functional analysis of the genes in the species which sustained the integrations and those retaining the same but native genes will help us to better understand our own evolution.

*Eugene D. Sverdlov*

# ABBREVIATIONS

| | |
|---|---|
| aa | amino acid |
| ADHD | attention deficit/hyperactivity disorder |
| AFLP | amplified fragment length polymorfism |
| ALV | Avian leukosis virus |
| *ATM* | ataxia telangiectasia mutated, gene responsible for ataxia telangiectasia |
| BAC | bacterial artificial chromosome |
| BES | bacterial artificial chromosome end sequence |
| BLAST | basic local alignment search tool |
| BLV | bovine leukemia virus |
| bp | base pair |
| *BRCA1 and 2* | genes responsible for hereditary breast cancer |
| CA | capsid protein surrounding the RNAs bound to nucleocapsid (NC) proteins |
| CGAP | Cancer Genome Anatomy Project |
| CGH | comparative genomic hybridization |
| CNS | central nervous system |
| *COX* | cytochrome *c* oxidase genes |
| CSF | cerebrospinal fluid |
| DD | differential display |
| DDRT-PCR | differential display reverse transcription PCR |
| DLBCL | diffuse large B-cell lymphoma |
| dUTPase | deoxyuridinetriphosphatase |
| EBV | Epstein-Barr virus |
| EM | electron microscopy |
| ERV | endogenous retroviral sequence |
| ES | embryonic stem cells |
| EST | expressed sequence tag |
| FeLV | feline leukemia virus |
| FISH | flourescence In Situ Hybridization |
| FV | friend virus |
| GaLV | gibbon ape leukemia virus |
| GDB | genome database |
| GC | germ cell |
| GCT | germ cell tumour |
| GEF | gene expression fingerprinting |
| GMS | genomic mismatch scanning; |
| GRE | glucocorticoid responsive element |
| HERV | human endogenous retroviral sequence, human endogenous retrovirus |
| HHV6 | human herpesvirus 6 |
| HHV7 | human herpesvirus 7 |
| HIV | Human Immunodeficiency Virus |
| HML | human MMTV-like sequence |

| | |
|---|---|
| Hsp90 | heat-shock protein 90 in *Drosophila* |
| HSV | Herpes Simplex virus |
| HTDV | human teratocarcinoma derived virus |
| HTLV | human T-lymphotropic virus, human T-cell leukemia virus |
| HVRI, HVRII | hypervariable regions I and II of mtDNA (sequences) |
| HYAL2 | hyaluronidase type 2 |
| IAP | intracisternal type A particle |
| IDDM | insulin dependent diabetes mellitus |
| IF | immunofluorescence |
| IFN | interferon |
| IgG | immunoglobulin G |
| IgM | immunoglobulin M |
| IN | integrase |
| Inr | initiator element |
| IRS-PCR | interspersed repetitive sequence PCR |
| ISU | immunosuppressive unit, a conserved sequence derived from p15E |
| JSRV | Jaagsiekte retrovirus |
| kb | kilobase pairs |
| L1 | LINE1, the most abundant LINE in mammalian genomes |
| LCR | locus control region |
| LINE | long interspersed nuclear element non-LTR retroelements encoding their own reverse transcriptase |
| LTR | long terminal repeat |
| MA | matrix protein,. matrix |
| MaLR | mammalian LTR retroelements. A large heterogeneous group of LTR retroelements found in mammals |
| Mb | megabase, million base pairs |
| MDV | Marek disease virus, a tumourigenic herpesvirus of turkeys |
| MHC | major histocompatibility complex |
| MHR | major homology region |
| MIR | mammalian interspersed repeat, class of SINEs |
| MLV | mouse (murine) leukemia virus |
| MMTV | mouse mammary tumour virus |
| MPMV | Mason-Pfizer monkey virus |
| MRCA | most recent common ancestor |
| MS | Multiple sclerosis |
| MSRV | Multiple sclerosis – associated retrovirus (retroviral element) |
| mtDNA | mitochondrial DNA |
| Mya | million years ago |
| Myr | million of years |
| MZ | monozygotic twins |
| NC | nucleocapsid proteins |
| ncRNAs | noncoding RNAs |
| Neu5Ac | N-acetyl-neuraminic acid |

| | |
|---|---|
| Neu5Gc | N-glycolyl-neuraminic acid |
| NK cells | natural killer cells |
| NMD | nonsense mediated decay |
| nr and htgs | nonredundant and high-throughput subset of GDB, correspondingly |
| NRE | negative regulatory sequence |
| nt | nucleotide |
| NWM | New World monkey |
| ODD | ordered differential display |
| ORF | open reading frame |
| OWM | Old World monkey |
| PA | polyacrylamide |
| PAGE | PA gel electrophoresis |
| PBL | human peripheral blood leukocytes |
| PBS | primer binding site |
| PCR | polymerase chain reaction |
| PERV | porcine endogenous retrovirus |
| PHA | phytohemagglutinin |
| PLZF | promyelocytic leukemia zinc finger |
| PML | promyelocytic leukemia |
| PPT | polypurine tract |
| PR | protease |
| PS | PCR suppression |
| PTN | pleiotrophin |
| RAPD | randomly amplified polymorphic DNA |
| RDA | representational difference analysis |
| RE(s) | retroelement(s) A transposable element that transposes via an RNA intermediate |
| READS | restriction endonucleolytic analysis of differentlly expressed sequences |
| REs | retroelements |
| RFLP | restriction fragment length polymorfism; |
| RIDGE | regions of increased gene expression |
| RLGS | restriction landmark genome scanning; |
| RNAi | RNA interference |
| RSV | rous sarcoma virus |
| RT | reverse transcriptase |
| RT-PCR | reverse transcription - PCR |
| SAGE | serial analysis of gene expression |
| SBH | sequencing by hybridization |
| SDD | systematic differential display |
| SDDIR | selective differential display of RNAs containing interspersed repeats |
| SH | subtractive hybridization |
| SINE | short interspersed nuclear element; non autonomous retroelement typically derived from a small functional RNA that has amplified in the genome by retrotransposition. |

| | |
|---|---|
| SLE | systemic lupus erythematosus |
| SNP | single nucleotide polymorphism |
| SRV | simian retrovirus |
| SSH | suppression subtractive hybridization |
| STR | short tandem repeat |
| SU | envelope surface glycoprotein, surface unit; outer portion of retroviral Env proteins |
| TDD | targeted differential display |
| TE | transposable element |
| TGDD | targeted genomic DD |
| TGF-β | transforming growth factor beta |
| TK | thymidin kinase |
| TM | transmembrane glycoprotein, inner portion of retroviral Env proteins |
| TPRT | target-site primed reverse transcription |
| 3'UTR | 3' untranslated region |
| 5'UTR | 5' untranslated region |
| VDA | variant detection array |
| VNTR | variable number of tandem repeats |
| VZV | Varicella-Zoster virus (mumps virus) |
| Vβ7TCR | T cell receptor containing a beta 7 variable chain |
| WGD | whole genome duplication |
| XRV | exogenous retrovirus |

# CHAPTER 1

# A Glance at Evolution through the Genomic Window

Eugene D. Sverdlov

*"Nothing in biology makes sense except in the light of evolution."*
Th. Dobzhansky

## Abstract

A large number of various genomes sequenced recently for the first time make it possible to analyze evolutionary changes at a whole genome level, unlike a single gene level. Intra- and interspecies comparisons of the sequenced genomes demonstrated that the organism's complexities did not directly correlate with the number of genes and suggested the importance of combinatorial interactions in cells and organisms as a major player in the complexity of live systems. They made it possible to reveal conserved and variable elements of the genomes and to suppose that tens of thousands of proteins are made of just about 1,500-2,000 discrete structural protein units called domains or modules. Different modular proteins are formed from these modules taken in different combinations, and this shuffling might play an extremely important role in the genesis of evolutionary novelties. The new domain architectures (defined as the linear arrangement of domains within a polypeptide) have emerged in evolution by shuffling, adding or deleting domains, resulting in new proteins composed of old parts. More complex organisms seem to contain more various protein architectures than the simpler ones. Whole genome comparisons allowed one to elucidate the role of gene duplications, gross genome rearrangements, transposable elements and other genomic changes in genome divergency, thus forming a solid basis for understanding genetic mechanisms of evolution. Whole genome analyses developed in parallel and interdependently with the development of new concepts of evolution such as evolutionary developmental biology (Evo-Devo) aimed at explaining how developmental processes and mechanisms become modified in evolution, and how these modifications produce changes in animal morphology and body plans. Among these new concepts can be found such fruitful notions as (i) a universal principle of modular organization at various levels of living systems, particular modules being changed and co-opted into new functions without affecting other modules, (ii) a concept of network-like organization of cellular regulatory systems with cis-regulatory elements of the genome functioning as major nodes of the networks, and a crucial evolutionary role of changes in the regulatory systems, (iii) an assumption of increase in functional load per regulatory gene with increasing the complexity of the organism, (iv) an idea of evolvability as a universal feature of the living entities, and a very important concept (v) that not only natural selection, but also internal developmental biases can form the basis for evolutionary changes.

*Retroviruses and Primate Genome Evolution*, edited by Eugene D. Sverdlov.
©2005 Eurekah.com.

This chapter considers new data and trends and supports the idea that transposable elements play a role of a major pacemaker in evolution being a "depot" of evolvability factors.

## Introduction

Jack London in his novel "The Sun Dog Trail" described the following scene:

*Sitka Charley smoked his pipe and gazed thoughtfully at the Police Gazette illustration on the wall. For half an hour he had been steadily regarding it… "Well?" I finally broke the silence.*

*He took the pipe from his mouth and said simply, "I do not understand."*

*"That picture-what does it mean? I do not understand."*

*I looked at the picture. A man, with a preposterously wicked face, his right hand pressed dramatically to his heart, was falling backward to the floor. Confronting him, with a face that was a composite of destroying angel and Adonis, was a man holding a smoking revolver.*

*"That picture is all end," he said. "It has no beginning."*

*"It is life," I said.*

*"Life has beginning," he objected…" Something happens in life. In pictures nothing happens. No, I do not understand pictures."*

*His disappointment was patent. I felt, also, that there was challenge in his attitude. He was bent upon compelling me to show him the wisdom of pictures. "Pictures are bits of life," I said. "We paint life as we see it. For instance, Charley, you are coming along the trail. It is night. You see a cabin. The window is lighted. You look through the window for one second, or for two seconds, you see something, and you go on your way. You saw maybe a man writing a letter. You saw something without beginning or end. Nothing happened. Yet it was a bit of life you saw."*

I think it is an exact description of our efforts to understand evolution. We see it as a momentary picture without beginning and end and try to understand life from the very beginning in all its diversity and in movement. And we try to animate the picture looking at it through different windows: through the window of paleontology, the window of phylogenetics, the window of developmental biology, and through the window of comparative genomics. In this chapter I will try to sketch what we see through the genomic window. It will mainly focus on human genome evolution, whereas the data on other genomes will be used for comparative purposes.

## Towards the Understanding of "The Mechanisms That Bring about Evolutionary Changes"

Dobzhansky in *Genetics and the Origin of Species*, first published in 1937 (citation from ref. 1) wrote: "The problem of evolution may be approached in two different ways. First, the sequence of the evolutionary events as they have actually taken place in the past history of various organisms may be traced. Second, the mechanisms that bring about evolutionary changes may be studied… ".

However difficult may appear to be the reconstruction of successive evolutionary events, the unraveling of the mechanisms leading to the morphological changes, which are then fixed due to natural selection and eventually lead to the emergence of new species is a much more difficult task. A fundamental question (further designated as Question 1) is what kind of the genomic changes are transformed into the phenotypic changes subject to natural selection and how these transformations are materialized?

DNA sequence variation is abundant in modern populations, yet the relationship between the phenotypic variation and the genomic variation producing it is extremely complex.[2] The main problems here involve, as a rule, many genes in creation of a function and a great gap in our understanding of the chain of events bringing about the conversion of the genetic information into the phenotype. A great variety of interdependent evolutionary changes could act cooperatively, sometimes within limited periods of evolution, and then cease to operate. Moreover, certain changes could accumulate in the genome first without any visible effect (dormant changes) and then suddenly manifest themselves due to a mutation in a single (or a few) "capacitor" gene (see below). Finally, an overwhelming majority of changes in the genome structure are most probably just neutral, or almost neutral, and play no role in the selection.

## *"A Relatively Small Number of Genetic Changes in Systems Controlling the Expression of Genes May Account for Major Organizational Differences between Human and Chimpanzees"*[3]

An attempt to answer Question 1 at least partially has been undertaken in the classical work of King and Wilson[3] on comparison of the chimpanzee (*Pan troglodytes*) and human (*Homo sapiens*) gene and protein primary structures. It was demonstrated that human proteins and genes are generally 99% identical to their chimpanzee counterparts. This remarkably low difference seemed to be too small to account for the evident dissimilarity of the organisms. Moreover, King and Wilson indicated that "Since the time that the ancestor of these two species lived the chimpanzee lineage has evolved slowly relative to the human lineage, in terms of anatomy and adaptive strategy". At the same time, the rates of molecular changes in proteins and genes were rather similar in these two species, and even close to the values for anatomically highly conservative species like frogs. The following two remarkable conclusions were drawn by King and Wilson: "The contrasts between organismal and molecular evolution indicate that the two processes are to a large extent independent of one another" and that "a relatively small number of genetic changes in systems controlling the expression of genes may account for major organizational differences between human and chimpanzees". These regulatory mutations may affect either *trans*-acting regulatory proteins, such as transcription factors, participants of the signal transduction pathways etc., or *cis*-acting sequences responsible e.g., for the regulation of the gene expression at the transcriptional or posttranscriptional levels, or even, as we understand now, non-protein regulatory molecules such as noncoding RNAs (ncRNAs, see below).

The regulatory character of the changes responsible for evolutionary progress is now widely accepted and the next question concerning the mechanisms of the appearance of new regulatory proteins and/or new *cis*-acting regulatory sequences is already being discussed.

## *Spate of Facts and Slow Progress to Real Knowledge of the Mechanisms of Evolutionary Changes*

*"...our ignorance of the laws of variation is profound."*
Darwin[4]

Gabriel Dover starts his seminal review[5] with a citation of the prominent evolutionist Richard Lewontin. "For many years population genetics was an immensely rich and powerful theory with virtually no suitable facts on which to operate. It was like a complex and exquisite machine, designed to process a raw material that no one had succeeded in mining....."

Quite suddenly the situation has changed. The mother-lode has been tapped and facts in profusion has been poured into the hoppers of this theory machine. And from the other end has issued nothing..... The machine can not transform into a finished product the great volume of raw material that has been provided... The entire relationship between the theory and facts needs to be reconsidered". Then Dover continues "...unless and until we uncover the 'rules of transformation' that connect 'genotype space' with 'phenotype space' then we can not seriously entertain, or be satisfied with, a gene based theory of evolution. How an individual phenotype emerges and reproduces from a given unique set of genes inherited from its sexual parents is the central question of evolutionary theory: all the rest is subsidiary". Both Lewontin and Dover describe the exact status of the modern theory of evolution. It operates now with a great number of facts concerning gene and genome structures, it tries to understand changes in the gene content and regulation in various species, and finally comes to a conclusion: 'Genome speaks biochemistry—not phenotype'.[6] This has been said right after the first complete genome sequence of a multicellular organism *C. elegans* has become available. Strong efforts of a consortium formed to inactivate all 19,099 genes of *C. elegans* led to an idea that inactivation of a great many genes of the animal yields either no obvious phenotype or early death. It was clearly understood that there is no quick and easy way to search for gene function even with this excellent model containing only 959 cells, 302 neurons and 97,000,000 bp forming about 19,000 genes.[7] Once again citing G.Dover,[5] we find that modern "Evolutionary genetics…in its current focus on DNA variation reduces phenotypes to symbols. Varying phenotypes, however, are the units of evolution, and if we want comprehensive theory of evolution we need to consider both the internal and external evolutionary forces that shape the development of phenotypes".

But however scanty is the information we have accumulated since Darwin wrote his bitter words cited in the epigraph, today we know much more than he could even dream of. We can be satisfied with the progress in understanding what is going on with the genome structure and, to a lesser extent, expression during evolution. Also, we considerably advanced with concepts on how the genomic information is transformed into phenomes at least at the very first biochemical stages of this process. And we move, though slowly, from traditional molecular biology to system-level understanding[8] and from 'one gene-one product' philosophy to modular cell biology and to genome-wide thinking. However, we do not understand phenotype and we do not understand evolution because we do not understand phenotype. We can only understand phenotype through its evolution and vice versa: Dobzhansky was right saying that "Nothing in biology makes sense except in the light of evolution". Comparison of phenotypic and genomic changes in a great variety of species is the only hope to answer Question 1.

## What Is Going on with Genomes during Evolution

Over the last decade massive information has accumulated on complete structures of the genomes starting from bacteria and finishing with an almost complete sequence of the human genome (leaving aside numerous viral genomes sequenced earlier). Some of the organisms with the genomes sequenced are listed in Table 1. These genomes together with the products of their expression form what could be called an integrated genome-information space. It opens enormous opportunities for comparisons aimed to reveal common and different features of various genomes and their functional organizations. In the following sections I will try to briefly outline what is emerging from such comparisons.

### The Complexity of the Organisms Does Not Correlate with the Number of Genes

An estimated total size of the human genome is 3.2 billions bp. Over half of the human DNA is occupied by repeated sequences of various types, and only 1.1% of the genome is spanned by exons, whereas 24% is introns, with 75% of the genome being intergenic DNA.

A comparison with some other sequenced genomes shown in Table 1 demonstrates that genome sizes increase with increasing complexity of organisms. It seems quite logical, though we remember the 'C paradox': the observed 40,000-fold variation in eukaryote haploid DNA content ('C value') is unrelated to organism complexity.[13] The problem is with the number of genes. The human genome is believed to contain 26,000[14] to 31,000[15] protein-encoding genes. In addition, several hundreds of genes are known to encode non-protein-coding RNAs. The number of coding genes in the human genome is thus only twice as large as in a worm *C. elegans*[7] (19,000) and approximately the same as in a plant, *Arabidopsis thaliana*[11] (about 26,000)

**Table 1. Some of the sequenced genomes with their characteristics***

| Organism | Genome Size (1,000 kb) | Gene Number Estimated |
|---|---|---|
| *Homo sapiens* (draft)[14, 15] | 2,900 (euchromatic part) | ~30 –40,000 |
| *Mus musculus*[9] | 2,500 (euchromatic part) | ~30, 000 |
| *Fugu rubripes*[10] | 365 | ~30 – 40,000 |
| *Arabidopsis thaliana*[11] | 125 | 25,498 |
| *Drosophila melanogaster*[12] | 120 | 13,600 |
| *Caenorhabditis elegans*[7] | 97 | 19,000 |
| *Saccharomyces cerevisiae*[12a] | 12.1 | 6,034 |
| *Escherichia coli* K-12[12b] | 4.6 | 4,288 |

* When this chapter was finished, a complete sequence of the human genome was reported: http://www.sanger.ac.uk/Info/Press/2003/030414.shtml.

and in fish *Fugu*.[10] Moreover, the genome of such a complex organism as *Drosophila melanogaster*[12] contains even fewer genes (~13,000) than a rather primitive worm.[7] If the estimated number of human genes is more or less correct (it is still being debated), then we have a new paradox—lack of correlation between the organism complexity and the gene number. Some authors called this new paradox 'N-paradox'.[16]

It is not so simple to give the definition to the organism complexity.[17-20] A common sense-based definition was suggested, for example, by David Baltimore[17] in his reflections on the appearance of the human genome draft sequence: "Understanding what does give us our complexity—our enormous behavioral repertoire, ability to produce conscious action, remarkable physical coordination (shared with other vertebrates), precisely tuned alterations in response to external variations of the environment, learning, memory...". To roughly evaluate the complexity I will use criteria (Table 2) based on the number of cells comprising the organism, on the number of neurons forming neural network, and on the number of cell types in the organism. The latter estimate has been used, in particular, by Raff and Kaufman.[21]

Table 2 demonstrates an enormous jump in complexity between *C. elegans* and *H. sapiens*. Then arises the question of how a modest increase in number of genes creates such a jump.

## The Number and Modifications of Proteins Encoded by Genes Do Not Explain Changes in Complexity

Estimates from the genes analyzed to date suggest that the average number of alternates spliced from the transcript of a single mammalian gene might be in the range of two to three or more.[14,15,23] With an estimate of about 30,000 genes, this would give us about 90,000 or more distinct proteins encoded by the human genome.[16,23,24] It was suggested that the extent of alternative splicing is higher in humans than in worm or *Drosophila*. However, this attractive explanation was recently called in question[23] and the extent of alternative splicing was shown to be likely similar in various animals, including invertebrates.

Another source of complexity can be at the protein level.[23,25] Proteins are much more complicated than nucleic acids: more than 200 different types of post-translational protein modifications are known. In addition, different proteins can be produced from one and the same gene due to alternative splicing (but see above), by varying translation start or stop sites, or by frameshifting due to which a different set of triplet codons in the mRNA is translated. All of these possibilities result in a proteome estimated to be an order of magnitude more complex than the genome.[25] Moreover, proteins respond to altered conditions by

*Table 2. Evaluations of the complexity of some organisms*

|                  | E. coli | S. cerevisiae | C. elegans | H. sapiens |
|------------------|---------|---------------|------------|------------|
| Cell number      | 1       | 1             | 959        | $10^{14}$  |
| Cell type number | 1       | 3–4*          | About 20   | 200**      |
| Neuron number    | 0       | 0             | 302        | $10^{10-11}$*** |

*Raff and Kaufmann,[21]  **Alberts et al,[22]  ***Koch and Laurent.[20]

translocation to different cellular compartments, by getting cleaved into pieces, and by changing their ability to bind other proteins, nucleic acids or low-molecular ligands. Important is also the ability of one protein to be involved in more than one process. Even minor alterations in the nature of protein-protein interactions, protein modifications, and localization can have dramatic effects on cellular physiology.[26] This increased complexity of proteome certainly contributes to organismal complexity but still seems to be not sufficient to be fully responsible for its enormous jump.

## Non-Coding RNAs and Epigenetic Mechanisms Might Be Partially Responsible for the Jump in Complexity

Yet another source of the increased complexity might be greater usage of ncRNAs for regulation.[15,27,28] Thousands of non-identified human genes may produce ncRNAs as their ultimate products.[15] Indeed, the analysis of mouse transcriptome[29] indicated that ncRNA is a major component of the transcriptome. The ncRNAs lack translated ORFs, they are often small and not polyadenylated and, accordingly, novel ncRNAs cannot readily be found by computational gene-finding techniques or experimental sequencing of cDNA or EST libraries and need special experiments to be revealed.[16] Their importance is now widely accepted[30] and their involvement in developmental processes has been demonstrated at least in model organisms.[31]

Epigenetic differences might also contribute to the greater complexity of mammalian genomes. Wide involvement of epigenetic mechanisms in gene expression regulation is now a common knowledge. Epigenetic modifications of mammalian DNA, such as methylation, are important for genome functioning in development and in adult organisms. DNA methylation is of central importance to genomic imprinting and other aspects of epigenetic control of gene expression, and methylation patterns are largely maintained during development in somatic lineages. The mammalian genome undergoes major reprogramming of methylation patterns in the germ cells and in the early embryo. Some of the factors that are involved both in maintenance and in reprogramming, such as methyltransferases, are being identified.[32] Epigenetic mechanisms may be quite different in different species. For example, *Drosophila*, *C. elegans* and yeast were long thought not to use methylation for their genome regulation. And although methylation was quite recently detected in *Drosophila*[33] and is suggested to have some regulatory function in development, it is still not known definitely whether DNA methylation has a functional role in *Drosophila*. In any case methylation features are quite distinct in fly and known mammalian systems. Epigenetic effects are known to regulate such important effects as dosage compensation by which the expression levels of sex-linked genes are appropriately altered in one sex to offset a difference in sex-chromosome number between males and females of heterogametic species. It was shown that different species use very different mechanisms to achieve such a compensation: the male X chromosome is hypertranscribed in drosophilid flies, both hermaphrodite X chromosomes are downregulated in the nematode *C. elegans*, and one of two X chromosomes is inactivated in mammalian females with the participation of ncRNAs. The *trans*-acting factors (proteins and ncRNAs) that have been shown to mediate dosage compensation are unrelated among the three lineages.[34]

## Multiple Combinatorial Interactions Can Be a Major Source of the Complexity

Many other possibilities were also suggested to explain the molecular basis of increased complexity, but the most powerful source of the complexity is probably a combinatorial use of the repertoire of regulatory factors. A single gene product can be involved in various processes, in particular during organism development.[35] This multiple involvement can be connected with the ability of modular promoters (enhancers) to interact with various combinations of transcription factors and thus to vary cell compartments and/or time of the gene expression during development.[5] The combinations of a protein with different partners form complexes with different features: a property known at least for different combinations of transcription factors, activators and co-activators capable of interacting with different enhancers or promotor regions switching various genes on and off.[36] Moreover, depending on the particular combination, the function of these factor complexes can be dramatically changed. For example, Dorsal transcription regulator in *Drosophila* is an activator, but in specific *cis*-regulatory regions it associates with two DNA binding proteins, Cut and Dead ringer (Dri), and is converted into a repressor through the recruitment of co-repressor Groucho.[36-38] The well known key participants of various cellular functions RB[39] and p53 [40] each contain potentials for interactions with tens of partners conferring new properties on the complexes formed and taking part in a multitude of cellular functions. The list of such multifunctional proteins is constantly growing. It is believed that over 2,000 transcription factors take part in cell-specific gene regulation.[41] The combinatorial use (Fig. 1) of these *trans*-regulators is an inexhaustible source of unique combinations ensuring the correct expression of each of the genomic genes. The vertebrate immune system is one more example of a biological system capable of generating a great repertoire of specific responses by using combinations of a few hundred different genes.[16]

To conclude the discussion of the interrelation between the complexity and *number* of genes I would like to mention after Venter et al[14] a speculation made by Haldane in 1937 that if the number of genes were too large, each zygote would carry too many new deleterious mutations thus making the population not able to maintain itself. In 1967 H. Muller calculated that the mammalian genome would contain a maximum of about 30,000 genes. Muller's estimate for *D. melanogaster* was 10,000 genes, compared to 13,000 derived from the annotated fly genome. Although these calculations may be too simplified and even not quite correct, they suggest that increasing complexity does not necessarily mean proportionally increasing gene number. However, it should inevitably lead to an increase in functional load at least on regulatory genes, as predicted by some evolution theories (reviewed in ref. 35) to explain one more paradox emerged in 1970s: multicellular organisms seem to use a highly conserved regulatory machinery for their functioning, while having different levels of complexity.

However, it should be stressed that all the speculations have important gaps. The genome may contain many additional, small genes expressed at relatively low levels which escape the detection by modern techniques.[9]

## Genes Common and Different among Various Species

In the previous paragraph I discussed the differences in the total gene number for various species. The next question is whether different species have mainly the same sets of genes just amplified in some of them, but not in the others. The scientific evidence accumulated

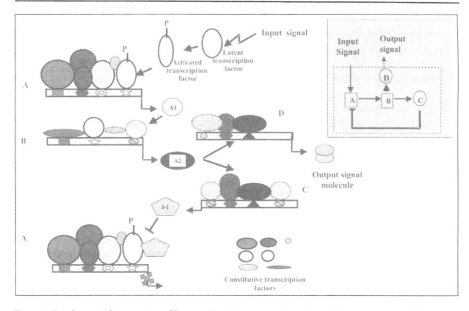

Figure 1. Combinatorial interactions of factors with their targets within cis--regulatory elements. A schematized functional module involves: (i) element A receiving an external signal and switching on synthesis of a transcriptional activator A1; (ii) element B accepting the A1 activator and switching on the next activator, A2; (iii) element C which responds to A2 activator and activates synthesis of I-1 Inhibitor which in turn inhibits the expression of element A at some step; (iv) element D which, in response to A2, directs synthesis of an output signal molecule connecting this functional module with another one. The constitutive transcription factors available in the cells participating in the regulatory events are shown as circles or ovals. The gene regulatory network representing this functional module as a formalized scheme is shown in grey box in the right upper corner.

over the past quarter of the century suggests that many essential mechanisms underlying similar functions of various organisms, from bacteria and yeasts to *Drosophila* and man, are highly conservative. In a review devoted to the problem of the molecular mechanisms underlying the evolution of greater biological complexity, Duboule and Wilkins[35] refer to the famous seminal article "Evolution and tinkering" by F. Jacob:[42] "What distinguishes a butterfly from a lion,... or a worm from a whale is much less a difference in chemical constituents than in the organization and distribution of these constituents". The authors raise a very important problem of to what extent phenotype diversity depends on new inventions of evolution and to what extent on the process of tinkering—reiterative usage of the same genes in different contexts. They left the final answer to this important question to the next 5-10 years, "...as comparative genomics broadens our knowledge about gene functions in different organisms". Five years have passed since the review was published in 1998, a great amount of structural information has been accumulated, and, though functional data has lagged far behind, one can try to look at the problem with this newly acquired knowledge. The most straightforward approach would be, of course, to compare total ensembles of proteins (proteomes) specified by various genomes.

Certainly, such an analysis will be incomplete due to difficulties in gene finding and functional classification of the proteins deduced from genomic sequences. Furthermore, as far as functional predictions are based on similarity to protein sequences with known functions, only basic biochemical functions can be assigned (rather than higher order cellular processes).[a] Besides, in such a way only about 60% of all deduced proteins can be assigned to certain broad functional categories, like "cytoskeletal structural protein", "ion channel", "transcription factor", or "cell adhesion".[14,15] In addition, a very serious problem is the identification of the "orthologs" for each of the human genes in other organisms under comparison. Orthologs, by definition, are the genes that appeared in the organisms by descent from a common ancestor. To analyze the evolutionary events, it is critical to separate orthologs from paralogs, that is the genes formed in a given species by duplication events in more than one copy (e.g., see refs. 43, 44). As a criterion, both human genome sequencing groups[14,15] used the highest sequence identity levels between pairs of cognate proteins in organisms under comparison. Since it is not a perfect criterion[b] of orthology, some of the derived evolutionary conclusions are probably not quite correct. However, despite all the limitations of such analyses, they allow one to draw some general conclusions and to get a deeper insight into the functional commonalties and diversity among eukaryotes.

Below I will confine myself to the human genome analysis of Venter et al,[14] but that of the Public consortium[15] is rather similar. In total, 2758 strict human-fly and 2031 human-worm orthologs were identified. 1523 were common to both sets, and they were defined as an evolutionarily conserved set of human proteins. Not surprisingly, the most basic cellular functions such as basic metabolism, transcription, translation, and DNA replication remained conservative since the divergence of single-celled yeast and bacteria. 60% of predicted human proteins display some sequence similarity to proteins from other species with sequenced genomes. About 40% of the human proteins show similarity with fruitfly or worm proteins. And 61% of fruitfly proteins, 43% of worm proteins and 46% of yeast proteins have sequence similarity to predicted human proteins.

The recently sequenced mouse genome demonstrated that approximately 99% of mouse genes had homologs in the human genome.[9] It is still unclear whether the remaining 1% predicted mouse genes do have no human homologs or this is just due to the unsequenced part of the human genome.

Very informative is the comparison of not the whole proteins but their discrete structural units, called domains (modules). The number of protein domains was estimated to be about 1,500-2,000, that by combining in different fashions can form tens of thousands of modular proteins. It appeared that only a relatively small proportion of various known protein domains have been invented in the vertebrate lineage, and that most domains trace at least as far back as a common animal ancestor: of 1,262 investigated domain and protein families only 94 (7%) representing 24 domain families and 70 protein families were 'vertebrate-specific'. The 94 vertebrate-specific families include defence and immunity proteins and proteins that function in the nervous system all of them contributing to important physiological differences between vertebrates and other eukaryotes. They emerged recently in evolution and/or were subject to rapid divergence.[15]

---

[a] It is important to differentiate what type of functions is under question. The term "function" is uncertain in itself. Just biochemical function like "protein kinase", "protein phosphatase" does not tell us much of the real functional role of an individual protein in the cell or in the organism: the same biochemical function can be used at different time during development, in different cellular compartments and in different signal transduction pathways (for review see ref. 45).

[b] See comment in ref. 46. Here a sober look at the problem of orthology identification is presented: this is an extremely difficult problem that can not be solved simply based on the highest sequence homology between two genes in two or more species. Therefore, all conclusions made using this criterion should be taken with caution.

The most prominent difference between humans and non-mammalian organisms with sequenced genomes is that human proteins contain more domains (modules) per protein and novel combinations of domains. Thus, new domain architectures, defined as the linear arrangement of domains within a polypeptide have been emerged in evolution. The human proteome set contains 1.8 times as many protein architectures as worm or fly and 5.8 times as many as yeast. [15] This difference is most prominent in the recent evolution of novel extracellular and transmembrane protein architectures in the human lineage. New architectures were created by shuffling, adding or deleting domains, resulting in new proteins composed of old parts: real tinkering took place in evolution.

Exon shuffling is a particular case of module shuffling, that frequently happened during evolution of various species, and represents a general mechanism for the origin of new genes. [5,47] One further source of vertebrate innovation is the expansion of protein families: about 60% of the families are more numerous in human than in any other non-mammalian organism with the sequenced genome. It means that gene duplication was among major evolutionary forces during vertebrate evolution (see below). Many of the expanded families are involved in distinctive aspects of vertebrate physiology. Compared with *C. elegans* and *Drosophila*, humans appear to have many proteins involved in cytoskeleton, defence and immunity, as well as transcription and translation. As to human and mouse proteomes comparison most of their protein or domain families have, as expected, similar sizes. However, many of the gene families have undergone differential expansion in at least one of the two genomes resulting in that some related protein families in the two proteomes significantly differed in their abundance. In particular mouse genome researchers[9] revealed numerous local gene family expansions that have occurred in the mouse lineage, most of them seemingly involving genes related to reproduction, immunity and olfaction.[9] Such gene family changes represent an insight into the aspects of physiology that have emerged since the last common ancestor.

Speaking of evolution, it is important to see what was going on with the molecules involved in development. Here also a great expansion of the developmentally important protein families is characteristic of the mammalian genomes. For example, the human genome encodes 30 fibroblast growth factors (FGFs), as opposed to two FGFs each in the fly and worm. It also encodes 42 transforming growth factors compared with nine and six in the fly and worm, respectively. These growth factors are involved in organogenesis, in particular that of the liver and the lung. A fly FGF protein, branchless, is involved in developing respiratory organs (tracheae) in embryos. Thus, developmental triggers of morphogenesis in vertebrates have evolved from related but simpler systems in invertebrates.

From point of view of the increase of functional load in the course of evolution, mammals also have many more proteins that are classified as falling into more than one functional category (426 in human versus 80 in worm and 57 in fly). Interestingly, 32% of them are transmembrane receptors.

Finally, the human and mouse organisms contain greater numbers of genes, domain and protein families, paralogues, multidomain proteins with multiple functions, and domain architectures than worm or fly. The corresponding potential increase in different protein—protein interactions is enormous and can form one of the major sources of the enormous increase in complexity.[48]

## Duplications As a Tool of Evolution

Thirty years ago Susumu Ohno (see for review ref. 49) put forth an extremely fruitful proposal that the appearance of new gene functions can be based on gene duplications. The duplications allow a gene to maintain the original function(s) whereas its copies can accumulate mutations and either lose the original function or gain a different one and be promoted by selection if this new function is advantageous. Duplications contribute to an important property of an organism—robustness which is a consequence of redundancy. Initially the duplicated genes would be fully redundant, but with time, one or the other member can diverge either in regulation of expression or in the biochemical function of the product encoded by the original gene.

Moreover, Ohno[49] suggested that great leaps in evolution could occur only if whole genomes were duplicated and that two rounds of whole genome duplication (WGD) occurred early in vertebrate history.

The whole human genome analysis could neither confirm nor disprove the WGD hypothesis.[14,15,51] However, there is evidence for such an ancient WGD event in the ancestry of yeast and several independent WGD events in the ancestry of mustard weed. Such ancient WGD events can be hard to detect because only a minority of the duplicated loci may be retained, with the result that the genes in duplicated segments cannot be aligned without many gaps. In addition, duplicated segments may be subsequently rearranged. For example, the ancient duplication in the yeast genome appears to have been followed by a loss of more than 90% of the newly duplicated genes.[50] The analysis of the whole genome sequences of many organisms has convincingly demonstrated that the duplications of the particular genome segments of various lengths were a common feature of all genomes investigated so far.

Interchromosomal or intrachromosomal segmental duplications in the human genome involve the transfer of 1-200 kb blocks of the genomic sequence to one or more locations in the genome. In total, the draft human genome sequence contains at least 3.3% segmental duplications. Such a high proportion of large duplications clearly distinguishes the human genome from such sequenced genomes, as those of fly and worm. As to the mouse genome, large regions of near-exact duplications in it seem to be less abundant than in the human one.[9] However, this conclusion is preliminary and needs further confirmation.

The duplicated copies are sometimes so similar (up to 99% identity) that paralogous recombination events can occur, giving rise to diseases such as Smith-Magenis syndrome or Charcot-Marie-Tooth syndrome 1A, respectively.[15] As judged by high sequence similarity and the absence in closely related species, some of such duplications appear to have arisen in very recent evolutionary time.

Three alternative outcomes in the evolution of duplicate genes can be envisaged:[52,53] (i) one copy may become silenced (nonfunctionalization); (ii) one copy may acquire a novel, beneficial function and become preserved by natural selection, with the other copy retaining the original function (neofunctionalization) or (iii) both copies may become partially compromised by mutation accumulation to the point at which their total capacity is reduced to the level of the single-copy ancestral gene (subfunctionalization). Most of the models predict that the usual fate of a duplicate-gene pair is the nonfunctionalization of one copy. Indeed, 50 to 92% of all duplicate genes are eventually lost over >50 Myr after the duplication. However, the frequency of duplicate-gene preservation following ancient polyploidization events, often suggested to be in the neighborhood of 30 to 50% over periods of tens to hundreds of millions of years, is unexpectedly high. This effect can be explained based on multifunctionality of eukaryotic genes. Each particular function can be independently affected by mutations that are not necessarily deleterious if two or more gene copies are available. The result of such partially degenerative mutations in one of two gene duplicates is that, being initially functionally identical, the duplicates evolve overlapping functions (subfunctionalization). Each gene can then perform at least one function the other gene can not, and the loss of such a gene might not be easily tolerated.

Lynch and Conery[54] analyzed duplicate genes in three completely sequenced eukaryotic genomes (*S. cerevisiae*, *D. melanogaster* and *C. elegans*) and in three others harboring hundreds or thousands of duplicate genes (*H. sapiens*, *M. musculus* and *A. thaliana*). Their evaluation of the half-life of duplicate genes ranges from 3 to 7 Myr for the taxa studied and also suggests that more than 90% of duplicates disappear before 50 million years have elapsed. They have also estimated the values of duplication rates ranging from 0.002 (fruit fly) to 0.02 (nematode) per gene per million years. This average duplication rate per gene is remarkably high. The authors point out that the rate of duplication of a gene is of the order of magnitude of the mutation rate per nucleotide site. In other words, as a source of variation gene duplications could be just as important as point mutations. Based partly on this finding, they argue that gene duplications and subsequent gene loss might have a significant role in speciation, because the loss of different gene copies in two isolated populations could cause a rapid accumulation of genetic differences.

One should keep in mind that the statistical approach of calculating the half-life of a gene provides only an averaged picture of genome evolution. There are many gene families that persisted perhaps much longer than the average life time. These conserved gene families deserve special consideration because they provide information on gene functions required for most living things.

Finally, the whole genome analysis confirms the general Ono's idea: although duplicate genes may only rarely evolve new functions, this phenomenon might play a significant role in the origin of new species.

### Genomes Experienced Multiple Rearrangements in Evolution, but Long Segments Remained Conserved among Species

About 342 conserved segments between mouse and human were identified so far,[9] and the minimal number of rearrangements needed to 'transform' the mouse genome into the human one was evaluated at 295. Assuming about 75 Myr of evolution from their common ancestor,[9] one can calculate an average estimated rate of about 2.0 chromosomal rearrangements fixed per 1 Myr. (Lander et al[15] obtained a lower value (1 rearrangement) due to less precise evaluation of the number of syntenic segments and higher species divergence time assumed). Most of the rearrangements occurred probably in the mouse genome. Among mammals, rodents may show unusually rapid chromosome alteration. By comparison, very few rearrangements have been observed among primates, as discussed in the review by Nadezdin and Sverdlov in this issue (see Chapter 4), and studies of a broader array of mammalian orders suggest an average rate of chromosome alteration of only about 0.2 rearrangements per Myr in these lineages.[15] Examination of the synteny between human chromosomes and those of mouse, chicken, pufferfish (*Fugu rubripes*), or zebrafish allows one to conclude that ages of many of the nearly chromosome-length duplications observed in humans are likely to be dated to the root of vertebrate divergence. Moreover, conserved segments have been observed in even more evolutionarily distant species such as fly and worm.[15]

With additional detailed genome maps of multiple species, it should be possible to determine whether this particular molecular clock is truly operating at a different rate in various branches of the evolutionary tree. Possibly, these data will eventually allow to reconstruct the karyotypes of common ancestors.

One more point deserves to be mentioned. The mouse genome sequencing allowed for the first time a direct alignment of two mammaliam genomes.[9] Approximately 40% of the human genome could be aligned to the mouse genome. These sequences seem to represent most of the orthologous sequences that remained in both lineages since the common ancestor, whereas the others were probably deleted in one or both genomes. Since the proportion of the mouse genes apparently having no homologs in the human genome (and vice versa) seems to be less than 1%, these deletions must have occurred mainly in nonfunctional parts of the genome. Assuming that the genome size of the last common ancestor of human and mice was about 2.9 billion base pairs and that 24.4% of the human genome (about 695 millions bp) consists of the repeats which appeared after the human and mouse lineages diverged, these newly emerged repeats probably compensated for the genome portion lost due to the deletions. Based on a similar assumption, the authors of the mouse genome sequence deduced that roughly 1,300 Mb of deletions partially compensated by repeats emergence occurred in the lineage leading to extant mouse from the human-mouse last common ancestor. However rough this estimate is, it demonstrates the scale of genome transformations during evolution.

### Transposable Elements (TEs) in the Human Genome

Repeat sequences are known to account for at least 50% and probably even more of the human genome. Among them are transposon-derived repeats, often referred to as interspersed repeats.[15] The genome contains DNA transposons which resemble bacterial transposons, have terminal inverted repeats and encode a transposase that binds near the inverted repeats and mediates mobility through a 'cut-and-paste' mechanism[15] (see also RepBase, http://www.girinst.org/~server/repbase.html). I will focus here on only a special class of the interspersed repeats that transpose through RNA intermediates—retrotransposones (for recent review see refs. 55-57).

LINEs (long interspersed nuclear elements), a large group of mobile elements known as the non-long terminal repeat (LTR) retrotransposons, are the most ancient interspersed repeats in eukaryotic genomes (see also Fig. 4 in Chapter 5). In humans they encode two open reading frames (ORFs) regulated by an internal polymerase II promoter. LINE1, or L1, is an abundant long interspersed nuclear element that has been amplified to a high copy number in mammalian genomes by retrotransposition. For example, in humans and mice there are over 100,000 L1 copies. Upon translation, a LINE RNA assembles with its own encoded proteins and moves to the nucleus, where the so called Target-site Primed Reverse Transcription (TPRT) and integration of the cDNA copy occurs.[55] Transposition-competent versions of L1 have been recently isolated and studied. L1 elements are 6 to 7 kb long and encode two proteins (ORF1p and ORF2p) necessary for retrotransposition. The endonuclease and reverse transcriptase activities, presumably required for TPRT, reside in ORF2p. The other protein, ORF1p, is also essential for retrotransposition, though its role is not fully understood. Reverse transcription frequently fails to proceed to the 5' end, resulting in many truncated, nonfunctional insertions. Indeed, most LINE-derived repeats are short, with an average size of 900 bp for all LINE1 copies, and a median size of 1,070 bp for copies of the currently active LINE1 element (L1Hs). New insertion sites are flanked by a small target site duplication of 7–20 bp. The LINE1 machinery is believed to be responsible for most reverse transcription in the genome, including the retrotransposition of the non-autonomous SINEs (short interspersed nuclear elements) and the creation of processed pseudogenes.[55] In addition to L1, two distantly related LINE1 families are found in the human genome: LINE2 and LINE3. However, only LINE1 is still active.[15]

SINEs are short (about 100–400 bp) elements, which harbor an internal polymerase III promoter and encode no proteins. These non-autonomous transposons are thought to use the LINE machinery for transposition.[15,55] The promoter regions of all known SINEs are derived from tRNA sequences, with the exception of a single monophyletic family of SINEs derived from the signal recognition particle component 7SL. This family includes the only active SINE in the human genome: the Alu element. By contrast, the mouse has both tRNA-derived and 7SL-derived SINEs. The human genome contains three distinct monophyletic families of SINEs: the active Alu, and the inactive MIR and Ther2/MIR3.

LTR retroposons are flanked by long terminal direct repeats that contain all of the necessary transcriptional regulatory elements. The autonomous elements (retrotransposons) contain *gag* and *pol* genes, which encode a protease, reverse transcriptase, RNAse H and integrase. Transposition occurs through the retroviral mechanism with reverse transcription occurring in a cytoplasmic virus-like particle, primed by a tRNA (in contrast to the nuclear location and chromosomal priming of LINEs). Although a variety of LTR retrotransposons exist, only the vertebrate-specific endogenous retroviruses (ERVs) appear to have been active in the mammalian genome. Most (85%) of the LTR retroposon-derived 'fossils' consist only of an isolated LTR, with the internal sequence having been lost by homologous recombination between the flanking LTRs.

### Intraspecific Variations in the Human Genome

Up to this point, all intergenomic comparisons were pertinent to interspecies and even interphyla levels. However, it seems probable that all evolutionary changes begin as intraspecific variations, which occur within particular populations living in particular environments.[2] Therefore, to understand the evolutionary origins of interspecific differences it is also necessary to investigate intraspecific genome variations (polymorphism) in coding and non-coding intergenic and regulatory genome regions and to try to correlate them with intraspecies phenotype changes.

Analysis of polymorphisms and their precise mapping in the human genome attracted great attention in view of their importance for identification and mapping of the loci involved in predisposition to human diseases. The data accumulated so far (reviewed in refs. 58-60) demonstrate enormous variability among individual genomes. The largest polymorphisms are due to microsatellites and single nucleotide polymorphisms (SNPs). The extent of intraspecific variability can be illustrated with an example of SNPs: on the average, each two homologous human chromosomes have one SNP per 1,000-1,200 bp of the DNA sequence. This enormous variability prompts a question of how it is possible to tolerate the genomic changes while retaining all the major phenotypic features, that is to remain humans, though rather different. The answer, probably, at least partially lies in special mechanisms which buffer the phenotypic expression of genotype variability.[61] A more detailed analysis of polymorphisms is presented in the Chapter 4 by Nadezdin and Sverdlov in this issue.

## Genome Structures and New Concepts of Evolution

A multitude of genomes are already sequenced and available for full-scale comparison. But are we really nearer to the solution of the question "what is life?" asked by many thinkers and, in particular, by a famous physicist Erwin Schroedinger.[62] The answer is "yes" and "no".

On the one hand, we have realized how conservative are a great many of genes and their protein products responsible for basic biochemical functions, thus confirming that all branches of life have the same biochemical background. On the other hand, the notorious questions still remain open: how is the genomic information converted into phenotypic traits (Question 1)? which of the differences among genomes are responsible for the phenotypic variations, and how and which of them were involved in evolutionary processes?

Below I will very schematically consider some of newly emerging hypotheses as to what genomic changes might be translated into phenotypic variations. More detail can be found in recent comprehensive reviews[2,35,63-67]

### *Not Only Natural Selection, but Also Internal Developmental Biases Can Form a Basis for Evolutionary Changes*

It is widely accepted that natural selection is the main mechanism responsible for adaptation of organisms to their environment. This is an 'external' mechanism acting as a fitness-based screen on the pre-emerging variety of the phenotypes that eventually appear due to mutations. The second widely accepted concept is that the major source of new phenotypes for natural selection is the mutations that cause changes in the development and thus lead to the appearance of new morphological features subject to the selection. This synthesis of developmental biology and evolution led to a new concept: Evo-Devo, based on the conviction that evolution cannot be understood without understanding the evolution of development, and how the process of development itself biases or constrains evolution.[65] Classical neo-Darwinism assigns the directive role solely to selection, while mutation just provides a supply of variation, being a stochastic non-biased process. The newly emerging concepts suggest that some mutations more readily produce phenotypic changes in certain directions than others, including the extreme case of some directions being apparently 'prohibited'. Such a state of affairs in general has been referred to as mutation bias or developmental bias ("evolution biased by development"[2]). Negative biases, both relative and absolute, constitute constraint, whereas positive biases have recently been termed developmental drive. Thus, these biases form the 'internal' aspect of evolution, including the ways in which developmental pathways evolve.[2] The constraints can be a consequence of increased developmental complexity at the expense of the increased load of biochemical functions per gene (see above). The involvement of a gene in a multitude of functions makes it less probable that it will be recruited into further function. According to this point of view, a complex developmental system can be expected to be more stable and less prone to changes than a simpler one. In other words, the complex system is more robust and inertial and is more determined to move along predetermined "ontogenic trajectory".[2] Such an effect has been termed "canalization" (for review see ref. 35). Due to these constraints most novelties will first be tested by an internal genetic selection to see whether they are compatible with the existing functional networks where the novelty is to be incorporated. The most recent analyses show strong interdependence of the evolutionary rates of the proteins involved in functional interactions in yeasts.[66] The connectivity of well-conserved proteins in the network is negatively correlated with their rate of evolution. Proteins with more interactions evolve more slowly not because they are more important to the organism, but because a greater proportion of the protein is directly involved in its function. At sites important for interaction between proteins, evolutionary changes may occur largely by coevolution, in which substitutions in one protein result in selection pressure for reciprocal changes in interacting partners and interacting proteins evolve at similar rates.

## *Modularity Is a Universal Characteristic of Organization of Biological Systems and Their Developmental Machinery*

An extremely important concept of the modern evolutionary theories is the concept of modularity (compartmentation) of the biological systems. According to Dover,[5] a module is defined as follows: "A module is an independent unit or process or function that may interact in a variety of combinatorial interactions with a variety of other units or processess or functions". A functional module[67] was defined as a discrete entity whose function is separable from those of other modules. A discrete biological function can only rarely be attributed to an individual molecule, it is always a result of the interplay of a variety of molecules forming a functional module.

An example of functional modules is signal transduction pathways integrating embryonic development. These pathways in various species constitute homologous modules consisting of many homologous species of interacting molecules. They are conserved through evolutionary time, though experience modification in the course of evolution. The signal transduction pathways demonstrate a very important property of functional modules—their overlapping in such a way that different modules use the same molecules for similar or different purposes. In case of signal pathways this feature was termed "crosstalk" (see for example ref. 68). Analyzing the signaling pathways integration development, Gilbert and Bolker[69] formulated four principles of "evo-devo" based on the functional module concept: (1) evolution depends on heritable changes in development, (2) development is modular such that different modules can change without affecting other modules, (3) modules can be co-opted (see below) into new functions, and (4) modules depend on intercellular communication. Functional modules need not be rigid, fixed structures; a given component may belong to different modules at different times. The higher-level properties of cells, such as their ability to integrate information from multiple sources, will hopefully be described by the pattern of connections among their functional modules.[67]

Recently a concept of genomic modularity (compartmentation) was proposed defined as the division of the cell's total genomic potential into partially independent subsets of expressed genes, as seen in cell differentiation.[70] In different such modules the subsets of genes are expressed in different combinations and at different levels. The pattern of the expression is determined by various transcription factors and enhancer binding proteins interacting with also modular *cis*-regulatory regions of the genes. As a result, certain constellations of genes are expressed during certain periods of time and at certain levels. This modularity of expression is also contingent on external signaling factors, which in metazoa are provided by other cells, and on internal factors (active transcription factors) brought forward in the cell lineage from previous stages of development.

Modular structures might facilitate evolutionary changes. Embedding particular functions in discrete modules allows the core function of a module to be robust to change, but allows for changes in the properties and functions of a cell (its phenotype) by altering the connections between different modules.

### The Increase in the Gene Functional Load Can Be Achieved by Various Means

The acquisition of new functions by the molecules involved in developmental pathways is suspected to cause important morphological novelties. Genes can be recruited to serve completely new functions in a new regulatory linkage, they can change their molecular specificity while remaining in the original (homologous) developmental program and can, at the same time, retain other functions.[64,71] At the gene level such an acquisition can be due to either changes in the gene coding potential and corresponding changes in the biochemical properties of the products encoded or changes in the gene regulation system affecting the temporal or spatial context of the gene expression (functional shift). Scenarios of functional shifts in evolution can be very different. Genes (modules) can get new functions after duplication or without it. Some features can emerge first as byproducts with no initial functions and then acquire a function, whereas the others can simply switch from one function to another. Finally, the genes (modules) having acquired new functions may or may not sustain the former functions. Ganfornina and Sanchez[71] proposed to use the term direct co-option (recruitment) for a functional shift of a gene itself without duplication, as opposite to co-option (recruitment) of a duplicated element usually considered as a classical way of functional diversification. Co-option is a central concept emerging from comparative studies of developmental genes and their interactions in different species. An extension of this concept is the idea that genes are co-opted not individually but as interacting cassettes or modules. Co-option embraces both biochemical function changes and regulatory changes. However, Duboule and Wilkins,[35] in their review remark that "...the primary source of developmental differences... will prove to be not unique gene products but rather the way that comparable, or the same, gene functions are differentially deployed in their development... Many so called heterochronic shifts, altering developmental programmes and morphologies involve no more than alteration in the times and cellular site at which particular regulatory molecules are expressed rather than alteration in those molecules themselves". In metazoan evolution, these processes have been brought under intercellular control regarding the time, place, and conditions of function. Regulatory processes have evolved greatly in metazoa. In gene expression, regulation can be at any of several levels: nuclear localization of a transcription factor, multimerization of a transcription factor with itself or with a cofactor or with an inhibitory protein, interaction and competition of different transcription factors for sites on DNA, and reversible or long-lasting modification of chromatin.[70] The co-option of a gene or a module to a new functional activity can be achieved by changes at any of these levels.

Vertebrates are among the species whose developmental evolution has, most probably, been deeply influenced by gene duplications (see above), though it is not a general rule for other phyla.[63] As compared to invertebrates, vertebrates have more *Hox* genes and many other developmentally important gene families due to duplication and reduplication of ancestral genes. It suggests that the genesis of the complex vertebrate body plan, with its novel cell types and organ systems, was enabled by the availability of extra genetic material which appeared due to duplications. This material could be gradually co-opted for playing new developmental roles.

### Regulatory Networks Determine Development and Evolution

Development unfolds as a succession of stages during which each next stage involves new subsets of cells each specified to implement a certain multitude of functions different both among newly appeared subsets and from cells existing at the previous stage. This specification occurs due to the expression of a given set of genes in the subsets of cells in each region of the developing animal. In turn, these changes in the transcriptional patterns are triggered by either external signals from other cells or by internal signals determined by maternal molecules distributed to particular cells with the egg cytoplasm and partitioned spatially during cleavage. The signals are received by *cis*-regulatory elements in the form of transcription factors interacting with specific target sites within the *cis*-elements. A *cis*-element thus form a node, to which a net of transcriptional factors and co-factors is tied at a certain time and within a certain set of the embryonic cells. All these interacting elements can be considered as a functional module (Fig. 1, insert) whose output is the switching on of a set of new regulatory genes. A multitude of such interacting modules forms regulatory networks. In each subset of the cells having appeared at each developmental stage its own regulatory network(s) is brought into being. The key role of the *cis*-nodes is due to their capacity to use "and" logic,[72] when two or more different transcription factors and co-factors, each present in a given spatial domain, must be bound to the node simultaneously in order to activate transcription (Fig. 1). A unique specificity of such a *cis*-node can be conferred by a certain combination of short specific sequences within it. To understand what makes a developmental process unfold as a very definite sequence of very definite stages one needs to identify the key inputs and outputs leading to the activation of certain *cis*-regulatory nodes in needed time and in needed place. Such a successful attempt has been recently reported for a particular case of a gene regulatory network involving over 40 genes that controls the specification of endoderm and mesoderm in the sea urchin embryo.[72] No doubt that the deciphering of such developmental regulatory networks will be the most important analysis of developmental mechanisms in future. It will require joint efforts of experimenters and computational scientists, because computational models are absolutely indispensable for the analysis of a wealth of data accumulated by now and for rational planning of future research.

When applied to the mechanisms of evolutionarily significant changes, the network approach will allow to perform a genome-wide analysis of how the developmental networks can be changed in evolution while preserving the functioning of the pre-existing systems. Surely, this is a tremendous problem considering that the work cited above has focused just on 40 genes involved in a comparatively simple stage of development, whereas thousands of genes participate in the whole process. New approaches using extensive techniques such as microchips will hopefully facilitate the network analyses by means of identification of the gene groups that are expressed coordinately and therefore probably involved in the same functional modules.[73]

### *Evolvability Is Probably a Universal Characteristic of Living Entities*

The discussion above was focused on the idea that conserved core processes constrain phenotypic variation, making most changes detrimental and acting as a barrier to evolution. The question is how this conservation of most of the core eukaryotic cell mechanisms can be reconciled with the observed great diversification among metazoa which certainly means their great capacity for phenotypic change. Do they have some cellular and developmental mechanisms enabling them to rapidly evolve? Attempts to answer this question led to a concept of evolvability (for recent reviews see refs. 70, 74).

Evolvability (evolutionary adaptability) was defined as the capacity of a lineage to evolve to generate heritable, selectable phenotypic variation.[70] It was hypothesized that multitudes of regulatory processes emerged during evolution to control the time, place, and other peculiarities of the conserved core processes have special properties relevant to evolutionary change. These properties depend on the modularity of the processes, weak linkage between separate modules, redundancy, and other features reducing the interdependence of organism's components. They allow to reduce constraints on change and to accumulate non-lethal variations.

As discussed above, functional redundancy is one of the consequences of gene duplication that protects old functions as new ones arise, hence reducing the lethality of mutation.

It has already been mentioned that modularity reduces the interdependence of processes, thereby reducing a chance of pleiotropic damage from mutation and facilitating phenotypic variation and evolutionary change. An important constituent of the evolvability concept is the idea of so called "weak linkage", that is a weak dependence of one process on another or its components. The concept of "weak linkage" implies the ability of modules to easily dissociate and to interact in various combinations, thus having a potential to function and respond in many contexts, rather than being precisely tailored to be effective in only one. Organization based on weak linkage facilitates a component's accommodation to novelty. After Kirshner and Gerhart[70] I will exemplify this idea with the eukaryotic transcriptional control organization, with its complex and highly conserved core components. The eukaryotic system recognizes many transcriptional signals from proteins bound at short conserved sequences of enhancer, promoter, silencer, or border elements. These short sequences form modules that can be met in different combinations within different regulatory sites governing expression of different genes. On the other hand, many factors interacting with these short *cis*-regulatory modules have limited affinity and rather low sequence specificity for these sites, and some bind with specificity only in the context of other proteins. Gene expression is controlled by many positively and negatively acting factors, whose own availability can be different in different compartments of embryos and differentiated cells. The different affinity of the factors and their different concentrations in different compartments may underlie the expression of different sets of genes in these compartments. This can be demonstrated by a recent example of regulation of transcription of pharyngeal genes by PHA-4 transcription factor, encoded by an organ-identity [c] gene *pha-4*,[75] (schematized in Fig. 2). PHA-4 controls pharynx formation in the *C. elegans* embryo. PHA-4 protein is present in nuclei of all pharyngeal cells throughout development, but its expression levels increase as development progresses. It was shown that PHA-4 directly regulates most of pharyngeal genes. Both early-acting regulatory genes and late-acting structural

---

[c] The process of organogenesis starts from setting up spatially defined regulatory domains that promote the differentiation programs. This mechanism enables the recognition and definition of regulatory fields as discrete territories of specific gene activities. The genes controlling the formation and identity of the various fields are defined as selector genes. A special class of selector genes are those field-specific genes that have the unique property of directing the formation of complex, specialized structures such as organs. These genes are usually necessary and often sufficient to establish organ identity.[76]

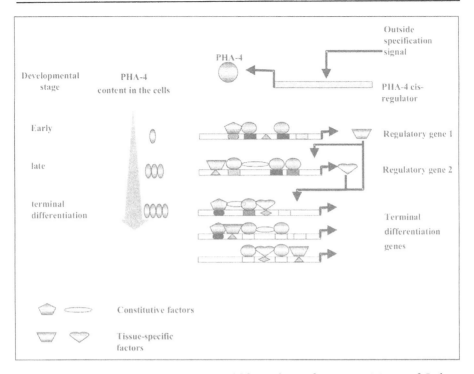

Figure 2. An affinity and factor concentration model for regulation of organogenesis in case of *C. elegans* pharynx. It assumes that (i) PHA-4 directly regulates a wide array of genes expressed at multiple stages of pharyngeal development, and (ii) the affinity of various versions of the consensus TRTTKRY target sequence for PHA-4 regulates the relative onset of the target genes expression. Consistent with this idea, PHA-4 is expressed at low levels at the ~50-cell stage of embryogenesis and its concentration in the cells increases with the progress of development. On the other hand, the genes expressed during early pharyngeal development carry high--affinity binding sites (black boxes), whereas those expressed at later stages harbor lower affinity sites (light grey boxes). The binding of PHA-4 with these latter targets is possible only at higher PHA-4 concentrations and possibly only in combination with other factors. In all cases PHA-4 most probably works as a constituent of a combinatorial regulatory complex containing various factors each bound to its own target and/or to other factors within the complex.[76,77]

genes involved in terminal differentiation contain in their *cis*-regulatory elements one or more PHA-4–binding sites with a TRTTKRY (R=A/G, K=T/G, Y=T/C) consensus sequence. But these different *cis*-acting nodes have different affinities to PHA-4 protein: those regulating early-acting genes contain high-affinity-sites, whereas late-acting genes have low affinity sites. Therefore, the regulation of pharynx formation can be realized in a very straightforward way: the timing of gene expression throughout the regulatory network is modulated through vary-ing affinities of PHA-4 for different target sites. When the concentration of PHA-4 is low (early developmental stages), it interacts efficiently with only high affinity nodes, but at later stages, when the concentration is increased, the late genes are switched on. This illustrates a principle of "weak linkage": PHA-4 is not determined to interact with only one strictly specific binding site, it can recognize many of them and thus function at various stages of development forming complexes with different regulatory nodes, possibly, in combination with various other factors and co-factors. It seems to be a rather common feature of organ-identity genes often acting both early and late in the process of organogenesis.[75,76] Various combinations of transcription factors and co-factors interacting with their targets form an extremely important basis of specificity and, at the same time, flexibility of the eukaryotic transcription regulatory machinery.

One more very recent example of such a flexibility[77] looks even more striking. The transcription factor T-beta is expressed in CD4, NK and in CD8 lymphocytes. All these cells produce also an important regulator of immune system development—interferon-γ (IFN-γ). But, paradoxically, T-bet is required for control of IFN-γ production only in CD4 and NK cells, but not in CD8 cells. This unexpected observation demonstrates that CD4 and CD8 T cells, although closely related and arisen from a common progenitor in thymus, have nevertheless evolved distinct mechanisms for transcriptional control of IFN-γ production. One and the same *cis*-regulatory region with its peculiar combination of the *cis*-regulatory modules is probably used by different combinations of *trans*-regulatory factors and/or co-factors to express the same gene. This divergence may have emerged to optimize the immune response of the organism to a diverse range of infectious invaders. This is again a demonstration of the "weak-linkage" principle: a gene is not destined to be expressed under control of only one strictly specific set of factors. Instead, different combinations of factors and cofactors can be used for its expression, depending on the cellular context.

The examples above demonstrate the importance of multiple inputs for the expression of metazoan genes at different times, places, and conditions, which is necessary to correctly respond to multiple signals. Weak linkage confers flexibility and robustness on transcriptional regulation and allows to add and subtract regulatory elements to eukaryotic genes thus increasing the evolvability of the system.

Another important contribution to evolvability may be provided by the buffering mechanisms.[61] These mechanisms probably enable a population to accumulate genetic variation while remaining phenotypically normal. But as soon as the buffering breaks down, phenotypic differences suddenly manifest themselves and become exposed for selection. The buffering mechanisms can serve as modulators of evolution and maintain a delicate balance between evolutionary stability and change. An example of such a buffer is a heat-shock protein Hsp90 in *Drosophila*.[78] This protein participates in signal transduction and is involved in several developmental pathways. When Hsp90 is impaired, phenotypic variants affecting nearly any adult structure emerge, with specific characteristics depending on the genetic background. These variants are produced by multiple, previously silent, genetic determinants and, after selectional enrichment become independent of the Hsp90 mutation. In such a way a promotion of evolutionary change in otherwise entrenched developmental processes may be implemented.

Various other aspects of evolvability have also been considered, but they are beyond the scope of this review, and the reader can learn more about them in very interesting and seminal reviews.[5,70,74] Summarizing this part, I should stress once more that to understand the mechanisms of evolution both internal and external evolutionary forces that shape the development of phenotypes have to be considered. The internal factors in particular involve the properties of the cellular and genetic organization. Genetic systems are redundant, modular and are readily modified by various processes like transposition, duplication, recombination, gene conversion and so on. These mechanisms contribute to creation and spread of novel combinations of modules, the major source of the evolution of phenotypes. As a result, yet more complex networks of genetic interactions and yet more functions for a given module emerge. Natural selection leads to a molecular coevolution between interacting modules and hence facilitates the establishment of biological novelties.

The details of the mechanisms reconciling increasing constraints with a capacity to produce evolutionary diversification will hopefully become clearer as more genomes are deciphered thus allowing more comprehensive inter- and intraspecial comparisons.

## How Could Transposable Elements Contribute to Evolution?

In this concluding paragraph I would like to briefly consider from the viewpoints above the evolutionary potentials of transposable elements, specifically retroelements (REs).

### *Long Co-Evolution of REs and Host Genomes Has Possibly Generated Mechanisms of Tolerant and Even Sometimes Mutually Beneficial Co-Existence*

Numerous data indicate that retroelements were repeatedly inserted/deleted during evolution, in particular that of higher primates.[79,55] It is widely accepted that the primary force for the origin and expansion of most transposons has been selection for their ability to create progeny and not a selective advantage for the host. However, to be not counterselected, REs probably had to evolve mechanisms of maximum harmlessness towards their host genomes so that hosts did not develop efficient mechanisms of RE-resistance. REs co-existed with their hosts for extremely long periods of evolution: for example, monophyletic LINE1 and Alu lineages are at least 150 and 80 Myr old, respectively. Some LTR retrotransposons, such as ERV-L and MaLRs, are over 100 Myr old.[15] Therefore, some compromise (co-adaptation) between host and REs has probably been elaborated.[56,57] Various mechanisms could function to protect the genome from excess transposon and retrotransposon activity,[55] among them being such well known general regulatory means as methylation and RNA interference (RNAi).

Moreover, the host cells sometimes exploit the ability of TEs to generate variations for their own benefit (see Chapter 7). Thus, REs evolve themselves in a permanent cross-talk with the constraints put by the host genome. Therefore, evolution of REs may possibly serve as an indicator of the rules that the host genome obeys in its own functioning and evolution, and forces to subordinate its "inhabitants".

Here one can ask what are the rules of REs evolution? The answers come partially from intra- and interspecial comparisons of the REs' families. The detailed comparisons will be presented in the review by Dixie Mager et al in Chapter 6, and here I will give just a list of what are, to my mind, the most essential generalizations from such comparisons.

1. Most interspersed repeats in the human genome predate the eutherian radiation. This indicates an extremely slow rate with which nonfunctional sequences are cleared from vertebrate genomes. The human genome contains a great amount of ancient transposon copies, whereas the transposons in the fly, nematode and mustard weed genomes,[15] as well as in the recently sequenced mouse genome[9] tend to be of more recent origin. This difference is most likely determined by different rates of the REs elimination through genomic deletion. The data available suggest that small deletions occur at a rate that is ~75-fold higher in flies than in mammals; the half-life of such nonfunctional DNA as pseudogenes is estimated at 12 Myr for flies and 800 Myr for mammals.[15]

2. In the human genome various families of REs went through periods of active retropositions and then lost their vital potential and became fossils providing a rich record of the genome evolutionary history, but inactive in further propagation. In this regard they were much like civilizations, flourishing for a while and then falling into decay and collapsing under pressure of various internal and external factors. For example, the most prolific elements ERV-L and MaLRs flourished for more than 100 Myr but apparently turned inactive about 40 Myr ago.[15] Only a single LTR retroposon family (HERV-K(HML-2)) is known to have transposed since our divergence from the chimpanzee some 4-7 Myr ago. After a burst of activity of Alus with maximum around 40 Myr ago, there would appear to have been a decline in their activity in the hominid lineage. The overall activity of all transposons has also declined over the past 35-50 Myr, and now almost all transposon sequences in the human genome are immobile, at least in normal conditions, though some background of

new retropositions, mainly L1 and Alu, in the human genome still exists. The major causative agents of endogenous genomic insertions are L1 retrotransposons of which there are
roughly 40–60 active copies per human genome.[15,55]

3. The substitution level within REs was an average of about 1.7-2 fold higher in mouse than
   human, in agreement with the generally higher rate of substitutions and recombinations in
   rodents (see above). The faster clock in mouse is also evident from the fact that the ancient
   LINE2 and MIR elements, which transposed before the mammalian radiation and are
   readily detectable in the human genome, cannot be readily identified in available mouse
   genomic sequence.[9,15]

4. Transposon activity in the mouse genome has not undergone the decline seen in humans
   and proceeds at a much higher rate. LTR retroposons are still alive in the mouse with such
   well known examples as IAP family and putatively active members of the ERV-L and MaLR
   families. Also LINE1 and a variety of SINEs are quite active, so that new spontaneous
   mutations are 60 times more likely to be caused by transposable elements in mouse than in
   humans.[15]

All these examples demonstrate that REs in the human genome seem to obey the general
rule: the rate of the evolution of the human genome constituents is among the slowest (still
debated, see for example ref. 80). The nucleotide substitution rate (per site, per year) in human
($\sim 1.2 \times 10^{-9}$) intronic sequences is less than that of Old World Monkeys ($\sim 1.8 \times 10^{-9}$), New World
Monkeys ($\sim 2.1 \times 10^{-9}$) and rats ($\sim 4.8 \times 10^{-9}$).[81] The slowdown in human, chimpanzee and gorilla
nucleotide substitution rates relative to those in other species is in correspondence with a similar slowdown in recombination and transposition rates discussed above. Especially striking is
the difference between the human and mouse genomes, which may mean some fundamental
difference between hominids and rodents, reflected in particular by the behavior of transposons.
One of the reasons may be different structure and dynamics of the populations of hominids
and rodents. Hominid populations are typically smaller than those of rodents and may undergo frequent bottlenecks. A difference in generation times may also play some role and,
moreover, it is very tempting to suggest that the hominids developed some constraints slowing
down all changes in the genome structure. What mechanisms could control such constraints?
Are they different and specific for substitutions, recombinations and retropositions, or there is
some common mechanism of monitoring of all kinds of changes? This will probably become
clearer as additional mammalian genomes are deciphered and the trasposition activity correlated
with their structural and functional features.

### Transposable Elements As a Depot of Evolvability Factors

Transposable elements can be considered as independent modules with a great regulatory
potential that often were and can be co-opted in various regulatory networks. Retroelements
can play a role of travelling carriers of a variety of control elements (see Chapter 7) for promoters,
transcriptional enhancers and silencers, poly(A) addition sequences, splice sites, non-coding
RNA for posttranscriptional regulation (e.g., through RNAi), DNA methylation, nucleosome
positioning, determinants of RNA stability or transport etc. The number of described cases
where retroelement sequences were shown to confer new traits to the host is constantly growing.[56,82,83] The data obtained for different species clearly demonstrate that insertions and deletions of TEs can alter not only the structure of genes and thereby their products but also their
regulation (see for review ref. 79). In addition, they can affect the stability of the genome by
introducing recombination hot spots. Moreover, if TEs have their own genes they can enrich
the genome with new genetic information like genes of reverse transcriptase or viral resistance.

LINE1 potential for genomic changes is especially well understood. In addition to the
above mentioned features, these REs can transduce the genes located downstream of the L1

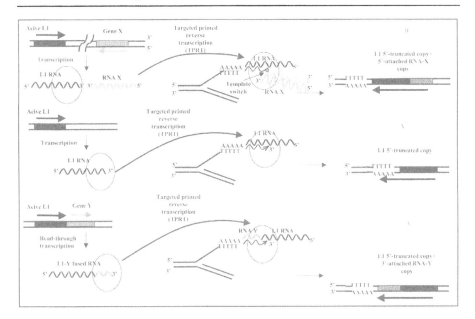

Figure 3. A schematic illustration of: A) a new 5'-truncated L1 copy integration in a new genomic site; B) formation of a chimeric retroelement containing cDNA of a gene transcript 5'-attached to 5'-truncated L1; C) gene transduction by means of an L1 element read--through into a neighboring gene area and reintegration of the chimeric cDNA into a new genomic position. In this case the gene transduced copy is attached to the L1 3'-terminus. In all cases the arrows indicate 5'→3' direction of the sense strand of the gene or L1, the grey circles symbolize the L1 reverse transcription/integration machinery, and Y-shaped elements in the center demonstrate the process of the target primed reverse transcription which is thought to include the introduction of a nick in the double--stranded DNA by L1 endonuclease, the formation of a duplex between poly-A tail of L1 RNA (or mRNA) and 3'-oligo(dT) stretch of the nick, and the reverse transcription of the RNA with the use of an oligo(dT) 3'-OH group for priming of the new DNA synthesis. In case (B) the template switch in the course of the reverse transcription is suggested to explain the existence of the genomic chimeric elements containing fusions of a 5'-proximal gene derivative and 5'-truncated L1 at the 3'-end.

elements by means of cotranscription of neighboring DNA, reverse transcription of the transcripts and reintegration of the cDNA into new genomic sites. In such a way the L1 machinery can also cause reverse transcription of various mRNAs, which typically results in nonfunctional processed pseudogenes but can, occasionally, give rise to functional processed genes[55] (see also http://www-ifi.uni-muenster.de/exapted-retrogenes/tables.html). Recently we have demonstrated a new possible mechanism of forming novel genetic elements. The mechanism is based on recombination between L1 RNA and cellular RNAs followed by reverse transcription and integration of a recombinant DNA copy using the L1 integration machinery.[84] Some of the L1 genome reshaping activities are illustrated in Figure 3.

As this book is devoted to the evolutionary aspects of endogenous retroviruses, I will use a couple of examples of the involvement of HERV in gene regulation modifications to demonstrate REs' evolutionary potential (a more detailed discussion is given in Chapter 7). HERVs are often located in the vicinity of genes.[79] Their LTRs represent regulatory cassettes with multiple short modules designed for retroviral expression regulation such as potential enhancer and hormone responsive elements, TATA-boxes and poly-A signal sequences. LTRs can modulate the expression of the nearby genes either directly as transcription regulatory signals or indirectly by means of chromatin remodeling at the sites of the integration. Moreover, when included in transcripts, they can participate in the regulation at the level of translation.[85] Several hundred genes in the human genome, for example, use transcriptional terminators donated by the LTRs.[15]

It was noticed that some of LTRs were located in the vicinity of the genes encoding Zn-finger or Zn-finger like proteins.[79] Many Zn-finger proteins are known to function as transcription regulators so the changes in their regulation can affect the expression of multitudes of other genes. Therefore, the embedding of an LTR in the vicinity of a regulatory gene can change its expression and thus trigger a new pattern of expression with many new genes involved. In the terms of the concepts discussed above, this can cause the emergence of a new *functional module*. This module can potentially be co-opted in some networks to implement new functions. However, because of constraints towards novelties in complex systems, these changes will have to be delicate enough to preserve the performance of the pre-existing systems.

A possible change in the tissue specificity of a functional module due to co-option of a retroviral (or other REs) regulatory system is shown in Figure 4. This scheme is in accord with recently described changes in the regulation of the human growth factor pleiotropin (PTN) gene.[86] A HERV insert in an intron of the gene between the 5'-untranslated and the coding region has occurred after the divergence of apes and Old World monkeys and therefore exists in human, chimpanzee, and gorillas but not in resus monkey.[87] This insert generated a phylogenetically new additional promoter with trophoblast-specific activity. It was recently shown that a retroviral Sp1 transcription factor binding site confers new traits on the PTN native promoter.[88] As a result, fusion transcripts between HERV and the open reading frame of PTN (HERV.PTN) were detected in all normal human trophoblast cell cultures as early as 9 weeks after gestation and in all term placenta tissues but not in other normal adult tissues. This trophoblast specificity is a new evolutionary acquisition: rodents lack it.[87] It adds a new trait to already existing ones: in both human and rodents the PTN is expressed from the phylogenetically common promotor of the PTN gene at highest level in the central nervous system during the perinatal period and in only a few adult tissues.[86,87] The formation of trophoblasts is one of the very early differentiation events in development, and the authors expressed a very cautious suggestion that trophoblast specific expression of the growth factor could be advantageous during phylogenesis. It may be of importance that many HERVs are expressed in embryonic tissues.[89] For example, HERV-R(ERV-3) is known to be primarily expressed in the placenta and is induced to high levels during differentiation of normal syncytiotrophoblasts.[90] These findings led Harris[89] to the hypothesis that one or more ancient trophoblastic ERVs could have played a role in the evolution and divergence of all placental mammals. In line with this hypothesis one can think that trophoblast specific expression of PTN in apes might also play a similar role in primate divergence.

Tissue-specific expression is a rather common characteristic of many REs and this property is possibly one of the ways of the co-adaptation of REs and host genomes that help to mitigate their deleterious effects.[56,83] Being integrated nearby genes, REs might bring these genes under their tissue-specific control. If this is the case, then the differences in integration sites of REs, and in particular HERV LTRs, among primate species could be expected to cause differences in tissue specificity of the neighboring genes. Quite a few of such LTR human specific integrations have been described in a recent research (see Chapters 6 and 8 in this book), and the genes that could be targets of possible regulatory impacts were identified. However, no proper comparative analysis of tissue specificity has yet been done.

The presence of REs, including HERVs and their LTRs, in primate genomes, their frequent location in the vicinity of genes, regulatory potential and differences in their integration sites among primate species make REs, and in particular HERVs, major candidates for being pacemakers of primate evolution. In keeping with King and Wilson[3] ideas, they could exert their evolutionary impact by changing gene expression patterns, especially during embryo de-

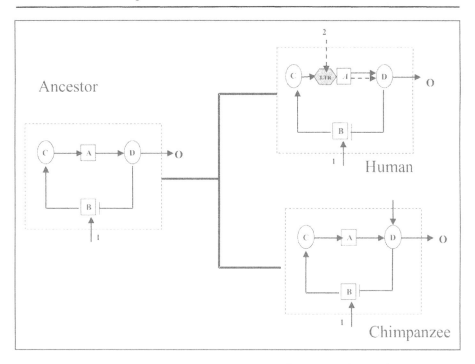

Figure 4. A possible scheme of a functional module evolution. The functional module with four cis-regulatory nodes (A, B, C and D) acts in a tissue specific mode in response to an external signal 1 and produces an output signal O. A co-option of an RE, in this case exemplified by LTR, in one of the descendant species (here human) confers to the module the capacity of responding to additional input signal 2 and thus functioning in a new tissue(s). The dashed arrows denote the pathway of the signal 2 transduction.

velopment and by co-opting newly formed functional domains into new functional networks. Speaking of primate and especially human evolution, REs integrations could affect the brain development having eventually caused a dramatic jump in human cognitive capacities.

But whether they did realize their evolutionary potential will hopefully be clearer when other genomes and especially primate genomes will be sequenced, compared to the human genome and become subject of functional genomics and proteomics. This is supposed to reveal genomic functional differences before and after REs' integrations and form a basis for conclusions on their involvement in evolution. However, we must confess that the real role of these invaders in evolution will never be disclosed unequivocally.

## Concluding Remarks: What about Genes That Make Us Humans?

Among the questions that the theory of evolution dreams to answer, one is, probably, most exiting: "Which of our genes make us human?"[91] There is a hope to get at least some initial momentum toward the solution of this problem through a comparison of the human genome with that of our closest relative, chimpanzee. This is how David Baltimore put it: "We wait with bated breath to see the chimpanzee genome. But knowing now how few genes humans have, I wonder if we will learn much about the origins of speech, the elaboration of the frontal lobes and the opposable thumb, the advent of upright posture, or the sources of abstract reasoning ability, from a simple genomic comparison of human and chimp. It seems likely that these features and abilities have mainly come from subtle changes—for example, in gene regulation, in the efficiency with which introns are spliced out of RNA, and in protein-protein interactions.... Another half-century of work by armies of biologists may be needed before this key step of evolution is fully elucidated".[17]

Since King and Wilson published their seminal work in 1975,[3] we learned that there are a lot of differences between the two species' genomes, however with only less than 2% sequence divergence. But we do not know whether they are relevant to the phenotypic differences or they resulted just from neutral changes accumulated over 5 million years after divergence of the two species. The question is exactly as Baltimore sees it: one can envisage myriads of subtle differences each contributing somehow to the phenome. We will meet a tremendously complex challenge, especially because all the history of human genetics is largely a history built on analysis of single-gene characters, and that history has influenced thinking in the genetics communities about genetic and molecular mechanisms forming phenotypes (traits). Only recently the community realized (though possibly not quite completely) how inefficient is the present theoretical and experimental armament even in the solution of such a comparatively simple task as identification of human genes responsible for predisposition to such well-known diseases as diabetes, asthma, rheumatoid arthritis, hypertension and the whole bunch of other complex diseases, each of which is just a particular case of one complex trait (see for review ref. 92). Moreover, even in the case of relatively simple, so called monogenic diseases, a closer look revealed a great complexity with multitudes of genes involved and epigenetic effects modifying the trait.[93] And although the application of 'one gene—one trait' approach enabled geneticists to characterize some of the genes responsible for monogenic diseases, these genes are rather rare. The problems faced in studying complex diseases show that 'complexity' is a subject that needs its own operating framework (for further discussion see ref. 94).

What kind of lessons give us all the data obtained from massive genome-based comparative evolutionary analyses, complex trait genetic architecture investigations, attempts to find the laws of converting genotypes into phenotypes and from many other efforts aimed at the understanding of evolutionary mechanisms?

In the concluding paragraphs of their paper, Venter et al[14] made an attempt to give a summary of how we should treat the data. In particular, they stress "...that in organisms with complex nervous systems, neither gene number, neuron number, nor number of cell types correlates in any meaningful manner with even simplistic measures of structural or behavioral complexity. Nor would they be expected to; this is the realm of nonlinearities and epigenesis.... Between humans and chimpanzees, the gene number, gene structures and functions, chromosomal and genomic organizations, and cell types and neuroanatomies are almost indistinguishable, yet the developmental modifications that predisposed human lineages to cortical expansion and development of the larynx, giving rise to language, culminated in a massive singularity that by even the simplest of criteria made humans more complex in a behavioral sense.

Simple examination of the number of neurons, cell types, or genes or of the genome size does not alone account for the differences in complexity that we observe. Rather, it is the interactions within and among these sets that result in such great variation.... Thus, there are no "good" genes or "bad" genes, but only networks that exist at various levels and at different connectivities, and at different states of sensitivity to perturbation".

From the Evo-Devo point of view, to understand differences between us and chimpanzee we will need to reveal the difference in developmental networks between the two species and to understand how the changes were derived from the homologous networks in their common ancestor. For these networks the genome provides only a program, and not a machine.[72] The machine is non-additively composed of many multi-modular systems implementing biochemical functions, conveying signals from inside and outside, monitoring the cell and its genome integrity and so on and so forth.

From all said above it follows inevitably that there are no single genes that make us humans. And when we think that we identified such a gene, it is most probably just one of many participants of an extremely complex network. A point mutation that has recently been found in a gene (*FoxP2*) encoding a transcription factor from affected members of a family with an autosomal dominant pattern of inheritance for specific speech and language impairment (SLI, the diagnosis given when language development goes awry in an otherwise normally developing child[95] ) exemplifies this point. It by no means indicates the gene encoding one of most important human-specific traits that has no equivalent in other primates-language. The phenotype is complex, and although SLI tends to run in families, such pedigrees are an exception rather than a rule. The gene found mutated is common to human and mouse. The fact that most people with SLI do not have this mutation means that SLI is an absolutely typical case of complex traits. However, this finding points to one of the components of a network having, probably, some upstream components affecting the *FoxP2* expression as well as downstream components influenced by FOXP2 transcription factor. An analysis of this network will help to identify target genes that play a role in development of the neural circuitry involved in language.

This example also presents one of the paradigms acceptable for identification of other components of this or other networks. It can be characterized as a genetic approach, that is the identification of mutations involved in the process of abnormal neurological development in a congenital disorder that affects high-level cognitive function, about which we know remarkably little.

This paradigm should be used in combination with the whole genome identification of differences between human and great apes, both in the genome structures (genotyping) and in expression patterns at transcriptome and proteome levels using, for example, microchip technology.

Parallel monitoring of thousands of genes provides a more global view of the system being studied than the classical one gene analysis. This approach again only identifies separate participants of the networks, but further analysis can give more systemic understanding of the differences, especially those involved in brain development and functioning.[96] This will be an important step towards an integrated picture of functional modules responsible for the whole complexity of brain and of the differences that can be considered candidates for an important role in speciation. Among these differences those in REs integrations may also help to identify some nodes of these networks.

## *Aknowledgements*

The work was partially supported by Physico-Chemical Biological Program of the Russian Academy of Sciences, and by the Russian Foundation for Basic Research 2006.200054 grant.

## References

1. Dobzhansky T. Genetics and the Origin of Species. New York: Columbia University Press, 1951.
2. Arthur W. The emerging conceptual framework of evolutionary developmental biology. Nature 2002; 415:757–764,
3. King MC, Wilson AC. Evolution at two levels in humans and chimpanzees. Science 1975; 188:107-116.
4. Darwin C. On the origin of species by means of natural selection or the preservation of favored races in the struggle for life. London: Murray 1859.
5. Dover G. How genomic and developmental dynamics affect evolutionary processes. BioEssays 2000; 22:1153-1159.
6. Plasterk RHA Hershey heaven and Caenorhabditis elegans. Nature Genet 1999; 21:63-67.
7. The C. elegans sequencing consortium genome sequence of the nematode C. elegans: A platform for investigating biology. Science 1998; 282:2012-2018.

8. Kitano H. Systems biology: a brief overview. Science 2002; 295:1662-1664.

9. Mouse genome sequencing consortium. Initial sequencing and comparative analysis of the mouse genome. Nature 2002; 420:520-562.

10. Aparicio S, Chapman J, Stupka S et al. Whole-genome shotgun assembly and analysis of the genome of Fugu rubripes. Science 2002; 23:1301-1310.

11. The Arabidopsis genome initiative. Analysis of the genome sequence of the flowering plant Arabidopsis thaliana. Nature 2000; 408:796-815.

12. Adams D, Celniker SE, Holt RA. The genome sequence of Drosophila melanogaster. Science 2000; 287:2185-2195.

12a. Goffeau A, Barrell BG, Bussey H et al. Life with 6000 genes. Science 1996; 274:546-567.

12b. Blattner FR, Plunkett III G, Bloch CA et al. The complete genome sequence of Escherichia coli K-12. Science 1997; 277:1453-1462.

13. Cavalier-Smith T. Nuclear volume control by nucleoskeletal DNA, selection for cell volume and cell growth rate, and the solution of the DNA C-value paradox. J Cell Sci 1978; 34:247-278.

14. Venter JC, Adams MD, Myers EW et al. The sequence of the human genome. Science 2001; 291:1304-1351.

15. Lander ES, Linton LM, Birren B et al. Initial sequencing and analysis of the human genome. Nature 2001; 409:860-921.

16. Claverie JM. What if there are only 30 000 human genes. Science 2001; 291:1255-1256.

17. Baltimore D. Our genome unveiled. Nature 2001; 409:814-816.

18. Goldenfeld N, Kadanoff LP. Simple lessons from complexity. Science 1999; 284:87-89.

19. Weng G, Bhalla US, Iyengar R. Complexity in biological signalling systems. Science 1999; 284:92-96.

20. Koch C, Laurent G. Complexity and the nervous system. Science 1999; 284:96-98

21. Raff RA, Kaufman TC. Embryos, genes, and evolution. New York: Macmillan, 1983.

22. Alberts B, Bray D, Lewis J et al. Molecular biology of the cell. New York-London: Garland Publishing Inc, 1983.

23. Brett D, Pospisil H, Valcárcel J et al. Alternative splicing and genome complexity. Nature Genet 2001; 30:29-30.

24. Galas DJ. Making sense of the sequence. Science 2001; 291:1257-1260.

25. Fields S. Proteomics in genomeland. Science 2001; 291:1221-1224.

26. McKinsey TA, Zhang CL, Lu J, Olson EN. Signal-dependent nuclear export of a histone deacetylase regulates muscle differentiation. Nature 2000; 408:106-111.

27. Hammond SM, Caudy AA, Hannon GJ. Post-transcriptional gene silencing by double-stranded RNA. Nature Rev Genet 2001; 2:110-119.

28. Eddy SR. Non-coding RNA genes and the modern RNA world. Nature Rev Genet 2001; 2:919-929.

29. The FANTOM consortium and the RIKEN genome exploration research group phase 1&II team, analysis of the mouse transcriptome based on functional annotation of 60 770 full-length cDNAs. Nature 2002; 420:563-573.

30. Storz G. An expanding universe of noncoding RNAs. Science 2002; 296:1260-1263.

31. Rougvie AE. Control of developmental timing in animals. Nature Rev Genet 2001; 2:690-701.

32. Reik W, Dean W. DNA methylation and mammalian epigenetics. Electrophoresis 2001; 22:2838-2843.

33. Lyko F. DNA methylation learns to fly. Trends Genet 2001; 17:169-172.

34. Marin I, Siegal ML, Baker BS. The evolution of dosage-compensation mechanisms. BioEsays 2000; 22:1106-1114.

35. Duboule D, Wilkins AS. The evolution of bricolage. Trends Genet 1998; 14:54-59.

36. Mannervik M, Nibu Y, Zhang H. Transcriptional coregulators in development. Science 1999; 284:606-609.

37. Valentine SA, Chen G, Shandala T et al. Dorsal-mediated repression requires the formation of a multiprotein repression complex at the ventral silencer. Mol Cell Biol 1998; 18:6584-6594.

38. Mannervik M. Corepressor proteins in Drosophila development. Curr Top Microbiol Immunol 2001; 254:79-100.

39. Hervig S, Strauss M. The retinoblastoma protein: a master regulator of cell cycle, differentiation and apoptosis. Eur J Biochem 1997; 246:581-601.

40 Lohrum MA, Vousden KH. Regulation and function of the p53-related proteins: Same family, different rules. Trends Cell Biol 2000; 10:197-202.

41. Brivanlou AH, Darnell JE Jr. Signal transduction and the control of gene expression. Science 2002; 295:813-818.

42. Jacob F. Evolution and tinkering. Science 1977; 196:1161-1166.

43. Chervitz SA, Aravind L, Sherlock G et al. Comparison of the complete protein sets of worm and yeast: orthology and divergence. Science 1998; 282:2022-2028.

44. Ouzounis C. Orthology: Another terminology muddle. Trends Genet 1999; 15:445.

45. Bork P, Dandekar T, Diaz-Lazcoz Y et al. Predicting function: from genes to genomes and back. J Mol Biol 1998; 283:707-725.

46. Theißen G. Orthology: Secret life of genes. Nature 2002; 415:741.

47. Long M. Evolution of novel genes. Curr Opin Genet Dev 2001; 11:673-680.

48. Rubin GM. Comparing species. Nature 2001; 409:820-821.

49. Ohno S. Gene duplication and the uniqueness of vertebrate genomes circa 1970-1999. Semin Cell Dev Biol 1999; 10:517-522.

50. Sankoff D. Gene and genome duplication. Curr Opin Genetics Dev 2001; 11:681-684

51. Hughes AL, da Silva J, Friedman R. Ancient genome duplications did not structure the human Hox-bearing chromosomes. Genome Res 2001; 11:771-780.

52. Massingham T, Davies LJ, Lio P. Analysing gene function after duplication. BioEssays 2001; 23:873-876.

53. Force A, Lynch M et al. Preservation of duplicate genes by complementary degenerative mutations. Genetics 1999; 151:1531-1545.

54. Lynch M, Conery J. The evolutionary fate and consequences of duplicate genes. Science 2000; 290:1151-1155.

55. Prak ETL, Kazazian HH. Mobile elements and the human genome. Nature Rev Genet 2000; 1:134-144.

56. Kidwell MG, Lish DR. Transposable elements and host genome evolution. Trends Ecol Evol 2000; 15:95-99.

57. Hurst GD, Werren JH. The role of selfish genetic elements in eukaryotic evolution, Nature Rev Genet 2001; 2:597-606.

58. Aquadro CF, DuMont VB, Reed FA. Genome-wide variation in the human and fruitfly: a comparison. Curr Opin Genetics Dev 2001; 11:627-633.

59. Stoneking M. Single nucleotide polymorphisms: from the evolutionary past. Nature 2001; 409:821-822.

60. Stephens JC, Schneider JA, Tanguay DA et al. Haplotype variation and linkage disequilibrium in 313 Human Genes. Science 2001; 293:489-493.

61. Rutherford SL. From genotype to phenotype: buffering mechanisms and the storage of genetic information. BioEssays 2000; 22:1095-1105

62. Schrödinger E. What is life? The physical aspect of the living cell. London: Cambridge University Press, 1944, 1955.

63. Holland PWH. The future of evolutionary developmental biology. Nature 1999; 402:41-44.

64. Eizinger A, Jungblut B, Sommer RJ. Evolutionary change in the functional specificity of genes. Trends Genet 1999; 15:197-202.

65. Raff R A. Evo-devo: the evolution of a new discipline. Nature Rev Genet 2000; 1:74-79.

66. Fraser HB, Hirsh AE, Steinmetz LM et al. Evolutionary rate in the protein interaction network. Science 2002; 296:750-752.

67. Hartwell LH, Hopfield JJ, Leibler S. From molecular to modular cell biology. Nature 1999; 402:C47-C52.

68. Pawson T, Saxton TM. Signalling networks—Do all roads lead to the same genes? Cell 2000, 97:675-678.

69. Gilbert SF, Bolker JA. Homologies of process and modular elements of embryonic construction. J Exp Zool 2001; 291:1-12.

70. Kirschner M, Gerhart J. Evolvability. Proc Natl Acad Sci USA 1998; 95: 8420- 8427.

71. Ganfornina MD, Sanchez D. Generation of evolutionary novelty by functional shift. BioEssays 1999; 21:432-439.

72. Davidson EH, Rast JP, Oliveri P et al. Regulatory network for development. Science 2002; 295:1669-1678.

73. Niiehrs C, Pollet N. Synexpression groups in eukaryotes. Nature 1999; 402:483-487.

74. West-Eberhard MJ. Evolution in the light of developmental and cell biology, and vice versa. Proc Natl Acad Sci USA 1998; 95:8417-8419.

75. Gaudet J, Mango S. E. Regulation of organogenesis by the Caenorhabditis elegans FoxA protein PHA-4. Science 2002; 295:821-825.

76. Arnon MI. Bringing order to organogenesis. Nature Genet 2002; 30:348-350.

77. Szabo SJ, Sullivan BM, Stemmann C et al. Distinct effects of T-bet in $T_H1$ lineage commitment and IFN-$\gamma$ production in CD4 and CD8 T cells. Science 2002; 295:338-342.

78. Rutherford SL, Lindquist S. Hsp90 as a capacitor for morphological evolution. Nature 1998; 396:336-342.

79. Sverdlov ED. Retroviruses and primate evolution. BioEssays 2000; 22:161-171.

80. Chen FC, Li WH. Genomic divergences between humans and other hominoids and the effective population size of the common ancestor of humans and chimpanzees. Am J Hum Genet 2001; 68:444-456.

81. Hacia JG. Genome of the apes. Trends Genet 2001; 17:637-645.

82. Britten RJ. Mobile elements inserted in the distant past have taken on important functions. Gene 1997; 205:177-182.

83. Kidwell MG, Lisch D. Transposable elements as sources of variation in animals and plants. Proc Natl Acad Sci USA 1997; 94:7704-7711.

84. Buzdin A, Ustyugova S, Gogvadze E et al. A new family of chimeric retrotranscripts formed by a full copy of U6 small nuclear RNA fused to the 5' terminus of L1. Genomics 2002; 80:402-406.

85. Kowalski PE, Mager DL. A human endogenous retrovirus suppresses translation of an associated fusion transcript, PLA2L. J Virol 1998; 72:6164-6168.

86. Schulte AM, Lai S, Kurtz A et al. Human trophoblast and choriocarcinoma expression of the growth factor pleiotrophin attributable to germ-line insertion of an endogenous retrovirus. Proc Natl Acad Sci USA 1996; 93:14759-14764.

87. Schulte AM, Wellstein A. Structure and phylogenetic analysis of an endogenous retrovirus inserted into the human growth factor gene pleiotrophin. J Virol 1998; 72:6065-6072.

88. Schulte AM, Malerczyk C, Cabal-Manzano R et al. Influence of the human endogenous retrovirus-like element HERV-E.PTN on expression of the growth factor pleiotrophin: a critical role of a retroviral Sp1-binding site. Oncogene 2000; 19:3988-3998.

89. Harris JR. Placental endogenous retrovirus (ERV): structural, functional, and evolutionary significance. BioEssays 1998; 20:307-316.

90. Boyd MT, Bax CM, Bax BE et al. The human endogenous retrovirus ERV-3 is upregulated in differentiating placental trophoblast cells. Virology 1993; 196:905-909.

91. Gibbons A. Which of our genes make us Human? Science 1998; 281:1432-1434.

92. Weissman S.M. Genetic bases for common polygenic diseases. Proc Natl Acad Sci USA 1995; 92:8543-8544.

93. Scriver CR, Waters PJ. Monogenic traits are not so simple. Lesson from phenylketonuria. Trends Genet 1999; 15:267-272.

94. Weiss KM, Terwilliger JD. How many diseases does it take to map a gene with SNPs? Nature Genet 2000; 26:151-157.

95. Bishop DVM. Putting language genes in perspective. Trends Genet 2002; 18:57-59.

96. Geschwind DH. Mice, microarrays, and the genetic diversity of the brain. Proc Natl Acad Sci USA 2000; 97:10676-10678.

# Complex Genome Comparisons:

## Problems and Approaches

Natalia E. Broude and Eugene D. Sverdlov

*Nothing is good or bad but by comparison.*
Thomas Fuller, 'Gnomologia'

## Abstract

Fast progress in human and other genome sequencing has created a foundation for global functional analysis of complex genomes. Genome-wide interindividual, population, and interspecies comparisons of genome variability and instability provide most valuable approaches for deciphering spatial and temporal regulatory networks. Importantly, comparative genomics allows also for studies of biological evolution. Genome-wide comparisons are also indispensable for understanding the genetic and environmental contributions to complex diseases afflicting modern society. In this review we discuss a variety of available approaches to genome-wide complex genome comparisons, their informativeness, advantages, and shortcomings. It is evident that new, more powerful methods are in critical demand for comprehensive comparisons of complex genomes.

## Introduction: Complex Genomes—The Ocean of Cryptic Information

First draft sequencing[a] of the three billion base human genome has recently been finished.[1,2] The sequence obtained as a result of enormous international efforts is just one of a huge number of variants united by a common motif: on average each of us differs at one per 1000 nucleotides.[3] This enormous variability is a major factor responsible for the peculiarities and uniqueness of each human being. The knowledge of DNA polymorphisms is paramount to answer the most important practical question: which sequence variations put borders between health and disease. Without this knowledge it is also impossible to address the fundamental biological questions such as: how far sequence diversions can extend and still maintain species rank, and what variability determines species differences?

However, despite all the achievements we still know too little about the human genome. Even such a crucial question as how many genes there are is very far from clear, saying nothing about functions of most human genes, and the complexity of the entire functional network programmed by the human genome only begins to be revealed and recognized.[4,5] Speaking of diseases, it is now appreciated that in reality, even monogenic diseases are complex, being

---

[a] When this chapter was finished a complete sequence of the genome was reported; http://www.sangcr.ac.uk//Info/Press/2003/030414.shtml

---

*Retroviruses and Primate Genome Evolution*, edited by Eugene D. Sverdlov.
©2005 Eurekah.com.

dependent on environment and on other genes modifying expression of the main causative gene.[6] And, naturally, we know much less about the mechanisms underlying complex diseases.

Oceans of cryptic information encoded by the genome can not be understood without uncovering differences in various genome structures in correlation with the corresponding differences in the phenotypes. When working with complex genomes and complex traits we meet acute shortage of experimental approaches. Many of the approaches, which were quite successful at deciphering 'one gene, one function' relationship, fail when applied to complex trait analyses. Unfortunately, there is no universal sequencing method for comparison of complex genomes and finding all the differences in one shot. However, a handful of different approaches of varying power have been developed. When used in a proper combination, they can give considerable information about similarities and differences among various genomes and their particular areas. Being by no means comprehensive, it still highlights many important aspects of genome variability and can serve as a solid basis for correlation of this variability with phenotype differences. In this review we give an outline of several techniques for comparative analyses of complex genomes and describe their advantages and limitations. We focus on methods allowing genome-wide comparisons, which can be applied to analysis of common allele variability (polymorphisms) as well as used for detection of de novo mutations. Special attention is devoted to suppression PCR, microarrays and subtractive hybridization. The methods analyzing mutations in rather small DNA fragments and known to be involved in disease etiology (e.g., multiple allele-specific PCR, mass spectrometry, mutation detection by single strand conformational polymorphism, or denaturing gradient gel electrophoresis) are beyond the scope of this review. The comparisons at a transcriptome level are discussed in Chapter 9.

## Classification of Comparative Techniques

Genomes can be compared at different levels, from cytogenetic analysis of chromosomes to direct comparisons of genomic sequences or their transcripts. In general, comparisons at the level of genomic variations were classified under the term 'genotype' approach.[7] Such methods are most straightforward and logical, however their application to complex genomes is hampered by tremendous complexity of mammalian genome and is feasible if only a certain fraction of the entire genome is exposed for comparison. In other words, some 'simplification' of genome is necessary for comparison.

An alternative to the genotype approach was termed the 'expression' approach.[7] It uses mainly cDNA comparisons and has its own advantages and limitations. The major advantage of expression approach is that cDNA itself represents a highly simplified fraction comprising only about 3-10% of the genome[8] which is expressed in a given cell type. Another advantage is that this fraction is one of the most functionally important, in which variations between individuals play an important role in phenotypic differences. On the other hand, only a small fraction of all genes is expressed in a given type of differentiated cells,[9-11] and one has to analyze about 200 different cell types of a mammalian organism to see the entire picture. Another problem with expression studies is related to varying copy numbers of different gene products in a given cell, which can vary from 0.3 to about 10,000 mRNA copies per cell.[11] The rare transcripts can be functionally very important, however they represent the fraction that is the most difficult to analyze. In this review, we will not consider the expression approach. Numerous reviews are available that summarize the studies aimed at either development or application of different approaches to gene expression studies (see Chapter 9 and refs. 12-15). Considerably less attention has been paid to the whole genome comparative analyses. As mentioned above, the genomes must be simplified prior to any comparative analysis. The methods used for genome simplification can be divided into two broad categories: (i) 'Pseudo-random' selection that uses genomic fractions selected more or less randomly. It should be noted, however, that there is no way to prepare truly random sets of genomic fragments due to biases of the

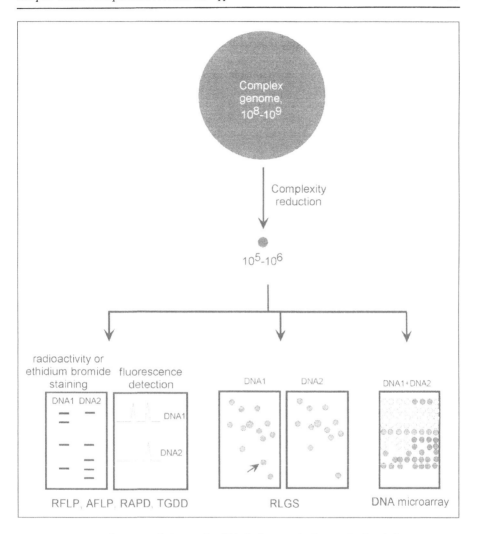

Figure 1. Summary of various techniques using "Display" approach. See text for description.

techniques used for such a selection, and (ii) hypothesis-driven directed selection that uses genomic fractions prepared according to a certain idea or hypothesis which predetermines the method of selection.

All methods for genomic differential analysis, both pseudo-random and directed hypothesis-driven, can be further conditionally divided into two groups. The first group includes methods that are mainly aimed at the detection of differences, whereas their isolation and identification must be done separately. This group can be defined as 'display' methods group (Fig. 1). The second group unites subtractive methods, which are specifically focused on isolation of differences, while all other fragments are discarded. This group includes also the genome mismatch scanning (GMS) method, which opposite to subtractive methods is aimed at isolation of sequences identical in two genomes. The major difference between two groups is that the display methods can simultaneously analyze as many samples as necessary, while the subtractive methods can analyze only two samples (or two groups of samples) at the time.

## Information Content and Resolution Power of Different Methods for Complex Genome Comparisons

Two characteristics are important in comparative display methods: information content and resolution power. The information content determines how large is the fraction of the genome, which is displayed for analysis, while the resolution power determines what type of differences can be detected by a method. It is clear that methods based on separation of genomic fragments are restricted by the resolution power of the separation technique. For example, the PCR-based fingerprinting methods with subsequent polyacrylamide gel electrophoresis (PAGE) display about 20 distinguishable fragments with an average length of 300 base pairs (bp). This allows analysis of about $6 \times 10^3$ bp or only $10^{-6}$ of the human genome. By running in parallel 30-40 different amplification sets, it is possible to increase the information content of one experiment up to $2 \times 10^5$ bp. At the same time, a single two-dimensional (2D) PAGE displays up to 2,000 DNA fragments with average size 300 bp, and analyzes about $6 \times 10^5$ bp. The increase in the amount of the displayed genome sequences is always accompanied by a decrease in resolution power. For example, 2D PAGE cannot detect single nucleotide polymorphisms (SNPs) and short insertions or deletions. In contrast, screening methods using short PCR-amplified fragments and PAGE would detect SNPs and short insertions or deletions, while long-range rearrangements could be missed.

Less evident are the information limits of techniques based solely on hybridization: DNA microarray hybridization and subtractive hybridization. In these cases, the limitations are determined by kinetic peculiarities of hybridization, on one hand, and by the content of each given genomic fragment in the mixture, on the other hand. The complexity of different genomes varies from $10^6$ bp in bacteria to $3 \times 10^9$ bp in humans and reaches $8 \times 10^{11}$ bp in some amphibians.[16] During hybridization in solution, as is the case in most subtractive hybridization methods, the annealing rate of complementary single-stranded DNA fragments follows second order reaction kinetics: it is proportional to the product of concentrations of the interacting single-stranded fragments. Under the same conditions, therefore, the annealing rate of bacterial genomic fragments will be $10^6$ fold higher than that of the single gene fragments obtained from the same mass amount of human genome. In the latter case, the hybridization rate will be too slow to achieve practical yields in reasonable time. This dictates the necessity of using only small portions of complex genomes for hybridization.

The similar kinetic problems are valid for hybridization of the whole-genome-derived fragments with DNA microarrays. Even though in this case kinetics follows the pseudo-first order, the hybridization rate of human genome fragments may still be 1,000 fold slower than that of bacterial fragments. In addition, there is another problem. To be detected as a positive signal, a certain amount of the fragment should hybridize with a probe on the microarray. Again, the relative amount of each fragment in the human genome is 1,000 times lower than in the bacterial genome, and this leads to a proportional decrease of the signal. Currently it is problematic to analyze the whole human genome on an array. Only certain simplified fractions, for example cDNAs, ESTs, or synthetic oligonucleotides, are usually immobilized on microarrays, and preamplified genomic fragments are used for hybridization, though recently an attempt was undertaken (see below) to use the entire genomic DNA for hybridization.[17]

These qualitative considerations show that whatever method is used to compare mammalian genomes, it is informative only when a relatively small fraction of the whole genome is put through the test. Different methodological approaches allow the reduction of complexity either randomly or sequence-specifically. Figure 1 schematically outlines genome complexity reduction and summarizes the display methods discussed below.

## Random Displays: From Fingerprinting to Flowthrough-Based Assays

Most methods for genome comparisons use electrophoresis for separation of the DNA fragments. However, the techniques that allow highlighting only a portion of DNA fragments from the entire multitude of genomic fragments vary, and use mainly three general approaches: (i) revealing a subset of DNA restriction fragments by hybridization with a probe representing a certain genome fraction; (ii) PCR amplification of genomic DNA restriction fragments using nonspecific adapters ligated to all genomic fragments. Since not all fragments are amplified with the same efficiency, only a subset of all possible fragments is amplified after multiple PCR cycles; additionally, the DNA fragments can be obtained by direct PCR amplification of a certain genome fraction with appropriate sets of primers. Below follows a description of the most frequently used techniques of the comparative display.

### *Restriction Landmark Genome Scanning (RLGS)*

This is a fingerprinting technology based on 2-dimensional (2D) gel electrophoresis. Radioactively labeled DNA fragments are prepared by sequential digestion of genomic DNA with three different restriction nucleases.[18-20] The first digestion uses a rare cutting restriction enzyme (usually NotI), and the fragments are radioactively labeled at the restriction site. After digestion with the second enzyme (usually 6 bp cutter), the fragments are resolved in an agarose gel. This is followed by a third restriction enzyme digestion, which is carried out in the gel. The agarose gel containing digested DNA is polymerized within the polyacrylamide gel, and the genomic fragments are resolved in a perpendicular direction (Fig. 2). 1,000-2,000 different loci can simultaneously be visualized by autoradiography on a single gel. NotI restriction enzyme is methylation sensitive, therefore, this approach enables studying

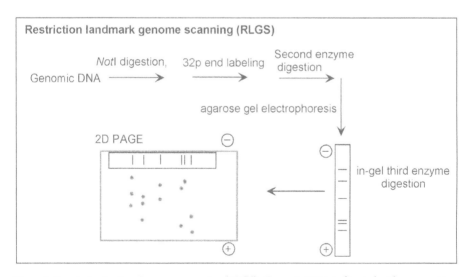

Figure 2. Restriction landmark genome scanning (RLGS). Genomic DNA is digested with a rare cutting restriction enzyme generating protruding ends, and the ends are filled in in the presence of radioactive dNTP. This is followed by a second restriction enzyme digestion (usually 6 bp cutter), and the products are separated by electrophoresis in an agarose gel. A third restriction enzyme (usually 5 or 4 bp cutter) is used to digest DNA fragments within the agarose gel. The agarose strip is polymerized within the acrylamide gel, and restriction fragments are separated by PAGE in the perpendicular dimension. Using human DNA, up to 2,500 distinct spots can be visualized.

DNA methylation. By analyzing spot intensity it is possible to determine whether the locus is diploid, haploid, or amplified.[21] The accuracy of RLGS is partially due to the fact that the DNA fragments are radioactively labeled before gel electrophoresis, and there is no need to perform Southern blotting and hybridization. Because RLGS uses NotI as a first restriction enzyme, it mostly targets CpG islands, which are often located near or within the transcriptional units.[22] This makes RLGS especially valuable in analysis of transcriptionally active genes.[21]

The RLGS method is relatively informative. Indeed, 2,000 resolvable restriction fragments allow surveying about $6 \times 10^5$ bp in a single gel. It is also discriminative, its application to DNA analysis from MZ twins discordant for schizophrenia revealed for the first time genomic differences.[23] The limitations of RLGS include the inability to detect small deletions, insertions and amplifications. It is also experimentally cumbersome, which hinders its application as a routine analytical method.

Another limitation of the RLSG and related fingerprinting techniques is their low sensitivity. To perform a complex genome fingerprinting, one needs 1-10 mkg of DNA, which translates into $10^5$-$10^6$ cells needed for DNA preparation. Since the invention of PCR, several methods have been developed to reduce the necessary amount of DNA by amplification of the whole set of restriction fragments, or by amplification of only a portion of these fragments. This allows analyses with less genomic DNA - down to the DNA from a single cell (see, e.g., ref. 24).

### Amplified Fragment Length Polymorphism (AFLP)

One possibility for whole-genome PCR amplification is to ligate all genomic restriction fragments to adapters, which will make possible simultaneous amplification of many fragments with adapter-specific primers. Several techniques have been published that use ligated adapters. Amplified fragment length polymorphism (AFLP) is one of the widely used methods.[25]

In AFLP, DNA is digested with two restriction enzymes, one rare and one frequent cutter. The restriction fragments are ligated to adapters, which are different for the different enzyme recognition sites. PCR with two primers, complementary to two different adapters, generates sets of amplified fragments, which consist mostly of the nonsymmetrical fragments, containing two different restriction sites at the ends.[25] Most likely, PCR suppression[26,27] inhibits amplification of the 'symmetrical' fragments with the same restriction sites at both ends because these have self-complementary ends (see below). The complexity of the amplicons is further regulated by the attachment of extra 3'-nucleotides to both primers.[25] These extra nucleotides match complementary genomic bases adjacent to the adapter, thus selecting for amplification only those fragments that contain the match. One extra nucleotide per each primer reduces complexity of the amplicons 16-fold, two extra nucleotides per each primer - 256-fold, etc. The AFLP method has been described for radioactive,[25] fluorescent[28] and chemiluminescent[29] detection. A microsatellite-specific variant of AFLP utilizes one AFLP-specific and one microsatellite-specific primer.[30]

The AFLP markers segregate in Mendelian ratios and are used for genetic linkage analyses.[31] The level of observed polymorphism between closely related genotypes is usually higher than for randomly amplified polymorphic DNA (RAPD, see below) and comparable to multilocus analysis with microsatellite markers.[32]

### Arbitrarily Primed PCR

Both RLGS and AFLP require preliminary digestion of DNA with restriction endonucleases. A technique that would allow simultaneous PCR amplification of a multitude of genomic fragments directly from undigested genomic DNA would be greatly advantageous. The original idea of direct DNA amplification is very simple: short oligonucleotides, used as PCR primers under nonstringent annealing temperatures, have a good chance to find complementary targets on both DNA strands not far from each other, thus, making possible amplification of DNA fragments between the two targeted sites. There should be enough coincidences so that a fingerprint with amplicons of different lengths can be obtained.

Arbitrarily primed PCR (AP-PCR),[33] randomly amplified polymorphic DNA (RAPD),[34] and DNA-amplified fragments,[35] are PCR-based methods using short arbitrary primers, usually 8-10 mers, and low stringency PCR. Genomic DNA is directly PCR amplified, and PCR products are separated on acrylamide or agarose gels. The polymorphic bands are revealed by their different mobility. This technique is simple, requires a small amount of DNA, and generates fingerprints containing from a few to several tens of fragments depending on the primer used. The well-recognized drawback of the original RAPD technique is its low reproducibility and, as a result, high extent of false polymorphisms.[36]

Since the RAPD outset in 1990, a number of modifications to the basic procedure have been published. Longer oligonucleotides have been proposed as primers in RAPD with low annealing temperature during the first cycles followed by higher stringency PCR.[37,38] Alternatively, short oligonucleotides with modified bases, which have stronger binding affinity, have been proposed.[39] Several attempts have been made to make RAPD a sequence-specific method. Primers designed for conserved promoter elements and sequences encoding protein motifs,[40] as well as primers containing di- and trinucleotide repeats[41] have been used for this purpose. Sequence analysis of amplified fragments, however, revealed low efficiency of such targeting since only 10-20% of the fragments contained targeted sequences.[42,43] This means that under the low stringency conditions of PCR, the primers work to a great extent as random primers and do not selectively amplify targeted sequences. Some improvements have been suggested to modulate the complexity of the amplified patterns.[44] For most RAPD markers it is impossible to distinguish heterozygous from homozygous genotypes. Nevertheless, Mendelian segregation has been demonstrated in some cases[34] and used to construct genetic maps in a variety of species.

Substantial improvements of the original RAPD methods were obtained by using automatic sequencing instruments and fluorescence detection,[45,46] which greatly increased sensitivity and throughput of the method.

One important result obtained by using RAPD is the detection of microsatellite instability in sporadic and hereditary nonpolyposis colon cancer.[47,48] This example demonstrates the utility of RAPD in the best way. The changes in the number of repeating units of hundreds of thousands of microsatellites in a variety of tumors were detected as multiple polymorphisms. In this case, the number of true polymorphisms was so high that it was not affected by the false positive signals common in RAPD.

All the techniques described above display pseudo-random sets of fragments. The fingerprint is very informative if one needs a 'yes' or 'no' answer to the question of whether the genomes under comparison are identical or different (e.g., in a case of forensic identification), or whether they are related (e.g., in a case of paternity interrogation or linkage studies). In such an approach, the location of the loci represented by the displayed bands is irrelevant and remains unknown. Additional experiments are needed to isolate and identify the differences. For other applications, it would be extremely useful to somehow predetermine the set of genomic sites used for the display. Such a goal is at least partially achieved using the targeted fragment selection methods discussed in the next section.

## Display Methods Targeted at the Flanks of Interspersed Repetitive Elements

In targeted analysis, genomic DNA fragments are selected sequence-specifically. The first example of targeted PCR amplification was interspersed repetitive sequence PCR (IRS-PCR). In IRS-PCR, specifically designed primers target the most conservative sequences in repeats (Alu or L1) and amplify sequences between two closely located repeats if they are in a tail-to-head position.[49,50] High-resolution gel electrophoresis of repeat-specific products generated from two or several different DNAs allows visualization of polymorphisms that segregate in Mendelian manner. Because the repeats differ substantially between species, IRS-PCR is species-specific, and sets of species-specific fragments can be obtained from hybrid cell lines containing heterologous chromosomes (human/rodent, for example). The gel analysis of IRS-PCR products provides a rapid method for identifying and monitoring the chromosomal content of somatic hybrid cells without conventional cytogenetic analysis. One of the widest applications of IRS-PCR is the construction of contigs and maps.[51-54] IRS-PCR reveals only those genomic fragments that are located between closely spaced repetitive elements so that the DNA fragment to be amplified is within the range of PCR amplification capacities. That puts strict limits on the inter-genomic comparison and allows analysis of only a small part of the genomes.

Several methods have been developed to display other types of short interspersed repeats.[41,55,56] Special attention has been paid to trinucleotide repeats due to their role in neurological diseases.[57] One of the methods is based on PCR suppression (PS)[26,27] and is called Targeted Genomic Differential Display method (TGDD).[58,59]

An outline of the PS principle is presented in Figure 3. Genomic DNA is digested with a restriction enzyme, and the resulting fragments are ligated to specially designed oligonucleotide adapters. The adapters are about 40 nucleotides (nt) long with a high GC content and 5'-protruding termini. After ligation and filling in the ends, each single-stranded DNA fragment acquires self-complementary ends, which enables formation of stem-loop structures during the PCR annealing step.

Formation of stem-loop structures makes PCR completely dependent on the target-primers (T-primers) complementary to the target sequences located in the single-stranded portions of the stem-loop structures. Indeed, the adapter-primers alone (A-primers) are inactive in PCR because of the formation of strong duplex stems by the self-complementary ends. The intra-molecular annealing of the complementary ends is kinetically preferable as compared to the annealing of shorter A-primers, and the intramolecular duplexes formed are more stable. Therefore, the A-primer cannot compete with the fragment ends for the same binding site and is not efficient in PCR. The inefficiency of the A-primers in PCR is called PCR suppression.[27]

In the presence of the A- and T-primers, however, PCR is efficient. Indeed, the T-primer anneals to its target and is used by DNA polymerase for initiation of the DNA synthesis. A newly synthesized PCR fragment has two termini, which are not self-complementary and, thus, cannot form stem-loop structures. This product is not subject to the PS effect and is efficiently amplified. As a result, only the fragments containing the target are exponentially amplified by PCR, while the fragments without the targets remain intact. In other words, this ensures high selectivity of PS PCR.

The PS effect allows selectively amplify a set of genomic DNA fragments containing repeats using a single primer targeting repetitive sequences. Therefore this method was called targeted genomic differential display (TGDD). The resulting PCR products contain a portion of the targeted repeat and one of the flanking sequences (Fig. 3). The complexity of the amplified fragments is modulated by 3'-anchoring of the A- and T-primers.[58-59] Sequence analysis of the PCR products revealed that more than 90% of the clones contained the targets.[58-59] This

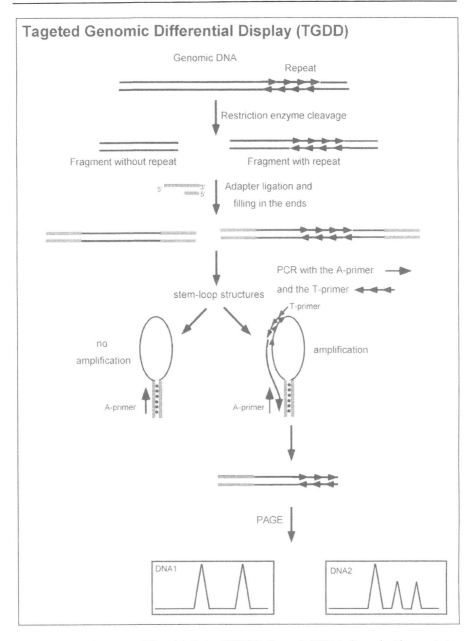

Figure 3. Targeted genomic differential display (TGDD). Genomic DNA is digested with a restriction enzyme and ligated with long GC-rich adapters. This produces strong base pairing of the self-complementary ends of single-stranded (ss) fragments, thus converting all ss DNA fragments to hairpin structures. PCR of such DNA fragments using only the primer corresponding to the adapter (A-primer) is suppressed. PCR is efficient, however, with two primers, one primer complementary to a target (T-primer) located within the ss loop of a hairpin structure, and one A-primer. If the T-primer is targeted at repeated sequences, multiple fragments are amplified. PAGE analysis reveals fingerprints specific for different DNAs and reflecting inter-individual polymorphisms.

high specificity of TGDD can be explained in part by a stem-loop shape of the target oligo-nucleotides, which provide enhanced specificity in hybridization with matched oligonucle-otides.[60-62]

Because the targeted repeats are highly polymorphic, this method displayed high discrimi-nation power. It was applied for analysis of differences in CAG-trinucleotide repeats and inser-tion sites of human endogenous retroviral long terminal repeats (LTRs) in DNAs of monozy-gotic (MZ) twins.[58-59] Both types of sequences in MZ twin DNAs showed rare, reproducible differences on the background of very similar patterns.

TGDD can be adapted for the display of any type of repeats. It can also be applied for methylation studies if the restriction enzyme used for DNA digestion is methylation sensitive. Recently, TGDD was combined with subtractive[63] and with array hybridization[64] to identify human specific integrations of endogenous retroviral LTRs. These enhanced methods were successfully applied for large-scale comparisons of repeat sequences in human and chimpanzee genomes. Such studies allowed isolation of new human-specific sites with integrated endog-enous retroviruses.

## DNA Microarrays As a New Tool to Display Genomic Differences

The DNA microarray technology allows automated high-throughput analyses of complex genomes. This technique is developing with an unparalleled speed, and the number of studies using this technology is rapidly increasing. Different aspects of microarray technique have been comprehensively discussed in several reviews (see e.g., refs. 65,66).[b] Here, we present a brief description of the basics of array technology in an attempt to place this technique into the context of other comparative genomic methods. Application of DNA microarrays to analyses of expression patterns of cells and tissues are briefly described in Chapter 9.

The DNA array approach is based upon physical hybridization of a DNA target with complementary oligo- or polynucleotide probes immobilized on a surface.[67] Thus, it is a vari-ant of filter dot-blot hybridization widely used in molecular biology for a long time. The cur-rent microarray technology is, however, much more sophisticated and enhanced with the use of modern methods of solid surface chemistry, microdispensing techniques, fluorescence de-tection, and computer software for microarray design, storing, and analyzing the data. Se-quencing by hybridization (SBH) was originally introduced as a method for fast high-throughput de novo sequencing.[68-70] The routine use of this technique for de novo sequencing is, however, hindered by many complications caused by kinetics and thermodynamics of oligonucleotide hybridization. As a result, the application of oligonucleotide microarrays has shifted to com-parative genomic studies, and sequencing is restricted mostly to resequencing.

One of the first successful examples of oligonucleotide microarray applications for muta-tion studies was the resequencing of the human mitochondrial genome (16.6 kb) in a large number of samples.[71] In this study, 135,000 unique 25-mers were put on a microarray. Analy-sis of mitochondrial DNA from 10 individuals revealed 505 polymorphisms, and sequencing was performed with 99% accuracy. This study demonstrated that the microarray approach could be much faster in polymorphisms detection than conventional sequencing methods.

An example of heterozygous DNA analysis was a study of mutations in the 3.43 kb exon 11 of the BRCA-1 breast and ovarian cancer gene.[72] This study (as well as human mitochon-drial DNA resequencing[71]) used so-called variant detection array (VDA). VDA is designed in such a way that a query base is centered in a middle position in a 25 mer, while all upstream and downstream flanking nucleotides are identical. A VDA contains all four variants of queried

---

[b] The most recent surveys of the DNA-microarray technique and its applications can be found in Supple-ment to Nature Genetics, 32, December 2002.

base, and the perfect match in a reference sequence generates a stronger signal relative to the three mismatches. Approximately 100,000 oligonucleotide probes were placed on a VDA, corresponding to the wild type and to the known mutations involved in cancer predisposition. The microarray was hybridized with an equimolar mixture of in vitro transcribed RNAs obtained from the preamplified PCR products from normal and cancer cells labeled with two different fluorescent labels. Cohybridization of a reference and test sequences labeled with different fluorophores had been developed earlier for so-called comparative genomic hybridization (CGH).[73,74] Direct comparison of the ratio of hybridization signal from two labels is applied to detect the mutation. Fourteen of fifteen known heterozygous mutations were correctly identified, and no false positive mutations were identified in 20 control samples. This study developed the strategy for analysis of heterozygous mutations and demonstrated the advantages and limitations of this technology.[72]

Later, the technique was refined substantially and was used for mutation screening in coding regions of several loci. In particular, 62 exons of the ATM gene responsible for ataxia telangiectasia,[75] and all 22 BRCA1 coding exons[76] were studied using two-color cohybridization experiments. One hundred forty nine chip designs, each containing 150,000 – 300,000 oligonucleotides, were used to screen 2.3 Mb of human genome sequence and 3,241 polymorphisms were detected.[3] These studies clearly demonstrated feasibility of DNA microarrays for large-scale screening for human variations. There is no doubt that many areas including population genetics and evolution studies will greatly benefit from this technique.[77]

Despite the impressive examples of oligonucleotide microarray applications, there are limitations to their routine use. The published data on SNP detection from Lander's group reported a 45% false positive rate using Affymetrix microarrays.[78] Therefore, an independent verification of new sequence variants is necessary. Insertions, deletions, and rearrangements are still a challenge for this technology.[76] One reason for a low fidelity of arrays is the dependence of the hybridization efficiency on the oligonucleotide base content and sequence: the GC-rich duplexes form much stronger hybrids than those with high AT content. This poses problems in signal discrimination between the mismatched GC-rich sequences and the perfectly matched AT-rich sequences. Another problem inherent to DNA chips is the effect of target DNA secondary structure. This could prevent hybridization of targets that form stable secondary structures. Hairpins, G-quartets, and bulged structures all pose severe problems in hybridization-based techniques.[76,79] Existence of repetitive elements is another unsolved problem for the DNA microarray approach. In addition to their tendency of forming various nonconventional secondary structures, it is difficult to discriminate repeats belonging to different loci.

Several approaches have been proposed to partially alleviate limitations of DNA microarrays. Chemical modifications of both the bases[80,81] and the backbone[82] to increase duplex stability have been suggested. Application of enzymatic reactions, such as ligation and primer extension, was suggested as another approach to partially overcome the drawbacks of array technology.[83-85] The high fidelity of the enzymatic reactions in discrimination of the matched sequences, in combination with enhanced stacking interactions, add new dimensions to conventional DNA array methods.[85-87] Double-probe arrays contain two different probes in every location of the microarray.[88] The presence of two probes allows enhanced hybridization of the target, if the target contains fragments complementary to both probe sequences. Hybridization of the physically linked sequences, even if they are separated, is preferable to hybridization of two separate fragments, each bearing one region of complementarity. New ideas to improve hybridization fidelity and mismatch discrimination also include application of an electric field to control and accelerate hybridization[89-91] and application of duplex DNA probes.[85,92]

Many efforts have been devoted to development of formats with long immobilized probes, cDNAs, PCR fragments, or BACs, to perform comparisons of complex genomes. It should be emphasized, however, that even in the case of the longest probes, it is far away from reaching full-genome coverage. Several examples of these studies are based on CGH and are aimed at detection of gross differences between genomes.[17,93,94] In the DNA array-CGH method, two genomic DNAs are differentially labeled with fluorophores and are cohybridized to the array. The hybridization pattern provides a locus-by-locus DNA copy number variation resulting from fluorescence ratios at arrayed DNA elements.[93-95]

The cDNA microarray with about 5,000 immobilized fragments was applied for quantitation of genomic differences DNA in cell lines with known chromosome copy number.[17] Breast cancer cell lines and tumors were also analyzed to detect known genomic deletions and amplifications. The authors were able to detect single-copy deletions with 85% sensitivity and 85% specificity. The technique allowed the detection of two-fold and smaller differences in DNA copy number (two-fold in XO versus XX, 1.5-fold in XXX versus XX), which is an indication of the ability to localize and identify new functionally important genes (e. g., tumor suppressor genes). The authors also detected the expected amplifications and deletions in MYC, ERB2, and TP53 loci in DNA from tumor cell lines. An important advantage of cDNA microarray-based CGH is that DNA copy number and gene expression can be characterized in parallel, thus making possible correlation between the genomic content and the functional fraction of the genome.

The ability of cDNA microarrays to scan complex genomes for DNA copy-number variations determined its applications in a variety of studies.[96-99] Most of them were aimed at correlation of gene copy number and gene expression levels in different cancers.

Several limitations, however, are inherent to array-based comparative genome hybridization. The hybridization interference between the homologous members of the gene families, between repetitive sequences and between some X-chromosomal genes and homologous autosomal sequences may lead to ambiguous results difficult for interpretation. Another principal limitation of this technique is its inability to analyze small genome aberrations like inversions and deletions. It should also be emphasized that, in all cases, the genome complexity is substantially reduced by the array design.

## Genomic Subtractive Hybridization

Two approaches to genome comparisons are based on hybridization of fragmented genomes in solution: subtractive hybridization (SH) and genomic mismatch scanning (GMS). Subtractive hybridization is used both for genomic and cDNAs comparisons, while GMS is applied exclusively to genomic studies. Both techniques are most efficient when applied to comparisons of the closely related genomes.

In subtractive hybridization (see also Chapter 9), both DNAs under comparison are digested to rather short fragments and mixed in a ratio such that one DNA is taken in great excess (it is defined as a driver DNA) over another (defined as a tracer DNA). The mixture is denatured and cooled to reanneal the denatured fragments. The homologous sequences in both tracer and driver DNA form heteroduplexes and, therefore, most of the tracer DNA is primarily hybridized with an excess driver DNA. At the same time, the sequences present in the tracer (defined as targets) but absent from the driver, do not form hybrids with the driver DNA and can be discriminated from the driver-tracer heteroduplexes. The reassociated fragments common for driver and tracer are discarded. The remaining DNA is enriched in target sequences and is cloned and analyzed. The enrichment of the library with the tracer is the ratio of the size of the initial genomic library, (which allows finding a target with a certain probability), to the size of the subtracted library (which allows finding the target with the same probability). The larger excess of the driver over the tracer, the higher is the enrichment of the subtracted library. Lamar and Palmer[100] were the first to apply subtractive hybridization to enrich a mouse library for Y chromosome fragments. Female mice DNA (driver) was randomly cut into fragments, while the male DNA (tracer) was cleaved with a restriction endonuclease MboI. Both DNAs were mixed, denatured and reannealed. The resulting double-stranded DNA included three types of molecules: driver-driver duplexes lacking MboI sticky ends; driver-tracer heteroduplexes, having only one MboI terminus; and reassociated tracer homoduplexes containing sticky MboI ends at both termini. Due to this specific feature, only the reassociated tracer can be selectively cloned into a pBR322 vector. The resulting library was enriched 40-fold with the clones containing Y chromosomal DNA. This work established two basic principles, which were used in all subsequent modifications, before and after the invention of PCR. (1) Hybridization of excess double-stranded driver with tracer was introduced to enrich the reannealed double-stranded tracer fractions in target sequences, and (2) different termini features characteristic of each type of duplexes were used for discrimination of the tracer homoduplexes from those of the driver and from the heteroduplexes. Detailed reviews of the techniques applied both for genomic and cDNA subtractions have been published.[101-103] Here, we will consider only the applications of the method to mammalian genomes and will focus mostly on PCR-based techniques, because they allow working with minimal amounts of DNA.

One of the widely used genomic subtraction schemes is called representational difference analysis (RDA).[104] RDA involves selective PCR amplification of the reannealed double-stranded tracer, while all other types of molecules (single-stranded, hybrid, and driver) remain unamplified. Selective amplification of the tracer was possible due to the specific PCR-adapters ligated to the tester,[105] which played the same role as MboI ends of the reannealed tracer in the work of Lamar and Palmer.[100]

Another important feature of RDA is the complexity reduction of driver and tracer DNAs before subtraction. To this end, both DNAs were cleaved separately with a restriction endonuclease. Each set of fragments was ligated to oligonucleotide adapters and PCR-amplified using adapter-primers. After multiple amplification cycles, the resulting driver and tracer contained only a fraction of the initial pool of fragments due to size-biased PCR amplification. The extent of simplification varied from 10- to 50-fold depending on the restriction enzyme used for cleavage. Most of the PCR products consisted of fragments shorter than 1 kb. The simplified, fragmented genomes, (or amplicons), were used for subtraction. The double-stranded amplified tracer was again subtracted with a new portion of driver, and the procedure was repeated. Each new round of subtraction produced additional enrichment in target sequences not only by virtue of the subtraction itself but also due to the higher reassociation rate of the enriched fragments resulting from second order kinetics. It should be noted that this kinetic component is equally efficient in all protocols that use repeated cycles of subtraction to enrich the double-stranded fraction.[104,106] Taken together, these advances allowed obtaining enrichment of the tracer 'greater than $5 \times 10^6$ fold from starting material, and about $4 \times 10^5$ fold from amplicons'.[104]

Since its introduction, RDA was successfully applied to identify a homozygous deletion in pancreatic carcinoma,[107] to clone probes that detect DNA loss and amplification in tumors,[102] and in many other comparisons of tumor and normal cell genomes. A new technique, genetically directed representational difference analysis (GDRDA), was developed for generating genetic markers linked to a trait of interest.[108] The application of RDA for the detection of losses of heterozygosity (LOH) in cancer is, probably, most fruitful, because in this case the subtraction is carried out using DNA from cancer cells against the DNA from healthy tissues of the same individual. This overcomes many problems related to general polymorphisms, focusing the subtraction only on those changes that occurred due to cancer development. Reviews of RDA techniques have been published.[109-111]

Despite many cases of successful application of RDA and other similar versions of subtraction, a number of problems remain to be solved. The evident drawback of genome simplification is that only a minor fraction of the genome (2-10%) is actually compared, while 90 - 98% remains beyond analysis. Another disadvantage of RDA is that the set of amplicons is not truly random, but rather represents a selection of fragments that could be amplified with the highest efficiency. This may introduce a systematic source of errors. In view of the known 'mosaic' structure of the human genome composed from blocks with different GC content,[112] the question is, whether amplicons equally represent different parts of the genome. The limitations of RDA applied to LOH were analyzed.[107] It was noticed, in particular, that due to the size selection in the process of multiple PCR cycles, DNA regions homozygously deleted in the driver can be missed, if the restriction fragments from the corresponding region of tracer DNA are too large to be amplified by PCR. Performing RDA with several restriction enzymes can solve this problem, but it considerably complicates the technique. All these 'cons' of RDA mean that the isolation of small genome differences might be a matter of luck. Point mutations, small deletions, as well as rearrangements (which do not cause losses or acquisitions of the genetic material, like inversions or chromosome translocations) also remain beyond the reach of this technique. Another problem common to all subtractive techniques is interference between different members of the same gene families that can cross-hybridize with the target, thus masking its deletion from the driver. The extent of such interference depends of the level of homology between the different members and the stringency of the hybridization conditions.

A version of subtraction using in-gel competitive reassociation,[113,114] avoids some shortcomings of subtraction in solution. The technique was developed to isolate orthologous restriction DNA fragments, in which molecular sizes differ in two genomic DNAs. In this technique, driver and tracer DNA are digested with the same restriction nuclease. Biotinylated tracer and dephosphorylated driver (the latter in large excess) are mixed and separated by gel electrophoresis. During electrophoresis, identical orthologous fragments of driver and tracer move with equal mobility and occupy the same positions in the gel, whereas the orthologous fragments of different size are physically separated in the gel. In-gel alkaline denaturation of DNA, followed by reassociation (in-gel competitive reassociation), generates driver/tracer heteroduplexes from identical counterparts, whereas tracer fragments differing from counterpart driver fragments form homoduplexes. DNA is eluted from the gel and ligated with an adapter. Only the tracer homoduplexes, which have ligated adapters at both termini, can be amplified by PCR after separation using biotin/streptavidin technology. By repeating these steps, Inoue et al., succeeded in preparing mouse and human subtracted libraries highly enriched with RFLP clones.[115] The developed approach likely allows avoiding cross-hybridization of paralogous gene family members, because many of DNA fragments belonging to different members of the same gene family are expected to produce restriction fragments of different lengths due to evolutionary divergency. This problem, however, needs further experimental investigation. It is reasonable to suggest that such a technique could easily identify products of large genomic rearrangements (inversions and/or translocations) that result in production of new genomic neighborhoods for the otherwise identical sequences. Indeed, the in-gel competitive reassociation was successfully used to clone DNA fragments with various structural changes from cancer tissues.[116] A similar idea, "RFLP subtraction," was described to isolate a large number of unique RFLPs in a single experiment.[117]

The genomic mismatch scanning (GMS) is also based on formation of heteroduplexes between restriction fragments derived from two genomes under comparison.[118] Opposite to subtractive hybridization, however, it is aimed at enrichment of identical sequences. Sequences that are identical-by-descent are of great importance for allele-sharing methods and are used for genetic dissection of complex traits.[119] Briefly, DNA fragments to be compared are obtained by restriction nuclease digestion. One of the samples is methylated with a DNA methylase. The methylated and nonmethylated fragments are hybridized and homoduplexes that are either completely methylated or completely un-methylated are removed with methylation sensitive restriction nucleases and exonucleases. The mismatch-containing heteroduplexes are also removed with a combination of Esherichia coli mismatch repair proteins and exonuclease. Mismatch-free homoduplexes are isolated and mapped. Originally applied to yeast genomic studies, GMS was used to scan human and mouse genomic DNA for identical sequences.[120-122] In the case of complex genomes, however, this method gives a rather low yield of identical fragments. The method did not find broad application in view of its experimental complexity.

## Conclusion

After completion of the Human Genome Project and with sequencing of growing number of other genomes, the next challenge for genomic studies is to gain a full understanding of the relevance of these enormous structures. Genomics alone is not able to assign functions to genes at the organismal level. As it was excellently observed, "Genome speaks biochemistry, not phenotype".[123] To reveal and understand the entire genomic network, the wide spectrum of approaches and methods will be used, from knock-out and knock-in techniques to in silico sequence comparisons.

Among other methods, comparative genomic studies within species and between species will play extremely important role.[124] Together with parallel expression profiling they will provide important information about the appearance of new and/or modification of old functions as well as about the correlation between genotype and phenotype.[125] Taken together, this information will reveal a global network of complex biochemical processes. To obtain and integrate this information the present day approaches need radical improvements to meet the coming demands of global functional analysis of cells, tissues, organs, and, eventually, organisms. It is likely that the future techniques will be enhanced by the achievements of many disciplines: chemistry, mathematics, physics, bioinformatics, and the emerging field of nano-engineering. New challenges will also drive the development of data management tools, which will use the achievements of such distant fields as meteorology, finance, and online management. Nevertheless, the future comparative methods will, most probably, be still based on the differential display, subtractive hybridization and microarray technologies described in this review, the methods, which highlight important genomic differences or similarities.

### *Acknowledgments*

The authors thank Charles R. Cantor, Vadim V. Demidov, and Chandran Sabanayagam for critical reading of the manuscript and valuable comments. We also apologize to all colleagues whose works were not cited due to a broad scope of this review and limited space. This work was partially supported by a grant from Packard Instrument Company, Inc., (NEB) and by the Russian Foundation for Fundamental Research (EDS).

## References

1. Venter JC, Adams MD, Myers EW et al. The sequence of the human genome. Science 2001; 291:1304-1351.
2. Lander ES, Linton LM, Birren B et al. Initial sequencing and analysis of the human genome. Nature 2001; 409:860-921.
3. Wang DG, Fan JB, Siao CJ et al. Large-scale identification, mapping, and genotyping of single-nucleotide polymorphisms in the human genome. Science 1998; 280:1077-1082.
4. Goldenfeld N, Kadanoff LP. Simple lessons from complexity. Science 1999; 284:87-89.
5. Koch C, Laurent G. Complexity and the nervous system. Science 1999; 284:96-98.

6. Scriver CR, Waters PJ. Monogenic traits are not simple: Lessons from phenylketonuria. Trends Genet 1999; 15:267-272.

7. Brown PO, Hartwell L. Genomics and human disease—variations on variation. Nature Genet 1998; 18:91-93.

8. Nelson PS, Hawkins V, Schummer M et al. Negative selection: A method for obtaining low-abundance cDNAs using high-density cDNA clone arrays. Genet Anal 1999; 15:209-215.

9. Tjian R, Maniatis T. Transcriptional activation: A complex puzzle with few easy pieces. Cell 1994; 77:5-8.

10. Sager R. Expression genetics in cancer: Shifting the focus from DNA to RNA. Proc Natl Acad Sci USA 1997; 94:952-955.

11. Velculescu VE, Madden SL, Zhang L et al. Analysis of human transcriptomes. Nature Genet 1999; 23:387-388.

12. Liang P, Pardee AB. Recent advances in differential display. Curr Opin Immunol 1995; 7:274-280.

13. Matz MV, Lukyanov SA. Different strategies of differential display: Areas of application. Nucleic Acids Res 1998; 26:5537-5543.

14. Ferea TL, Brown PO. Observing the living genome. Curr Opin Genet Dev 1999; 9:715-722.

15. Green CD, Simons JF, Taillon BE et al. Open systems: Panoramic views of gene expression. J Immunol Methods 2001; 250:67-79.

16. Cantor CR, Smith CL. Genomics: The science and technology behind the human genome project. NY: John Wiley and Sons, 1999:26.

17. Pollack JR, Perou CM, Alizadeh AA et al. Genome-wide analysis of DNA copy-number changes using cDNA microarrays. Nature Genet 1999; 23:41-46.

18. Hatada I, Hayashizaki Y, Hirotsune S et al. A genomic scanning method for higher organisms using restriction sites as landmarks. Proc Natl Acad Sci USA 1991; 88:9523-9527.

19. Hirotsune S, Hatada I, Komatsubara H et al. New approach for detection of amplification in cancer DNA using restriction landmark genomic scanning. Cancer Res 1992; 52:3642-3647.

20. Hayashizaki Y, Hirotsune S, Okazaki Y et al. Restriction landmark genomic scanning method and its various applications. Electrophoresis 1993; 14:251-258.

21. Watanabe S, Kawai J, Hirotsune S et al. Accessibility to tissue-specific genes from methylation profiles of mouse brain genomic DNA. Electrophoresis 1995; 16:218-226.

22. Lindsay S, Bird AP. Use of restriction enzymes to detect potential gene sequences in mammalian DNA. Nature 1987; 327:336-338.

23. Tsujita T, Niikawa N, Yamashita H et al. Genomic discordance between monozygotic twins discordant for schizophrenia. Am J Psychiatry 1998; 155:422-424.

24. Cheung VG, Nelson SF. Whole genome amplification using a degenerate oligonucleotide primer allows hundreds of genotypes to be performed on less than one nanogram of genomic DNA. Proc Natl Acad Sci USA 1996; 93:14676-14679.

25. Vos P, Hogers R, Bleeker M et al. AFLP: A new technique for DNA fingerprinting. Nucleic Acids Res 1995; 23:4407-4414.

26. Launer GA, Lukyanov KA, Tarabykin VS. Simple method for cDNA amplification starting from small amount of total RNA. Mol Gen Microbiol Virusol 1994; 6:38-41.

27. Siebert PD, Chenchik A, Kellog DE et al. An improved PCR method for walking in uncloned genomic DNA. Nucleic Acids Res 1995; 23:1087-1088.

28. Roman BL, Pham VN, Bennett PE et al. Nonradioisotopic AFLP method using PCR primers fluorescently labeled with Cy5. Biotechniques 1999; 26:236-238.

29. Lin JJ, Ma J, Kuo J. Chemiluminescent detection of AFLP markers. Biotechniques 1999; 26:344-348.

30. Witsenboer H, Vogel J, Michelmore RW. Identification, genetic localization, and allelic diversity of selectively amplified microsatellite polymorphic loci in lettuce and wild relatives. Genome 1997; 40:923-936.

31. Liu Z, Nichols A, Li P et al. Inheritance and usefulness of AFLP markers in channel catfish (Ictalurus punctatus), blue catfish (I. furcatus), and their F1, F2, and backcross hybrids. Mol Gen Genet 1998; 258:260–268.

32. Mackill DJ, Zhang Z, Redona ED et al. Level of polymorphism and genetic mapping of AFLP markers in rice. Genome 1996; 39:969-977.

33. Welsh J, McClelland M. Fingerprinting genomes using PCR with arbitrary primers. Nucleic Acids Res 1990; 18:7213-7118.

34. Williams JG, Kubelik AR, Livak KJ. DNA polymorphisms amplified by arbilrary primers are useful as genetic markers. Nucleic Acids Res 1990; 18:6531-6535.

35. Caetano-Anolles G, Bassam BJ, Gresshoff PM. DNA amplification fingerprinting using very short arbitrary oligonucleotide primers. Biotechnology 1991; 9:553-557.

36. Ellsworth DL, Rittenhouse KD, Honeycutt RL. Artifactual variation in randomly amplified polymorphic DNA banding patterns. Biotechniques 1993; 14:214-227.

37. Zhao S, Ooi SL, Pardee AB et al. New primer strategy improves precision of differential display. Biotechniques 1995; 18:842-846,848,885.

38. Gillings M, Holley M. Amplification of anonymous DNA fragments using pairs of long primers generates reproducible DNA fingerprints that are sensitive to genetic variation. Electrophoresis 1997; 18:1512-1518.

39. Lebedev Y, Akopyants N, Azhikina T et al. Oligonucleotides containing 2-aminoadenine and 5-methylcytosine are more effective as primers for PCR amplification than their nonmodified counterparts. Genet Anal 1996; 13:15-21.

40. Birkenmeier EH, Schneider U, Thurston SJ. Fingerprinting genomes by use of PCR with primers that encode protein motifs or contain sequences that regulate gene expression. Mamm Genome 1992; 3:537-545.

41. Zietkiewicz E, Rafalski A, Labuda D. Genome fingerprinting by simple sequence repeat (SSR)-anchored polymerase chain reaction amplification. Genomics 1994; 20:176-183.

42. Weising K, Atkinson RG, Gardner RC. Genomic fingerprinting by microsatellite-primed PCR: A critical evaluation. PCR Methods 1995; 4:249-255.

43. Donohue PJ, Hsu DK, Winkles JA. Differential display using random hexamer-primed cDNA, motif primers, and agarose gel electrophoresis. In: Liang P, Pardee A, eds. Differential display: Methods and protocols. Methods in Molecular Biology, v 85. Humana Press, 1997:25-35.

44. Caetano-Anolles G, Bassam BJ, Gresshoff PM. DNA amplification fingerprinting using very short arbitrary oligonucleotide primers. Biotechnology (NY) 1994; 9:553-557.

45. Ito T, Kito K, Adati N et al. Fluorescent differential display: Arbitrarily primed RT-PCR fingerprinting on an automated DNA sequencer. FEBS Lett 1994; 351:231-236.

46. Yasuda J, Kashiwabara H, Kawakami K et al. Detection of microsatellite instability in cancers by arbitrarily primed-PCR fingerprinting using a fluorescently labeled primer (FAP-PCR). Biol Chem 1996; 377:563-570.

47. Peinado MA, Malkhosyan S, Velasquez A. et al. Isolation and characterization of allelic losses and gains in colorectal tumors by arbitrarily primed polymerase chain reaction. Proc Natl Acad Sci USA 1992; 89:10065-10069.

48. Perucho M. Cancer of the microsatellite mutator phenotype. Biol Chem 1996; 377:675-84.

49. Nelson DL, Ledbetter SA, Corbo L et al. Alu polymerase chain reaction: A method for rapid isolation of human-specific sequences from complex DNA sources. Proc Natl Acad Sci USA 1989; 86:6686-6690.

50. Ledbetter SA, Nelson DL, Warren ST. Rapid isolation of DNA probes within specific chromosome regions by interspersed repetitive sequence polymerase chain reaction. Genomics 1990; 6:475-481.

51. Cox RD, Copeland NG, Jenkins NA et al. Interspersed repetitive element polymerase chain reaction product mapping using a mouse interspecific backcross. Genomics 1991; 10:375-384.

52. Hunter KW, Ontiveros SD, Watson ML et al. Rapid and efficient construction of yeast artificial chromosome contigs in the mouse genome with interspersed repetitive sequence PCR (IRS-PCR): Generation of a 5-cM, >5 megabase contig on mouse chromosome 1. Mamm Genome 1994; 5:597-607.

53. Hunter K. Application of interspersed repetitive sequence polymerase chain reaction for construction of yeast artificial chromosome contigs. Methods 1997; 13:327-335.

54. Himmelbauer H, Schalkwyk LC, Lehrach H. Interspersed repetitive sequence (IRS)-PCR for typing of whole genome radiation hybrid panels. Nucleic Acids Res 2000; 28:e7.

55. Lench NJ, Norris A, Bailey A et al. Vectorette PCR isolation of microsatellite repeat sequences using anchored dinucleotide repeat primers. Nucleic Acids Res 1996; 24:2190-2191.

56. Phan J, Reue K, Peterfy M. MS-IRS PCR: A simple method for the isolation of microsatellites. Biotechniques 2000; 28:18-20.

57. Richards RI, Sutherland GR. Dynamic mutation: Possible mechanisms and significance in human disease. Trends Biochem Sci 1997; 22:432-436.

58. Broude NE, Storm N, Malpel S et al. PCR based targeted genomic and cDNA differential display. Genet Anal 1999; 15:51-63.

59. Lavrentieva I, Broude NE, Lebedev Y et al. High polymorphism level in genomic sequences flanking human endogenous retrovirus LTRs insertion sites. FEBS Lett 1999; 443:341-347.

60. Roberts RW, Crothers DM. Specificity and stringency in DNA triplex formation. Proc Natl Acad Sci USA 1991; 88:9397-9401.

61. Bonnet G, Tyag S, Libchaber A et al. Thermodynamic basis of the enhanced specificity of structured DNA probes. Proc Natl Acad Sci USA 1999; 96:6171-6176.

62. Broude NE. Stem-loop DNA oligonucleotide probes: A robust tool for molecular biology and biotechnology. Trends Biotechnol 2002; 20:249-256.

63. Buzdin A, Khodosevich K, Mamedov I et al. A technique for genome-wide identification of differences in the interspersed repeats integrations between closely related genomes and its application to detection of human-specific integrations of HERV-K LTRs. Genomics 2002; 79:413-422.

64. Mamedov I, Batrak A, Buzdin A et al. Genome-wide comparison of differences in the integration sites of interspersed repeats between closely related genomes. Nucleic Acids Res 2002; 30:e71.

65. Hacia JG, Brody LC, Collins FS. Applications of DNA chips for genomic analysis. Mol Psychiatry 1998; 3:483-492.

66. Watson SJ, Meng F, Thompson RC et al. The "chip" as a specific genetic tool. Biol Psychiatry 2000; 48:1147-1156.

67. Southern EM. Detection of specific sequences among DNA fragments separated by gel electrophoresis. J Mol Biol. 1975; 98:503-517.

68. Bains W, Smith GC. A novel method for nucleic acid sequence determination. J Theor Biol 1988; 135:303-307.

69. Lysov YuP, Florent'ev VL, Khorlin AA et al. Determination of the nucleotide sequence of DNA using hybridization with oligonucleotides. A new method. Dokl Akad Nauk SSSR 1988; 303:1508-1511.

70. Drmanac R, Labat I, Brukner I et al. Sequencing of megabase plus DNA by hybridization: Theory of the method. Genomics 1989; 4:114-128.

71. Chee M, Yang R, Hubbell E et al. Accessing genetic information with high-density DNA arrays. Science 1996; 274: 610-614.

72. Hacia JG, Brody LC, Chee MS et al. Detection of heterozygous mutations in BRCA1 using high density oligonucleotide arrays and two-color fluorescence analysis. Nature Genet 1996; 14:441-447.

73. Kallionemi A, Kallioniemi OP, Sudar D et al. Comparative genomic hybridization for molecular cytogenetic analysis of solid tumors. Science 1992; 258:818-821.

74. Forozan F, Karhu R, Kononen J et al. Genome screening by comparative genomic hybridization. Trends Genet 1997; 13:405-409.

75. Hacia JG, Sun B, Hunt N et al. Strategies for mutational analysis of the large multiexon ATM gene using high-density oligonucleotide arrays. Genome Res 1998; 86:1245-1258.

76. Hacia JG. Resequencing and mutational analysis using oligonucleotide microarrays. Nature Genet 1999; 21:42-47.

77. Chakravarti A. Population genetics-making sense out of sequence. Nature Genet 1999; 21:56-60.

78. Cargill M, Altshuler D, Ireland J. Characterization of single-nucleotide polymorphisms in coding regions of human genes. Nature Genet 1999; 22:231-238.

79. Southern E, Mir K, Shchepinov M. Molecular interactions on microarrays. Nature Genet 1999; 21:5-9.

80. Nguyen HK, Bonfils E, Auffray P et al. The stability of duplexes involving AT and/or G4EtC base pairs is not dependent on their AT/G4EtC ratio content. Implication for DNA sequencing by hybridization. Nucleic Acids Res 1998; 26:4249-4258.

81. Hacia JG, Woski SA, Fidanza J et al. Enhanced high-density oligonucleotide array-based sequence analysis using modified nucleoside triphosphates. Nucleic Acids Res 1998; 26:4975-4982.

82. Weiler J, Gausepohl H, Hauser N et al. Hybridisation based DNA screening on peptide nucleic acid (PNA) oligomer arrays. Nucleic Acids Res 1997; 25:2792-2799.

83. Broude NE, SanoT, Smith CL et al. Enhanced DNA sequencing by hybridization. Proc Natl Acad Sci USA 1994; 91:3072-3076.

84. Dubiley S, Kirillov E, Lysov Y et al. Fractionation, phosphorylation and ligation on oligonucleotide microchips to enhance sequencing by hybridization. Nucleic Acids Res 1997; 25:2259-2265.

85. Gunderson KL, Huang XC, Morris MS. Mutation detection by ligation to complete n-mer DNA arrays. Genome Res 1998; 8:1142-1153.

86. Yershov G, Barsky V, Belgovskiy A et al. DNA analysis and diagnostics on oligonucleotide microchips. Proc Natl Acad Sci USA 1996; 93:4913-4918.

87. Parinov S, Barsky V, Yershov G et al. DNA sequencing by hybridization to microchip octa-and decanucleotides extended by stacked pentanucleotides. Nucleic Acids Res 1996; 24:2998-3004.

88. Gentalen E, Chee M. A novel method for determining linkage between DNA sequences: Hybridization to paired probe arrays. Nucleic Acids Res 1999; 27:1485-1491.

89. Edman CF, Raymond DE, Wu DJ et al. Electric field directed nucleic acid hybridization on microchips. Nucleic Acids Res 1997; 25:4907-4914.

90. Sosnowski RG, Tu E, Butler WF et al. Rapid determination of single base mismatch mutations in DNA hybrids by direct electric field control. Proc Natl Acad Sci USA 1997; 94:1119-1123.

91. Gilles PN, Wu DJ, Foster CB et al. Single nucleotide polymorphic discrimination by an electronic dot blot assay on semiconductor microchips. Nature Biotechnol 1999; 17:365-370.

92. Broude NE, Woodward K, Cavallo R et al. DNA microarrays with stem-loop DNA probes: Preparation and applications. Nucleic Acids Res 2001; 29:e92.

93. Pinkel D, Segraves R, Sudar D. High resolution analysis of DNA copy number variation using comparative genomic hybridization to microarrays. Nature Genet 1998; 20:207-211.

94. Geschwind DH, Gregg J, Boone K et al. Klinefelter's syndrome as a model of anomalous cerebral laterality: Testing gene dosage in the X chromosome pseudoautosomal region using a DNA microarray. Dev Genet 1998; 23:215-229.

95. Solinas-Toldo S, Lampel S, Stilgenbauer S et al. Matrix-based comparative genomic hybridization: Biochips to screen for genomic imbalances. Genes Chromosom Cancer 1997; 20:399-407.

96. Hodgson G, Hager JH, Volik S et al. Genome scanning with array CGH delineates regional alterations in mouse islet carcinomas. Nature Genet 2001; 29:459-464.

97. Veltman JA, Schoenmakers EF, Eussen BH et al. High-throughput analysis of subtelomeric chromosome rearrangements by use of array-based comparative genomic hybridization. Am J Hum Genet 2002; 70:1269-1276.

98. Cai WW, Mao JH, Chow CW et al. Genome-wide detection of chromosomal imbalances in tumors using BAC microarrays. Nature Biotechnol 2002; 20:393-396.

99. Pollack JR, Sorlie T, Perou CM et al. Microarray analysis reveals a major direct role of DNA copy number alteration in the transcriptional program of human breast tumors. Proc Natl Acad Sci USA 2002; 99:12963-12968.

100. Lamar EE, Palmer E. Y-encoded, species-specific DNA in mice: Evidence that the Y chromosome exists in two polymorphic forms in inbred strains. Cell 1984; 37:171-177.

101. Ermolaeva OD, Sverdlov ED. Subtractive hybridization, a technique for extraction of DNA sequences distinguishing two closely related genomes: Critical analysis. Genet Anal 1996; 13:49-58.

102. Lisitsyn NA, Lisitsina NM, Dalbagni G et al. Comparative genomic analysis of tumors: Detection of DNA losses and amplification. Proc Natl Acad Sci USA 1995; 92:151-155.

103. Sverdlov ED. Subtractive hybridization - a technique for extracting DNA sequences, discriminating between two closely related genomes. Mol Gen Microbiol Virusol 1993; 6:3-12.

104. Lisitsyn NA, Lisitsyna NM, Wigler M. Cloning the differences between two complex genomes. Science 1993; 259:946-951.

105. Lisitsyn NA, Rosenberg MV, Launer GA et al. A method for isolation of sequences missing in one of two related genomes. Mol Gen Microbiol Virusol 1993; 3:26-29.

106. Straus D, Ausubel FM. Genomic subtraction for cloning DNA corresponding to deletion mutations. Proc. Natl Acad Sci USA 1990; 87:1889-1893.

107. Schutte M, da Costa LT, Moskaluk CA et al. Isolation of YAC insert sequences by representational difference analysis. Nucleic Acids Res 1995; 23;4127-4133.

108. Lisitsyn NA, Segre JA, Kusumi K et al. Direct isolation of polymorphic markers linked to a trait by genetically directed representational difference analysis. Nature Genet 1994; 6:57-63.

109. Brown PO. Genome scanning methods. Curr Opin Genet Dev 1990; 4:366-373.

110. Lisitsyn N, Wigler M. Representational differences analysis in detection of genetic lesions in cancer. Methods Enzymol 1995; 254:291-304.

111. Jonsson JJ, Weissman SM. From mutation mapping to phenotype cloning. Proc Natl Acad Sci USA 1995; 92:83-85.

112. Bernardi G. The isochore organization of the human genome. Annu Rev Genet 1989; 23:637-61.

113. Yokota H, Oishi M. Differential cloning of genomic DNA: Cloning of DNA with an altered primary structure by in-gel competitive reassociation. Proc Natl Acad Sci USA 1990; 87:6398-402.

114. Yokota H, Amano S, Yamane T et al. A differential cloning procedure of complex genomic DNA fragments. Anal Biochem 1994; 219:131-138.

115. Inoue S, Kiyama R, Oishi M. Construction of highly extensive polymorphic DNA libraries by in-gel competitive reassociation procedure. Genomics 1996; 31:271-276.

116. Ohki R, Oishi M, Kiyama R et al. A whole-genome analysis of allelic changes in renal cell carcinoma by in-gel competitive reassociation. Mol Carcinog 1998; 22:158-66.

117. Rosenberg M, Przybylska M, Straus D. "RFLP subtraction": A method for making libraries of polymorphic markers. Proc Natl Acad Sci USA 1994; 91:6113-61117.

118. Nelson SF, McCusker JH, Sander MA et al. Genomic mismatch scanning: A new approach to genetic linkage mapping. Nature Genet 1993; 4:11-18.

119. Lander ES, Schork NJ. Genetic dissection of complex traits. Science 1994; 265:2037–2048.

120. Mirzayans F, Mears AJ, Guo SW et al. Identification of the human chromosomal region containing the iridogoniodysgenesis anomaly locus by genomic-mismatch scanning. Am J Hum Genet 1997; 61:111-119.

121. McAllister L, Penland L, Brown PO. Enrichment for loci identical-by-descent between pairs of mouse or human genomes by genomic mismatch scanning. Genomics 1998; 47:7-11.

122. Cheung VG, Nelson SF. Genomic mismatch scanning identifies human genomic DNA shared identical by descent. Genomics 1998; 47:1-6.

123. Plasterk RH. Hershey heaven and Caenorhabditis elegans. Nature Genet 1999; 21:63-64.

124. Fields S, Kohara Y, Lockhart DJ. Functional genomics. Proc Natl Acad Sci USA 1999; 96:8825-8826.

125. Enard W, Przeworski M, Fisher SE et al. Molecular evolution of FOXP2, a gene involved in speech and language. Nature 2002; 418:869-872.

# CHAPTER 3

# A Brief Introduction to Primate Evolution

Hans Zischler, Christian Roos and Gerhard Hunsmann

## Abstract

Functional aspects of genetic information as well as the diversity of different genomes will be major biomedical issues in postgenomic research. Genomic information from numerous organisms is closing in at an ever-increasing rate. Complete genomes are available from different pro- and eukaryotic taxa including that of the mouse—the most prominent eutherian model organism—and humans. Meaningful comparative analyses in character evolution, irrespective on which level this is carried out require an undisputed phylogenetic framework to ultimatively decide between homology and analogy. Though of outmost importance for evolutionary research in humans, an undisputed phylogenetic framework linking the mouse with primate-related eutherians, nonhuman primates and humans is still missing and research results on this issue are often controversially discussed. We herein review the actual status of the investigations on primate phylogeny. An emphasis is given to the divergence of nonhuman primates, relevant interpretations of the fossil record and molecular, including retropositional evidence. Whereas a congruent view is coming up concerning the phylogenetic relationships among primate taxa at a higher taxonomic level, e.g., the primate infraorders, there is still considerable debate on primate origins or very recent splits in primate evolution. For the latter this is mostly due to an incomplete taxon sampling, since primate material from at least more than 200 currently recognized species that is suitable for molecular research is not easily accessible. Obtaining more clarity about primate origins is to a large degree hampered by the sparseness of the critical fossil record. In as far as both molecular and fossil evidence is available for a certain splitting, the impression is that many interpretations based on the two completely different molecular and fossil-based approaches are remarkably compatible.

## Introduction

Nonhuman primates are the closest relatives of *Homo sapiens*. Needless to mention, that the phylogenetic history of primates is part of our own evolutionary history, and it is therefore of outmost importance to elucidate the phylogenetic affiliations of primates to other eutherians, among nonhuman primates and to learn more about our own history taking place after the chimpanzee split off.

Basically, an undisputed phylogenetic framework is necessary to discriminate between homology and analogy in character evolution, irrespective of whether one looks at morphological, physiological, behavioral or molecular characters. This phylogenetic framework is more important in times when incoming sequence data of both man and eutherian model taxa represent a first platform for a thorough comparison of DNA sequence changes over evolutionary time scales.

*Retroviruses and Primate Genome Evolution*, edited by Eugene D. Sverdlov.
©2005 Eurekah.com.

The effect of these sequence changes on human transcriptomes and proteomes, the increase in transcriptome and proteome complexity on the lineage to humans by mechanisms like alternative splicings, RNA editings, RNA interference, epigenetic and posttranscriptional modifications, just to quote a few, will be the main targets of future biological research.

It is hoped that knowledge about all these phenomena that shape the spatio-temporal pattern of transcriptomes and proteomes during the development of an organism will ultimately tell us what makes us humans.

Apparently sequence changes taking place on the lineage to humans are therefore the main target of scientific interest. However it has to be kept in mind that all primate taxa represent different realizations of the evolutionary process starting from a common origin, the ancestor of all living primates. The different mechanisms and molecular principles acting on basically the same starting material, the genome of the primate ancestor, will tell us a lot of how and by what mechanisms genomes, transcriptomes and proteomes can be shaped.

Thus primate evolutionary research is not only an interesting biological and zoological topic with countless aspects per se it is rather an irreplaceable complement to the studies of human evolution that is also inherently linked to the emergence of human disease phenotypes.

The review is not intended to give a summary over the genomic changes that occurred during primate divergence and on the lineage to anatomically modern humans. Though we will mention and discuss a few general trends of primate genome evolution this review rather focuses on the evolution of primates and their phylogenetic classification. Both fossil as well as molecular evidence will be discussed, and, to mention this already in the beginning, the overall congruence of the phylogenetic interpretations based on the two sources of data is remarkable.

## The Problem of Primate Definition

Primates constitute a eutherian order and, concerning the pure number of different species, primates represent one of the most successful eutherian orders. With at least 230 currently recognized species[1] primates are merely surpassed in species numbers by the rodents, chiropterans, insectivores and carnivores. As a general trend in recent primate research it is getting more and more evident, that species diversity in living primates has been greatly underestimated hitherto. Due to intensified field work, the continuous advent of primate sequence data and the application of genetic distances to obtain a better taxonomic classification it is getting obvious that many primate genera are taxonomically too lumped. Interestingly it could be found upon comparing molecular data that the mostly morphology-based classification of mammals of large body size tends to be oversplit, whereas that of small mammals has an excess of lumping. This might be due to an increased difficulty in finding diagnostic characters in the classification of small animals.[2] Apparently this applies to primates too and indeed several authorities in primate research propose an extension of the primate species list with a number of currently recognized extant species going well beyond 350 (see ref. 3). Thus the near future is likely to generate an impression of an increase in primate biodiversity, which at first sight seems to be at odds with the fact that there is a continuing and ever increasing threat of extinction to many primate taxa, due to habitat destruction and hunting.

The definition of a primate did not change significantly since the summary given by Le Gros Clark in 1959 which was extended by Napier and Napier.[4]

Primates are characterized by a generalized limb structure with primitive pentadactyly and an enhanced mobility of the digits, especially of the pollux and hallux. Flat nails replace sharp and compressed claws and sensitive tactile pads on the digits were developed. The snout progressively shortens during primate evolution whereas the visual system develops, with an invention of binocular vision. In parallel the olfactory apparatus gets reduced. Primate dentition is hallmarked by the loss of some elements of the primitive mammalian dentition, whereas a simple molar cusp pattern is preserved. The brain and especially the cerebral cortex underwent a size increase. Postnatal life periods are getting extended and there is a progressive development of truncal uprightness leading to a facultative bipedalism.

Though the list of primate features is remarkable, the definition of what makes a primate and separates it from other eutherians is problematic and complex. At first no unique morphological, physiological or behavioral characteristic defines a primate except some still to determine molecular traits. It is rather a collection of shared characteristics and trends, and the listing mentioned above does not contain an undisputed synapomorphy that might define a clade. Moreover most of these characteristics represent retentions of ancestral features. Many "typical primate" features are behavioral, or depend on soft tissue anatomy, thus identifying a fossil and deciding whether this fossil represents a primate or not is difficult. In fact, there are no specific morphological features that can fossilise and unify all primates. Earliest primates are therefore only marginally different from other orders of placental mammals. This together with the scanty fossil record of early primates represents a major problem for studying the origin of primates.

The origin and remarkable diversification of primates is usually linked to three different theories, the arboreal theory, the second being the visual predation theory, and finally the angiosperm radiation theory. Though pros and cons were raised for all these theories, the most popular one, which most primatologists probably favour, is the visual predation theory. In that orbital convergence, grasping hands and feet, and reduced claws are considered to represent an adaptation to nocturnal foraging for fruit and insects. To be better able to move through the environment, the taxa thus evolved full stereoscopic vision which was achieved in that early primates underwent orbital convergence. In parallel to the increase of the visual capacities, the olfactory system was reduced, aptly mirrored by reductions of the snout and the brain areas responsible for processing the olfactory signals.

All these gross anatomical changes are shaped through genomic changes during primate evolution. Thus trends are recognizable in that the olfactory receptor genes, primates possess more than 1000 copies of these genes in their genomes, are getting successively nonfunctional on the lineage to humans (reviewed in ref. 5). The increase in frequency of nonfunctional pseudogenes might be due to a reduced selective pressure.

## Linking Primates to Other Eutherian Orders

Growing evidence based on considerable sequence information from both nuclear and mitochondrial DNA (mtDNA) suggests that eutherian orders can be partitioned into four major groups the Laurasiatheria, Xenarthra, Afrotheria and the Euarchontoglires.[6-8] Primates constitute a member of the latter group clustering together with the rodents, lagomorphs, tree shrews and flying lemurs. A large degree of congruence between the interpretations based on mtDNA sequences and nuclear DNA combined with mtDNAsequences can be made out for the assembling in the major groups. Thus humans and mice as the prime mammalian model organism, both taxa for that already complete sequence information in draft version exists are confined to one major eutherian group. Whereas both the rodents and the lagomorphs each constitute a monophyletic group, with the rodents and lagomorphs being sister groups, the phylogenetic affiliations among the remaining members of the Euarchontoglires, the primates, flying lemurs and tree shrews remain contradictory.

Since the formulation of the concept of the superorder Archonta that was first proposed by Gregory[9] and revived in the last decades[10] in which, traditionally, primates, bats (Chiroptera), flying lemurs (Dermoptera), and tree shrews (Scandentia) are grouped together in various constellations, the question which eutherian order shares a common ancestry with primates exclusive of all other mammalians is still not adequately answered.

The concept of the superorder Archonta, is mainly based on anatomical evidence obtained from extant species.[10] Neither paleontological data[11] nor most interpretations of molecular data[12-14] support this concept.

Though the monophyly of the four taxa is partially refuted by both molecular[15-17] and palaeontological evidence[11] it is common sense that the primate origin lies somewhere among archontan representatives (for overview see ref. 18).

## Fossil Evidence of Primate Origins and Early Evolution

The early history of the primate evolution between the late Cretaceous and the end of the Eocene and the inter-relationships among extant and fossil primates and eutherians is continuing to be an issue of lively debate.

Extant primates may coalesce to two species of the archaic primate genus *Purgatorius* living in the late Cretaceous and early Paleocene of Eastern Montana at least 65 million years ago (Mya).[19] However, the respective sampling level of fossils, some fragmented jaws and teeth, is far to low, to represent a basis for drawing meaningful conclusions on early primate evolution. Indeed Tavaré and coworkers[20] proposed that no more than 7% all primate species that ever existed are known from fossils. Taking this into account they estimate, applying a new statistical method based on an estimate of species preservation derived from a model of the diversification pattern, that the most recent common ancestor (MRCA) of all primates existed about 81.5 Mya. This date is considerably earlier than previous estimations (see Chapter 4). Based on *Purgatorius*, fossil evidence points to a geographic origin of primates in North America and Eurasia.

The first major adaptive radiation from ancestral primates was formerly thought to be represented by the Plesiadapiformes or "archaic primates". However this view has been shaken by new fossil evidence and analyses. It is now getting evident that this group is not monophyletic and various phylogenetic constellations of the different plesiadapiform families, including Micromomyidae, Saxonellidae, Carpolestidae, Plesiadapidae, Picrodontidae and Paromomyidae and extant members of the Archonta superorder have been proposed (reviewed in ref. 21). Obviously the discovery of new fossils has to be awaited to get a more reliable picture how archaic primates and extant nonprimates are phylogenetically linked to primates of modern aspect. However molecular data obtained from extant taxa can at least help to establish falsifiable theories about the phylogenetic affiliations between the primate and nonprimate crown groups thus directing the development of evolutionary hypotheses related to the archaic primates.

## Candidates for the Next Living Relatives to Primates

The two eutherian orders of Dermoptera and Scandentia, both belonging to the "Euarchonta" are currently discussed as potential extant sister groups to primates. Among the members of the former Archonta superorder, bats are currently recognized as the least likely candidate for the primate's sister. Molecular evidence, such as tree reconstructions obtained from the complete mtDNA sequences[22] or interpretations based on nuclear and combined nuclear and mtDNA datasets reveal phylogenetic affiliations of bats to ferungulates (carnivores, perissodactyls, artiodactyls, and cetaceans) with overwhelming support.[17] Moreover, the supposed close phylogenetic relationship between Dermoptera and Chiroptera, suggested by various neontological, morphological analyses, is not supported by fossil evidence.[11] Whereas the earlier interpretation regarded members of the Plesiadapiformes to constitute the first adaptive radiation from ancestral primates, it becomes evident, that at least some genera of the Plesiadapiformes might be more directly linked to Dermoptera. Beard[23] observed similarities in phalangeal morphology of paromomyids and extant taxa, suggesting that the former possessed a gliding membrane (patagium) as extant flying lemurs do. Therefore, it seems conceivable that numerous morphological characters formerly recognized as shared derived characters and linking bats and flying lemurs are largely due to convergences that appeared with the evolution of gliding or flight. On the other hand, flying lemurs share postcranial features with extinct Paromomyidae, the archaic radiation of the order Primates.[23-24] As a consequence the term Primatomorpha was coined to describe the phylogenetic relatedness of primates and dermopterans. However, Kay et al[11] conclude from craniodental and postcranial synapomorphies that a clade comprising plesiadapiforms—and with them the living dermopterans—and primates seems to be unfounded and propose the term 'archaic dermopterans' for the extinct plesiadapiforms.

Flying lemurs are therefore, beside the Scandentia representatives, prime candidates for being the extant descendants of the branching event in eutherian evolution that is closest to the time point when primates of modern aspect diverged.

Both the Philippine and Indonesian flying lemurs (*Cynocephalus volans* and *C. variegatus*), two parapatric species with a narrow geographic distribution restricted to Southeast Asia constitute the mammalian order of Dermoptera.

Sequence data from flying lemurs are at hand, however, the phylogenetic results obtained from them are contradictory though different taxonomic samplings render comparisons between the different interpretations difficult. In these dermopterans are variously linked with either Scandentia, Anthropoidea, Primates in general[25] or Chiroptera.

Unexpectedly to primatologists was the close affiliation of dermopterans to the anthropoid primates, that was proposed in various studies of combined mtDNA and nuclear datasets[6-8] or complete mtDNA information alone.[26] In the latter the term "Dermosimii" was even coined to express the close evolutionary relationship of anthropoid primates and the dermopterans that renders the primates a paraphyletic grouping. However, this interpretation is not supported by e.g., retropositional evidence. Multilocus analyses, i.e., the presence of short interspersed nuclear elements (SINE)-Alu elements in all primate taxa and their absence in dermopterans, or single locus retropositional evidence[27-28] clearly speak against primate paraphyly. Instead it was proposed that the unexpected positioning of the dermopterans is rather due to a similarity of mtDNA base composition than reflecting true phylogenetic relationship. This effect is assumed to be so strong that even combined mtDNA and nuclear datasets might be affected by this phenomenon.[27-28] It is therefore of outmost importance that more molecular data, especially nuclear information is obtained from dermopterans to finally settle the problem of its affiliation to primates.

The other candidate for a close phylogenetic relationship to primates is currently recognized in the members of the eutherian order Scandentia. Extant representatives of this order are comprised in the family Tupaiidae with two subfamilies Tupaiinae and Ptilocercinae. The Tupaiidae or tree shrews—a potentially misleading name since they are not uniformly arboreal—represent a small radiation currently consisting of 18 recognized extant species, geographically distributed in South and Southeast Asia.[29] Tree shrews are small, omnivorous animals with a fairly unspecialised mammal-type skull and claws on each digit. Superficially regarded they look very much like squirrels in fact the Malay word "tupai" designates both a squirrel and tupaia.

Their uncertain phylogenetic affiliation is reflected in the various systematic classifications attributed to this taxon in the past. Before being included in the mammalian order of Scandentia, the Tupaiidae were recognized as the deepest primate split, after they were classified as being a member of the former insectivore subgroup Menotyphla.[29] Now the characteristics shared between tree-shrews and primates have been classified as primitive amongst placental mammals.[29] The problem with placing tree-shrews into the order of primates is that tree-shrews and primates do not share any synapomorphies separating them from the other mammals.[29] It has to be noted that the possible inclusion of *Tupaia* in the order of primates does not only represent a taxonomic problem, the more important question is if tupaias, or alternatively the dermopterans, and primates share an ancestral stock exclusive to all other mammals. This phylogenetic uncertainty is mainly due to the fact that members of this group e.g., tupaia show a complex mixture of plesiomorphic and apomorphic morphological characters,[29-30] a lack of fossil information pertaining to the Scandentia, and the long independent evolutionary history of Scandentia.[30] Nevertheless, the tupaia is currently generally recognized as an intermediate between primates and other eutherian orders. Molecular evidence in favour of a sister group-relationship to primates is confined to some nuclear DNA datasets e.g., MHC-DRB introns or combined nuclear and mtDNA datasets.[31] For the latter different samplings and questionable phylogenetic affiliations among the Euarchonta, e.g., the above mentioned sister group relationship of anthropoid primates and dermopterans, preclude a final and satisfying answer to the problem of primate-eutherian affiliations. mtDNA as well as a large body of nuclear data point away from a close phylogenetic link between primates and tupaias, however different methodological or mtDNA inherent problems could be theoretically raised that might weaken these interpretations.

At the end of this paragraph we have to conclude that reliable evidence in favour of Scandentia or Dermoptera being the primates sister taxon (or both taxa being the sister group to primates) is still missing and a firm phylogenetic hypothesis, on which large scale comparative molecular work can be based, still requires further scrutiny. More sequences brought together with a meaningful taxonomic sampling as well as e.g., retropositional evidence are required to settle the problem of linking primates to eutherian model organisms like the mouse.

## The Major Groups of Primates

Definitely less disagreement concerns the major primate groups and their phylogenetic affiliations among each other. Probably the most challenging problem involving the different primate infraorders pertains to the position of tarsier, a group of primates with a narrow geographic distribution in Southeast Asia. However recently established retropositional evidence[32-33] has settled this problem from a molecular cladistic perspective.

In the following the phylogenetic affiliations of the members of each major group of primates, that split off from the lineage leading to humans will be briefly discussed together with the still numerous problems e.g., concerning various biogeographical aspects. We will deal with this in a linear fashion following the succession of the major branching events in primate evolution (Fig. 1).

### *Strepsirrhini*

The deepest split in primate evolution separates the strepsirrhines from the other primate taxa. Strepsirrhines constitute a remarkable diverse group of primates, with about 50 extant species (data combined from refs. 1, 34) and thus including ca. 20% of all living primate species. Indeed, the number of strepsirrhines should be already expanded by nearly 20, to include the subfossil lemurs that went extinct over the past few thousand years. The heyday of these animals lasted until about 1,500 years ago, when humans from Africa and the Malay archipelago colonized Madagascar. Subsequently anthropogenic habitat modifications and hunting set an end to these subfossil lemurs that were among the largest primates ever existing with e.g., *Archaeoindri* reaching the size of a gorilla.

An even further expansion of the list of strepsirrhine species is to be expected for the near future. At first, beside rediscoveries of taxa, that were assumed to be extinct, there is still an enormous potential of future identifications of additional species during fieldwork programs, especially nocturnal strepsirrhines.[35-36] Yoder and coworkers[37] could confirm this assumption for the genus *Microcebus*, a member of the Malagasy Cheirogaleid family. Their phylogenetic

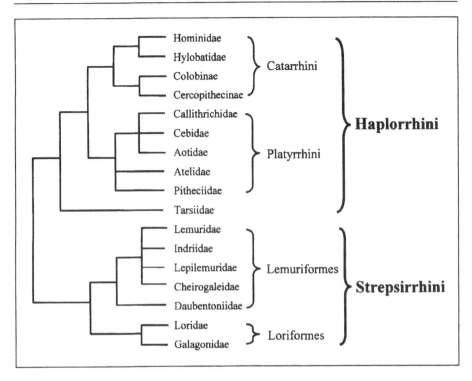

Figure 1. Summary of the infraordinal phylogenetic affiliations of the major primate groups as discussed in the text. Polytomies are displayed for branchings that are only insufficiently supported by hitherto available data.

analyses of mtDNA sequence data led to the description of three species that are new to science, and the resurrection of two others from synonymy, thus doubling the number of recognized mouse lemur species. These molecular data have been corroborated from the morphological site and were further supported by evidence of reproductive isolation in sympatry.[37] Secondly, with the advent sequence information of a taxonomically broad strepsirrhine species sample the phenomenon of "taxonomic lumping" (see ref. 2) gets apparent for these primates too.

Strepsirrhines are commonly divided in two infraorders (Fig. 1): the Lemuriformes, geographically exclusively restricted to Madagascar and the Loriformes found on two continents, Africa and Asia. Current species distribution and accepted palaeocontinental reconstructions suggest an African origin of strepsirrhines with two subsequent migrations to Madagascar and Asia. The colonization of Madagascar was accompanied by a remarkable adaptive radiation resulting in the broad diversification of extant Lemuriformes. At least 33 extant lemuriform species classified in the families of Lemuridae, Lepilemuridae, Indriidae, Cheirogaleidae and Daubentoniidae are currently known to inhabit Madagascar. This is the more surprising considering the small area lemurs inhabit on the Malagasy island. A long standing question, still not settled, is how often Madagascar was colonized by lemuriform ancestors. Since there is virtually no fossil record of lemuriform ancestors (except for the subfossil lemurs) available, both the timing and the way of crossing the channel of Mozambique from Africa to Madagascar is still enigmatic. Even considering an underestimation of the time of primate origin and splitting events, a controversially discussed issue supported by both palaeontological and molecular evidence, the crossing of the huge water barrier is hard to imagine. It is the more unconceivable, that such an event could be repeated several times in the history of Madagascar primates. However, this could not be ruled out on the basis of phylogenetic reconstructions on morphological characters. At first cheirogaleids, mostly due to their peculiar blood supply of the brain, were formerly grouped together with Loriformes. Interpreting such a tree topology on the basis of the geographical distribution would lend support to at least two independent colonizations of Madagascar. Though this Cheirogaleid-Loriformes affiliation can be now firmly excluded e.g., by several molecular analyses, another problem is still existing linked to the position of the aye-aye (*Daubentonia madagascariensis*). The aye-aye both in morphological as well as molecular examinations is not placed in a monophylum together with the other lemuriforms with conclusive evidence. Alternative topologies e.g., being basal to all strepsirrhines, joined with the Indriidae or not fully resolved with respect to other strepsirrhines[38] were proposed. Combined molecular and morphological evidence[39] is in favour of a monophyly of Malagasy lemurs, however there is still need to settle this discussion e.g., by retropositional evidence. In addition to this monophyly problem, there is a continuing debate about the phylogenetic affiliations of the classical Malagasy lemur families of the Cheirogaleidae, Indriidae, Lemuridae and Lepilemuridae. As mentioned above an adaptive radiation took place shortly after the lemur ancestors colonized Madagascar and most likely *Daubentonia* split off. This radiation resulted in consecutive splitting events taking place in a short time span and leading to the extant members of the above mentioned families. Obviously lineage sorting phenomena, the unequal distribution of polymorphisms into progeny lineages, are the possible consequence of such a radiation-like diversification. This hampers the establishing of a reliable phylogeny irrespective of what characters are used and it can be envisaged that still a considerable amount of sequence data will be necessary to resolve this question with certainty.

A further phylogenetic puzzle that needs to be mentioned pertains to the Loriformes - group of strepsirrhines, comprising the Galagonidae (bush babies) and Loridae that is distributed over Asia and Africa. Further to the still unsettled discussions about the phylogenetic affiliations among the African Galagonidae members, different Loridae taxa display two profoundly different morphotypes on each continent. Thus gracile morphotypes are found e.g., for the African *Arctocebus* and Asian *Loris*, whereas the African potto (*Perodicticus*) and the Asian slow loris (*Nycticebus*) display the more compact morphotype. The phylogenetic affiliations among these taxa and to the bush babies are still not sufficiently resolved. Further molecular data are urgently required to determine their phylogenetic relationships and decide whether the evolutionary affiliation follows the biogeography of these animals or is expectable by the general morphotype of the taxa in question.

## *Tarsiiformes*

Regarding the intraordinal taxonomic level, the most striking cladistic problem in evolutionary primatology pertains to the phylogenetic relations of *Tarsius* to other extant primates. This is e.g., reflected in the different taxonomic nomenclatures in that *Tarsius* are either combined with lemurs and lorises in a group called prosimians or in that *Tarsius* together with anthropoid primates forms the group of Haplorrhini (see Fig. 1) that is separated from the strepsirrhines. Moreover this problem is both exemplary of a conflict between neontological and palaeontological data and an incongruence between nuclear and mitochondrial DNA analyses. On one side evidence exists mainly based on morphological, neontological data which allows to group *Tarsius* in the haplorrhine suborder being a sister group to the Anthropoidea (Platyrrhini and Catarrhini). On the other side analyses of fossil records favour alternative evolutionary tree topologies which either place *Tarsius* as a sister group to the Strepsirrhini ("adapid-theory"), or which shows *Tarsius* to branch off before the Anthropoidea-Strepsirrhini split or give rise to a polychotomous branching involving these three taxa.[19] This conflict is mostly due to the fact that *Tarsius* represents the only surviving genus of a formerly diverse group of Eocene Tarsiiformes. Currently at least five species, *T. bancanus*, *T. dianae*, *T. pumilus*, *T. spectrum and T. syrichta* are recognized, two further taxa were recently proposed to be separated as distinct species as well (*T. pelengensis* and *T. sangirensis*).

Fossils from the early Oligocene of Egypt and Miocene of Thailand are potentially referable to the family of Tarsiidae. The oldest fossil assigned to this family was from middle Eocene in China, showing the antiquity of this group.

Fossils of the second family belonging to the Tarsiiformes, the Omomyidae have a geological range of early Eocene to late Oligocene in North America and early Eocene to early Oligocene in Eurasia. This diverse group contains 30 known genera, emphasizing the diversity of extinct Tarsiiformes[39] (reviewed in ref. 40). Beside the acquisition of autapomorphies in their long independent evolutionary history, it is therefore unlikely that the only living *Tarsius* species fully represent the diversity of all tarsiiform primates[29] Molecular sequence data obtained both from mitochondrial and nuclear DNA could not resolve this problem adequately. This is partly due to a lack of congruence between mitochondrial and nuclear DNA based phylogenies. Moreover, these data need to be interpreted with caution since evidence exists that mitochondrial sequence evolution might either deviate from a purely neutral model of evolution on the lineage to simians after the strepsirrhines branched off[41] or, as alternative explanation, might be subjected to a strong directional mutation pressure, resulting in a remarkable base compositional plasticity of primate mitochondrial genomes.[28-29] However this major cladistic problem could be solved on the basis of retropositional evidence. Thus, transpositions of Alu sequences, representing the most abundant primate SINE, were evaluated as molecular cladistic markers to analyze the phylogenetic affiliations among the primate infra orders. Altogether 118 human loci, containing intronic Alu elements, were PCR analysed for the presence of Alu sequences at orthologous sites in each infra order. In this way three Alu transpositions, which can be regarded as shared derived molecular characters linking tarsiers and anthropoid primates could be revealed[32] underpinning the monophyly of haplorrhine primates (Anthropoidea and Tarsioidea) from a novel perspective.

### Platyrrhini

Together with the catarrhines (Old World monkeys and hominoids), the platyrrhines or New World monkeys constitute the group of higher primates, also designated as simian or anthropoid primates. Platyrrhines are enormously diverse from anatomical and behavioral aspects, mating strategies and social systems, locomotion and feeding adaptations. In this context it has to be taken into account that these animals, contrasting to the other major primate groups of strepsirrhine and catarrhine primates, inhabit just one land mass, Central and South America. Classically, altogether 16 New World monkey genera are recognised in that about 50 extant and extinct species are subsumed.[1] As for all primate groups, these numbers are in flux due to a recognition that the formerly valid taxonomy might be oversplit or too lumped. E.g., on one site growing evidence suggests that a formerly recognized genus like *Cebuella* cannot be maintained and including *C. pygmaea* into the genus *Callithrix* is justified on several grounds. On the other site however, the introduction of the new Atelid genus *Oreonax* is taking the only recently appreciated divergence in the genus *Lagothrix* better into account.[3,35]

Thus from the pure number of species platyrrhines are about as diverse as the strepsirrhines or Old World monkeys. However, as a sharp contrast to the catarrhines, New World monkey phylogeny appears not to be characterised by a relatively well defined succession of splitting events leading to the extant taxa. Fossil evidence and molecular data rather point towards the fact that current platyrrhine diversity is the result of a radiation that took place after the platyrrhine ancestors entered the New World. Together with the scarce fossil record pertaining to the

New World monkeys, which renders it extremely difficult to define primitive features in turn lading to serious difficulties for the determination of shared derived markers, it is comprehensible that the phylogenetic affiliation among many platyrrhine taxa is still very controversially discussed. Platyrrhine classification obviously reflects that New World monkey phylogeny is strongly linked to adaptations, e.g., shifts in dietary possibilities and examples of parallel evolution on different platyrrhine lineages are numerous. There is a general agreement, that platyrrhines can be partitioned into five major groups: First the Callitrichidae or marmosets and tamarins that comprise the genera *Callithrix, Saguinus, Leontopithecus* and *Callimico*. The phylogenetic affiliations among the callitrichine genera and especially their relation to the other New World monkeys have always been a controversially discussed subject. Callithrichids, the "dwarfs" among the simian primates that bear claws at all digits except for the big toe and give regularly birth to twins, were for long considered to represent the primitive ancestral simian state.[42] However this opinion is getting less popular and evidence exists that e.g., the above mentioned traits are rather derived specializations,[9,43-45] which separate them from all other New World monkeys. *Callimico goeldii*, giving birth to singletons, was formerly considered an intermediate between the callitrichines and the other platyrrhines. Nowadays, mainly based on genetic data, they are firmly placed amidst the callitrichines as sister to *Callithrix*.

The question for the next living relative to the callitrichines is one of the most controversially discussed issues in evolutionary primatology and closely linked to the question of the positions of the representatives of two other major platyrrhine groups the Cebidae (*Cebus* and *Saimiri)* and Aotidae (*Aotus*) in the New World monkey tree. As the fourth and fifth major platyrrhine group the Pitheciidae comprising the genera *Cacajao, Callicebus, Chiropotes* and *Pithecia* and the Atelidae with the genera *Alouatta, Ateles, Brachyteles* and *Lagothrix* have to be mentioned. The monophyly of these two groups is undisputed, however the general phylogenetic relation to the other groups is not satisfactorily resolved, a problem that mainly hinges to the questionable phylogenetic affiliations of *Saimiri, Cebus* and *Aotus*. Although most authorities agree that the Cebidae and Aotidae are more closely affiliated to the Callithrichidae than the Atelidae and Pitheciidae are, it is to expect that still a considerable amount of data have to be collected to give a final answer to the problem of New World monkey phylogeny. The latter is above all due to the radiation-like divergence of New World monkeys.

Another major phylogenetic puzzle pertaining to the New World monkeys is related to their origin. Though several features like skin histology, hair follicle arrangement, immunological and cytogenetic findings do not support a common origin of extant platyrrhine taxa, numerous morphological and anatomical features and mainly molecular data including retropositional evidence[46] are clearly in favour of a platyrrhine monophyly. However the exact link of the platyrrhine ancestors to the simian primate stock outside America and how Central or South America could be colonized is still not understood, mainly due to the scarce fossil record. For at least 65-80 Myr South America is geographically isolated from other land masses. Three waves of immigration of animals from other continents into America are currently discussed, with a second wave in that rodents and primates colonized America. A third wave of colonization took place after the formation of the isthmus of Panama that was completed in the range of 3-4 Mya enabling a biotic exchange between the Americas. Irrespective of how platyrrhines managed to enter South America it is obvious that this colonization was linked to a surmounting of huge water barriers. Two major alternatives were formerly discussed in that New World monkeys were directly related to ancestral simians from Africa, which harbours a rich fossil record of early anthropoid primates in the e.g., Fayum deposits that are well over 31 Mya. However, only catarrhine anthropoids and more primitive omomyiids, tarsiers and lorises can be traced there. As a next alternative an invasion of Central and South America from North America was postulated where more archaic adapid and omomyid primates, but no recognizable anthropoids lived from at least 56 Mya. This discussion was further complicated by the description of a simian fossil in China, dated at around 40 Mya that displayed a dentition that is more primitive as compared to the Fayum fossils. This mouse sized animal, *Eosimias centennicus*[47] corroborates the hypothesis that the anthropoid clade originated on the Asian landmass at least 40 Mya. This simian origin hypothesis is further corroborated by the fact that many primitive anthropoids could be found in Asia and in that the closest extant relative to simians, *Tarsius*, is nowadays geographically restricted to Asia. From these observations four main colonization scenarios are currently discussed all taking place in the Eocene. A route with an origin in Asia via Australia, Antarctica and South America is proposed as well as an Africa-Antarctica-South America invasion scenario. As a third possibility a transoceanic migration on "floating islands" also designated as the rafting theory was proposed too. As a last alternative the simian ancestor was of North American origin from where colonization of Asia, via the Bering Strait, and South America started. Though pros and cons for all these hypotheses are discussed extensively, the mystery of the colonization of the New World by the primates is still not resolved and has to await a more dense sampling of fossils. As the platyrrhine ancestors arrived in South America it appears that there were several consecutive waves of expansions perhaps accompanied by a merging of the formerly more patchy tropical forest in huge areas. These expansion waves extended the geographic distribution of New World monkeys even into the Caribbean islands (Cuba, Haiti) and in direction to the Atlantic coast.

## Catarrhini

Whereas platyrrhines apparently never dwelled on the ground, a major success of the other simian primate group, the catarrhines, is that they were able to invade the grasslands.

The primate group of the catarrhines comprises the Old World monkeys or Cercopithecoidea (Cercopithecinae and Colobinae) and the hominoids including the gibbons, orang-utans, gorillas, chimpanzees, bonobos and humans. Nowadays, the geographic distribution of the catarrhines, with the exception of humans, is restricted to Africa and Asia, however fossil record indicates that apes and monkeys lived in Eurasia from about 5 to 17 Mya. Combined evidence from fossil record and molecular data allowed the formulation of a synthetic view on catarrhine evolution in that the branching pattern as well as the timing based upon both fossils and molecular data are remarkably congruent.[48] Taken together the earliest catarrhine fossil *Aegyptopithecus* of 34 Mya suggests an African origin of catarrhines. It is not decided whether *Aegyptopithecus* truly represents the stem lineage leading to extant catarrhines or if it represents its sister. This poses a problem for the dating of the Hominoidae-Cercopithecoidea-split, however there is consensus that this took place before 20 Mya. The repeatedly cited catarrhine divergence date of 25 – 30 Mya seems reasonable in this context, however fossil evidence for the time span close to the catarrhine divergence would be of great help for a more reliable time estimation.

For the major Cercopithecoidea split, separating the Colobinae and Cercopithecinae, molecular data result in a time estimate of about 14 Mya, being congruent to the fossil evidence as obtained from *Victoriapithecus*. However it has to be stated that this time period is underrepresented in the fossil record.

Within the Cercopithecinae, two tribes, the Papionini (*Macaca, Papio, Theropithecus, Lophocebus, Mandrillus, Cercocebus*) and the Cercopithecini (*Cercopithecus, Chlorocebus, Erythrocebus, Allenopithecus, Miopithecus*) are recognized. In contrast to the mainly resolved phylogenetic relationships among the Papionini, the branching pattern among the Cercopithecini genera is still disputed. However, the paraphyly of mangabeys within the Papionini tribe, with *Lophocebus* closest related to *Papio* and *Theropithecus*, and *Cercocebus* as next relative to *Mandrillus* is mainly the result of recent genetic studies.[49] However, many further phylogenetic problems within Cercopithecinae genera as e.g., in macaques exist due to radiation. For this group there is now consensus that the only extant African macaque species, *M. sylvanus* represents the sister to the remaining Asian macaque representatives. On the other hand we are far from appreciating the full genetic divergence of many of these taxa, which still requires molecular analyses to give a good estimate on species number and geographic distribution. As in Cercopithecinae, Colobinae are also divided in to two tribes, the African Colobini and the Asian Presbytini. Molecular data favor a monophyly of both African and Asian colobines. This bears important biogeographic consequences but is not supported when analyzing morphological characters. However the interpretations based on the hitherto available data are not sufficiently supported which is mainly caused by the fact that a representative species sampling of both the African and Asian colobines is extremely difficult to achieve due to the accessibility of tissue material from these animals. Thus from molecular data the most parsimonious explanation for the colobine origin is that this took place in Africa with a dispersion into Eurasia less than 10 Mya.

Given an African origin of the MRCA of all extant catarrhines and including the fossil evidence Stewart and Disotell proposed a most parsimonious dispersal scenario of the extant hominoids out of Africa. A speciation into lesser and great apes took place subsequently in Eurasia followed by a back-dispersal into Africa by a great ape species, which eventually gave rise to the lineage leading to the African great apes nowadays represented by gorillas, chimpanzees and bonobos. It has to be noted that this scenario was initially proposed purely on neontological data whereas the recently introduced synthesis of fossil and molecular data lends support from a palaeontological site too. The dates for the separation of gibbons (lesser ape) and great apes are still controversially discussed with different arguments based on fossil and molecular evidence. However it has to be taken into account that datings based on mtDNA data[38] might be hampered by the plasticity of the primate mtDNA composition.[27] It was even proposed that this phenomenon might even lead to wrong tree topologies created on the basis of primate mtDNA information. Thus an adequate correction for this would be of outmost importance when applied to estimate branching dates in primate evolution. As a general point and already stated for the other primate groups, also Old World monkey and hominoid genetic diversity is greatly underestimated leading to taxonomic lumping. One prominent example for the hominoids is the genetic divergence of the lesser apes.[50] Corroborated by molecular sequence data and by chromosomal analyses gibbons are now regarded as being partitioned into four different genera representing the *Nomascus*, *Bunopithecus*, *Hylobates* and *Symphalangus* group. Further work, mainly on the molecular level, will be required to reliably assign the phylogenetic positions to the thirteen currently recognized lesser ape species in these genera and the branching pattern among them. However also this will be hampered by the restricted accessibility of these animals and the threat of extinction they are faced with. As for so many primate taxa we are currently in a situation that we begin to appreciate the full diversity of these animals in times when we have to seriously fear that they will be extinct in the near future. Another example of a better incorporation of the actual genetic divergence into hominoid taxonomy might be represented by the elevation of the formerly orang-utan subspecies inhabiting Borneo and Sumatra to species level. More extensive field work combined with molecular analyses is urgently required which will bring us a better view on genetic differentiation on all primate species including the great apes. It is of outmost importance that these endeavors ultimately have to serve for the conservation of these animals.

### African Apes

Genomic relationship of African great apes and humans can be shortly summarized as follows: DNA sequence analyses and cytogenetics have indicated striking similarities between gorillas, chimpanzees and humans.[5] Due to the former lack of sequence data from the African great apes, other than man sequence comparisons have largely been restricted to hybridization-melting curve analysis of nonrepetitive DNA. Such data indicate a divergence date for human and chimpanzee separation (5.8 – 7.1 Mya) and a point in time for the branching off of the gorilla lineage (8.3 – 10.1 Mya). According to direct comparison of several orthologous loci the common ancestor of chimpanzees and humans is estimated to have existed about 4.6 to 6.2 Mya (see also Chapter 4). Moreover, DNA and amino acid sequence similarities indicate a human-chimpanzee sister group relationship with gorilla as the closest relative. As calculated with parsimony analysis the effective population size of the human-chimpanzee ancestors was 52,000 to 90,000. The significantly smaller number of early human individuals of about 10,000 represents a dramatic reduction in number of the early human population[5] (see also Chapters 1 and 4). Comparison of the Xq13.3 locus on the X chromosome[51] reveals that great apes have substantially larger sequence variability than humans compatible with an estimated early chimpanzee population size of 35,000 (see also Chapter 4). By comparison of this locus from common (*Pan troglodytes*) and pygmy (bonobo, *P. paniscus*) chimpanzee the divergence time of their lineages was estimated at 0.93 Mya conflicting with data from mtDNA (2.5 Mya) and the ß-globin locus (2.78 Mya). The high variability of the mtDNA D-loop is frequently used to determine interspecies genetic variability. Bonobo, common chimpanzee and gorilla retain an up to 10 times higher mitochondrial variation than all humans do. Moreover, the same technique can also discriminate three gorilla subspecies (western lowland, eastern lowland and mountain gorilla) as well as two subspecies of the common chimpanzee (*P. troglodytes troglodytes* and *P. troglodytes schweinfurthii*).[5]

The genetic similarity between humans and chimpanzees is striking, given the major phenotypic differences between these taxa. Solely based on this, it would not be justified to separate chimpanzees and humans into two different genera. However, interspecies genomic sequence divergence due to single-base pair mutations in specific genes is only one mechanism generating differences of nuclear DNA. Small changes in gene regulatory regions or major chromosomal rearrangements by breakage and relegation as well as segmental duplications, and the position of interspersed repeat sequences are additional important forces possibly altering gene function and regulation, which consequently result in major phenotypic changes.[52] Human cells contain a diploid set of 46 chromosomes. The other great apes have one additional chromosome pair because the human chromosome 2 was formed by fusion of two ancestral chromosomes. The fluorescence in situ hybridization (FISH) technique has allowed fine mapping of inversions on several human chromosomes when compared to chimpanzee chromosomes (for more detail see Chapter 4).

Segmental gene duplications were identified by FISH technique and bioinformatic tools. They are enriched 3 to 5 fold in pericentromeric and subtelomeric regions of all human chromosomes and may add up to 5 to 10 % of the human genomic sequence.[52] Large blocks (1 to over 200 kb) of genes were duplicated and moved by transposition within the same or to another chromosome probably by a two-step mechanism of duplication followed by transposition. Minisatellite-like repeats often flanking integration sites suggest a targeted integration. Moreover, segmental duplication more likely occurs at nonrandom chromosomal sites and with fluctuating probability over time.[52-53] It has been hypothesized that this partial polyploidization followed by positive selection is a major prerequisite for speciation. Several differences in such duplications have been identified between the human and other hominoid genomes e.g., in the RhD blood group gene, the high-affinity IgG receptor gene, and the melanin-concentrating hormone locus. A duplication of a survival motor neuron gene (*SMN2*) is found in the human but not in the chimpanzee lineage, but the chimpanzee carries more *SMN* genes. The functional relevance of most segmental gene duplications is not yet understood. However, the ability of trichromatic colour vision of different primate taxa was obviously achieved by the duplication of opsin genes.

Another major factor shaping primate genomes is represented by interspersed repeats derived from transposable elements that constitute up to 50 % of the human genome.[54] The major groups of these sequences are DNA transposons, long terminal repeat (LTR) transposons, long interspersed nuclear elements (LINEs) and short interspersed nuclear elements (SINEs). The most prolific members of the latter group are 300 – 400 bp Alu repeats. Over one million of Alus have been identified in the human genome comprising about 10 % of its sequence. The typical dimeric Alu repeats are primate specific. After dimerization from so-called free monomers they expanded in number in a wave like fashion during primate divergence. Some of them still display retropositional activity. The African great apes have 99.8 % of the Alu repeats in common.[54] However, the integration of interspersed repeat elements may lead to altered gene function and regulation (see Chapter 7). Thereby relatively minor genetic changes could lead to substantial phenotypic differences.

## Evolution of Anatomically Modern Humans

Numerous bone fragments and partial skeletons belonging to various hominid species have been found predominantly along the East African Rift Valley. Due to their anatomical features, these specimens are considered not to belong to direct ancestors of extant chimpanzees but rather to predecessors of humans, the hominids. Their age can be determined with satisfactory precision by modern physical and biological dating techniques. However, in several instances the exact position of these respective species along the path of human evolution is still a matter of debate.

Recently, a fossil skull from a taxon named *Sahelanthropus chadensis* has been discovered in Northern Chad and dated by faunal evidence to be 6 to 7 Myr old.[55] It was proposed to represent the hitherto earliest taxon pertaining to the lineage leading to anatomically modern humans after it separated from the chimpanzee/bonobo lineage.

These findings challenge the East African origin of the hominid clade (see also Chapter 4). Its earliest members include *Orrorin tugenensis* (6 Myr old) and *Ardipithecus ramidus* (4.4 – 5.8 Myr old) and younger specimens of *Australopithecus anamensis* (3.9 – 4.1 Myr old), *Kenyanthropus* (3.5 Myr old) and *Australopithecus afarensis* (3.6 – 2.7 Myr old) from Ethiopia and Kenya. The famous Lucy skeleton (3.6 Myr old) belongs to this latter species as well as footprints of similar age discovered at Laetoli, Tanzania. So obviously, australopithecins had already developed a bipedal gait and roamed a wide range of the African savanna at that time. Three sidelines known as *Paranthropus aethiopicus*, *P. boisei* and *P. robustus* branched off from *A. afarensis* by about 2.5 Mya and lived in the area of Tanzania, Kenya and Ethiopia up to 1 Mya. They must have shared the habitat with the first two hominoid species for perhaps 2 Myr. *Homo habilis* probably originated about 3 Mya from *A. africanus* populating Malawi and later Southern Africa while *H. rudolfensis* (2.5 – 1.8 Mya) was preceded by *A. garhi* in Ethiopia. *H. rudolfensis* is the likely ancestor of *H. ergaster* and subsequent *H. erectus* (2 – 0.5 Mya) while *H. habilis* vanished 1.5 Mya. Early *H. ergaster* or late *H. rudolfensis* were able to use tools and began to move out of Africa into Eurasia less than 2 Mya. Respective fossils found in China, on Java, in ice-free Southern Spain, Georgia and Israel dated to be 1.5 – 1.8 Myr old, document this. The European variant of *H. erectus* exemplified by *H. heidelbergensis* (0.7 Myr old) developed into the *H. neanderthalensis* about 0.5 Mya. At that time, the archaic *H. sapiens* developed from *H. erectus* in Africa into the anatomically modern human, *H. sapiens*. When moving out of Africa in a second wave they met with the Neanderthals in the Near East about 80,000 years ago. This *H. sapiens* gradually spread all over the globe living beside or together with the Neanderthals for about 50,000 years before the latter disappeared. Local populations isolated by distance diversified from one another. However, modern man developed from a single evolutionary lineage.

Ancient DNA analysis of Neanderthal specimens supports the idea of an African origin of anatomically modern humans, disproving alternative scenarios with multiple geographic origins and yielded no evidence for gene flow between *H. neanderthalensis* and anatomically modern humans.[56]

Numerous excellent discussions are available on the genetic evidence pertaining to the origin, historical demography and geographic distribution of anatomically modern humans. A recently introduced interpretation of genetic data as obtained from extant populations relies on haplotype trees. To construct such trees sequence data from conserved nonrecombinational coding or noncoding regions of mtDNA, Y- and X-chromosomal DNA, as well as from several autosomal haplotypes derived from populations in Europe, Asia and Africa are compared by statistical parsimony. Such trees derived for individual haplotypes are either identical or topologically compatible. Results of such analyses support the fossil evidence that the human lineage originated in Africa from where it began spreading out to Eurasia about 1.7 Mya. Comparison of individual haplotype trees identify clustered changes in the mtDNA and Y-DNA as well as in the autosomal loci examined compatible with a second more recent out-of-Africa expansion about 0.08 to 0.15 Mya. Interestingly, this latter movement and possibly the former was not a replacement but is rather characterized by interbreeding.[57] More recent out-of-Asia migrations e.g., of the Amerindians cannot yet be reliably dated with confidence by this technique.

## Concluding Remarks

The ongoing detailed analysis is revealing an unpredictable degree of evolutionary plasticity at least for the nonhuman primate and human genome. A more extensive knowledge of the underlying mechanisms generating this diversity and selecting for specific genetic traits will be essential to finally understand the phenotypic differences between humans and its closest relatives and to reconstruct the path of their evolution.

## References

1. Rowe N. The pictorial guide to the living primates. Charlestown: Pogonias Press, 1996.
2. Castresana J. Cytochrome b phylogeny and the taxonomy of great apes and mammals. Mol Biol Evol 2001; 18:465-471.
3. Geissmann T. Vergleichende Primatologie. Berlin: Springer Verlag, 2002.
4. Napier JR, Napier PH. A handbook of living primates. London: Academic Press, 1967.
5. Hacia JG. Genome of the apes. Trends Genet 2001; 17:637-645.
6. Madsen O, Scally M, Douady CJ et al. Parallel adaptive radiations in two major clades of placental mammals. Nature 2001; 409:610–614.
7. Murphy WJ, Eizirik E, Johnson WE et al. Molecular phylogenetics and the origins of placental mammals. Nature 2001; 409:614–618.
8. Murphy WJ, Eizirik E, O'Brien SJ et al. Resolution of the early placental mammal radiation using bayesian phylogenetics. Science 2001; 294:2348–2351.
9. Gregory WK.The orders of mammals. Bull Am Mus Nat Hist 1910; 27:1–524.
10. Novacek MJ. Mammalian phylogeny: Shaking the tree. Nature 1992; 356:121–125.
11. Kay RF, Thorington RW, Houde P. Eocene plesiadapiform shows affinities with flying lemurs not primates. Nature 1990; 345:342–344.
12. Bailey W J, Slightom JL, Goodman M. Rejection of the "flying primate" hypothesis by phylogenetic evidence from the -globin gene. Science 1992; 256:86–89.
13. Graur D, Duret L, Gouy M. Phylogenetic position of the order Lagomorpha (rabbits, hares and allies). Nature 1996; 379:333–335.
14. Porter AP, Goodman M, Stanhope MJ. Evidence on mammalian phylogeny from sequences of exon 28 of the von Willebrand factor gen. Mol Phylogenet Evol 1996; 5:89–101.
15. Lin YH, Penny D. Implications for bat evolution from two new complete mitochondrial genomes. Mol Biol Evol 2001; 18:684–688.
16. Schmitz J, Ohme M, Zischler H. The complete mitochondrial genome of Tupaia belangeri and the phylo-genetic affiliation of Scandentia to other eutherian orders. Mol Biol Evol 2000; 17:1334-1343.

17. Teeling EC, Scally M, Kao DJ et al. Molecular evidence regarding the origin of echolocation and flight in bats. Nature 2000; 403:188-192.

18. Fleagle JC, Macphee RD. Primates and their relatives in phylogenetic perspective. New York: Plenum Press, 1993.

19. Shoshani J, Groves CP, Simons EL et al. Primate phylogeny: Morphological vs. molecular results. Mol Phylogenet Evol 1996; 5:102-154.

20. Tavare S, Marshall CR, Will O et al. Using the fossil record to estimate the age of the last common ancestor of extant primates. Nature 2002; 416:726-729.

21. Martin RD. Primate origins: Plugging the gaps. Nature 1993; 363:223-234.

22. Pumo DE, Finamore PS, Franek WR et al. Complete mitochondrial genome of a neotropical fruit bat, Artibeus jamaicensis, and a new hypothesis of the relationships of bats to other eutherian mammals. J Mol Evol 1998; 47:709-717.

23. Beard KC. Gliding behaviour and palaeoecology of the alleged primate family Paromomyidae (Mammalia, Dermoptera). Nature 1990; 345:340-341.

24. Beard KC. Origin and evolution of gliding in early Cenozoic Dermoptera (Mammalia, Primatomorpha). In: Fleagle JG, Macphee RDE eds. Primates and Their Relatives in Phylogenetic Perspective. New York: Plenum Press, 1993:63-90.

25. Killian JK, Buckley TR, Stewart N et al. Marsupials and eutherians reunited: Genetic evidence for the Theria hypothesis of mammalian evolution. Mamm Genome 2001; 12:513-517.

26. Arnason U, Adegoke JA, Bodin K et al. Mammalian mitogenomic relationships and the root of the eutherian tree. Proc Natl Acad Sci 2002; 99:8151-8156.

27. Schmitz J, Ohme M, Zischler H. The complete mitochondrial sequence of Tarsius bancanus: Evidence for an extensive nucleotide compositional plasticity of primate mitochondrial DNA. Mol Biol Evol 2002; 19:544-553.

28. Schmitz J, Ohme M, Suryobroto B et al. The Colugo (Cynocephalus variegatus, Dermoptera): The primates' gliding sister? Mol Biol Evol 2002; 19:2308-2312.

29. Martin RD. Primate origin and evolution: A phylogenetic reconstruction. London: Chapman Hall, 1990.

30. Starck D. Vergleichende Anatomie der Wirbeltiere auf evolutionsbiologischer Grundlage. Berlin-New York: Springer-verlag, 1978.

31. Kupfermann H, Satta Y, Takahata N et al. Evolution of Mhc-DRB introns: Implications for the origin of primates. J Mol Evol 1999; 48:663-674.

32. Schmitz J, Ohme M, Zischler H. SINE insertions in cladistic analyses and the phylogenic affiliations of Tarsius bancanus to other primates. Genetics 2001; 157:777-784.

33. Kuryshev VY, Skryabin BV, Kremerskothen J et al. Birth of a gene: Locus of neuronal BC200 snmRNA in three prosimians and human BC200 pseudogenes as archives of change in the Anthropoidea lineage. J Mol Biol 2001; 309:1049-1066.

34. Martin RD. Origins, diversity and relationships of lemurs. Int J Primatol 2000; 21:1021-1048.

35. Groves CP. Primate taxonomy (Smithsonian Series in Comparative Evolutionary Biology). Washington, London: Smitsonian Institution Press, 2001.

36. Tattersall I, Schwartz JH. Phylogeny and nomenclature in the "Lemur-group" of malagasy strepsirrhine primates. Anthropol Pap Am Mus Nat Hist 1991; 69:1-18.

37. Yoder AD, Rasoloarison RM, Goodman SM et al. Remarkable species diversity in malagasy mouse lemurs (primates, Microcebus). Proc Natl Acad Sci USA 2000; 97:11325-11333.

38. Arnason U, Gullberg A, Janke A. Molecular timing of primate divergences as estimated by two nonprimate calibration points. J Mol Evol 1998; 47:718-727.

39. Yoder AD, Cartmill M, Ruvolo M et al. Ancient single origin for malagasy primates. Proc Natl Acad Sci USA 1996; 93:5122-5126.

40. Gunnell GF, Rose KD. Tarsiiformes: Evolutionary history and adaptation. The Primate fossil record. Hartwig WC, ed. New York: Cambridge Univ Press, 2002.

41. Andrews TD, Jermiin LS, Easteal S. Accelerated evolution of cytochrome b in simian primates: Adaptive evolution in concert with other mitochondrial proteins? J Mol Evol 1998; 47:249-257.

42. Hershkovitz P. Living new world monkeys (Platyrrhini) volume 1: With an introduction to the Primates. Chicago: University of Chicago Press, 1977.

43. Ford SM. Callitrichids as phyletic dwarfs, and the place of the Callitrichidae in Platyrrhini. Primates 1980;21:31-43.

44. Rosenberger AL. Systematics:the higher taxa. In:Coimbra Filho AF, Mittermeier RA. eds. Ecology and behavior of neotropical primates, vol.1. Rio de Janeiro: Academia Brasiliera de Ciencias, 1981:9-27.

45. Garber PA, Rosenberger AL, Norconk M. Marmoset misconceptions. In: Norconk MA, Rosenberger AL, Garber PA, eds. Adaptive radiations of neotropical primates. New York: Plenum press, 1996:87-95.

46. Singer SS, Schmitz J, Schwiegk C et al. Molecular cladistic markers in new world monkey phylogeny (Platyrrhini, Primates). Mol Phylogenet Evol 2003; 26:490-501.

47. Gebo DL, Dagosto M, Beard KC et al. The oldest known anthropoid postcranial fossils and the early evolution of higher primates. Nature 2000; 404:276-278.

48. Stewart CB, Disotell TR. Primate evolution—in and out of Africa. Curr Biol 1998; 13:R582-588.

49. Page SL, Goodman M. Catarrhine phylogeny: Noncoding DNA evidence for a diphyletic origin of the mangabeys and for a human-chimpanzee clade. Mol Phylogenet Evol 2001; 18:14-25.

50. Roos C, Geissmann T. Molecular phylogeny of the major hylobatid divisions. Mol Phylogenet Evol 2001; 19:486-494.

51. Kaessmann H, Wiebe V, Weiss G et al. Great ape DNA sequences reveal a reduced diversity and an expansion in humans. Nature Genet 2001; 27:155-156.

52. Samonte RV, Eichler EE. Segmental duplications and the evolution of the primate genome. Nature Rev Genet 2002; 3:65-72.

53. Bailey JA, Gu Z, Clark RA et al. Recent segmental duplications in the human genome. Science 2002; 297:1003-1007.

54. Nekrutenko A, Li WH. Transposable elements are found in a large number of human protein-coding genes. Trends Genet 2001; 17:619-621.

55. Brunet M, Guy F, Pilbeam D et al. A new hominid from the upper Miocene of Chad, Central Africa. Nature 2002; 418:145-151.

56. Krings M, Stone A, Schmitz RW et al. Neandertal DNA sequences and the origin of modern humans. Cell 1997; 90:19-30.

57. Templeton AR. Out of Africa again and again. Nature 2002; 416:45-51.

# CHAPTER 4

# How Different Is the Human Genome from the Genomes of the Great Apes?

Eugene V. Nadezhdin and Eugene D. Sverdlov

## Abstract

During Hominoid evolution a lot of sequence and chromosomal organization differences between highly related genomes of human and the African great apes were accumulated. Some of them certainly form a genetic basis for recently evolved, specifically human traits such as brain size of at least 600 cubic centimeters, extended period of childhood growth and development, possession of language, enhanced cognitive capacity, and many others. The human genome sequencing revealed its characteristic features, and the ongoing sequencing of the chimpanzee genome continuously widens the possibilities of large-scale systematic comparison of the two genomes. Such a comparison provides information on all genomic differences, both evolutionary significant and neutral, fixed in evolution by chance. Singling out functionally significant differences from this mess is a rather challenging task. Here we review known general and some particular differences between human and chimpanzee genome organization. Most probably, there is little or no hope to find the genes that make us humans since hundreds or even thousands of genes were involved in the divergence of the two species. The divergence might be caused by changes in gene regulation, by modifications of protein biochemical functions, by gene duplications, losses and acquisitions.

## Introduction

The first undisputed primates appeared in the fossil record about 55 million years ago (Mya) in the Eocene (for more detail see Chapter 3). They were most probably furry and small, weighing from about 200 gram to about 4 kilogram.[1] By the Miocene Epoch, apes had evolved from monkeys, and, in the late Miocene, an evolutionary line leading to hominids (human-like primates) had finally branched off. This hominid line included our direct ancestors. Modern primates include about 200 species, in particular the Hominoidea superfamily (Fig. 1): gibbons and siamangs (*Hylobatidae*), orangutans (*Pongidae*), gorilla, chimpanzee, bonobo (*Panidae*), and humans (*Hominidae*).[a] These Hominoidea are remarkably similar and at the same time dramatically different. They are different not only in their appearance but also in such characteristics as behavior and resistance to various diseases, including cancer and AIDS. Among

---

[a] This is only one possible classification, which will be used in this review. For discussion of others see for example ref. 2.

---

*Retroviruses and Primate Genome Evolution*, edited by Eugene D. Sverdlov.
©2005 Eurekah.com.

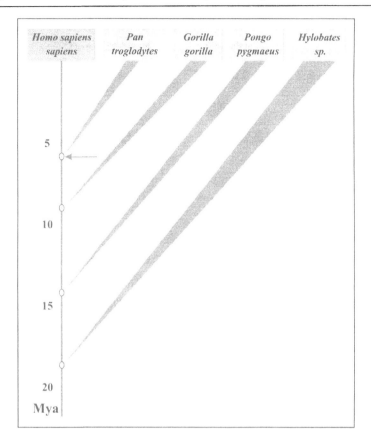

Figure 1. Evolutionary tree of the Hominoid superfamily.

them most closely related are humans and chimpanzees. Many lines of evidence indicate that humans and chimps had a common ancestor, and a great challenge is to reconstruct its genetic architecture and then to understand the ways of its transformation into two closely related, but different architectures of human and chimpanzee. What events caused their divergence in evolution? What genes and regulatory systems were involved in branching hominids off from their closest relatives, chimpanzees and bonobo and then in their proceeding to the Homo genus crowned with extant *Homo sapiens*, that is humans with their brain size of at least 600 cubic centimeters, extended period of childhood growth and development, possession of language and many other human specific traits?[3] One can imagine a dramatic history of our ancestors' severe struggle for existence with unfriendly environment, climate changes, floods and glaciations prior to the Homo genus had been shaped and then progressed further with extinction of some branches and survival of others. During this history each genomic locus could have its own destiny, so that in general the history of genes and the history of species and populations are different histories.[4] The great trial and error game of nature with the random choice of the material for further construction of the genomic architecture system put a great challenge to the researchers who try to unravel the genetic reasons of speciation. Unfortunately, the data available do not provide enough material to make any confident conclusions. Moreover, there is no even agreement on the structure of the phylogenetic tree describing relations among the whole set of the known extinct human relatives that populated the Earth in the past. Figure 2

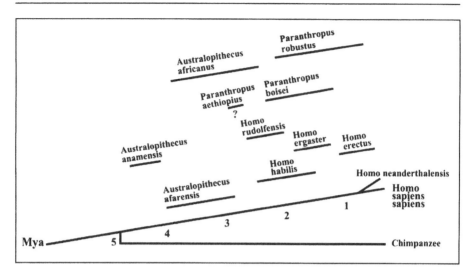

Figure 2. Schematic presentation of hominid evolution (modified from ref. 3). It should be, however, remembered that many of the elements of evolution are still debatable, and some species are not proven to be separate rather than simply a result of distortion of fossils or normal variation of another species phenotype.[140]

(see the discussion in Chapter 3) is a schematic presentation of the information available at this point, which is extremely incomplete and contradictory.

The genomic changes which caused speciation had occurred, most probably, in many genomic loci common for all the species representatives. One can say that these changes form a species specific genetic architecture—a certain number and organization of the genetic peculiarities characteristic for a given species. In parallel, a great many of random genomic changes, neutral or almost neutral for natural selection, had also been accumulated. Many of them might be fixed in species quite randomly, whereas the others varied from individual to individual, thus being polymorphic. For this reason, a vast majority of sequence changes that had occurred during evolution did not directly affect phenotypes. Direct genome-wide comparisons will allow one to quantitatively evaluate the divergence between species and intraspecies polymorphism. This kind of comparison enable researchers to construct phylogenetic trees and to deduce the genetic histories of species branching and population formation and migration. This is essentially a hypothesis-free approach. It would also allow to discard intraspecies polymorphisms as candidates for species specific changes.

However, such a wide approach will not allow to single out from the mass of randomly fixed neutral mutations and polymorphisms the positively selected mutations that defined the genetic basis of being human. Revealing the genes that make us humans will most probably use hypothesis driven approaches, when the research starts from the hypotheses aimed at explaining why and how the most significant human features, such as language or cognitive capacities, could emerge. In this approach, only particular loci will be taken for interspecies comparison. This last line of research will be undoubtedly stimulated by new information on genome-wide comparisons.

Obviously, the chances to reconstruct the succession of the genetic events which had occurred during millions of years of evolution are negligible. But what we can hope to gain as a result of such a comparative research is the deeper understanding of the mechanisms governing the modern genome functioning and the differences in spatial-temporal networks of the events determining the rules of existence of different species representatives. If this goal is achieved, then we will make a great bound toward understanding of what is life in general, what are its nuances regarding our presently flourishing but however endangered species, and how these nuances could evolve.

## First Human—Chimpanzee Molecular Comparisons: Changes in Regulatory Systems Are Most Important for Speciation

Chimpanzee (*Pan troglodytes*) and human (*Homo sapiens*) have been thoroughly compared both as organisms and using biochemical and genetic data available, and since a classical work of King and Wilson[5] it is widely accepted that human proteins and genes are basically 99% identical to their chimpanzee counterparts. This remarkably low difference allowed King and Wilson to conclude that "a relatively small number of genetic changes in systems controlling the expression of genes may account for major organizational differences between human and chimpanzees".[5] It is now evident that the regulatory systems involve not only cis-acting genomic sequences (promoters, enhancers etc.), but also numerous trans-acting factors and cofactors forming complicated networks. Using new, recently developed efficient techniques of structural analysis of proteins and nucleic acids, a large number of new structures were compared having confirmed that homologous, orthologous sequences of human and chimpanzee are indeed at least 98.5% identical. However, quite a number of qualitative differences between the two genomes were also found up to the absence of some chimpanzee DNA stretches from the human genome, and vice versa. The differences can be classified under several groups:

1. Clearly evident differences in chromosome organization.[6-13]
2. Markedly different relative copy numbers, locations, and functional status of individual genes or pseudogenes within multigene families.[13-18]
3. Differences in structures of gene coding regions leading sometimes to serious distinctions between humans and great apes with regard to the presence/absence of gene expression products.[19]
4. Differences in gene expression regulation resulting in different levels of certain mRNAs in different tissues. These differences are present both at the transcriptional and posttranscriptional level, such as alternative splicing.[20]
5. Differences in the number and distribution of interspersed and tandem repeats, segmented duplications and especially of different transposable elements (TEs) that occupy about 35% of the human genome.[21] Numerous data evidence that these DNA sequences were repeatedly inserted/deleted during higher primate evolution.[22-25]
6. General differences in single nucleotide polymorphisms (SNP) and short tandem repeats and satellites' lengths and distribution.

Each of these differences in itself, or much more probably their combinations might affect regulation and be responsible for some events important for speciation. However, at this point it is impossible to single out those genomic variations that led to phenotypic differences important enough to be selected by natural selection.

The rapid accumulation of data in this field is accompanied by periodic reevaluation of the conclusions made from the data obtained earlier, and new conclusions are quite often substantially different from the former ones. Sometimes even very fundamental concepts have to be revised. For example, it is widely accepted that noncoding regions of the genomes can be used for more accurate tracing the history of evolution because they are not directly subject to natural selection. Now, when a great variety of noncoding RNA (ncRNAs) are discovered (see for review refs. 26, 27) and shown to participate in the regulation of genome functions,[28,29] it is becoming more and more evident that this conception is not fully correct.

In view of the above, we will first discuss large scale comparisons of the genomic sequences aimed at the evaluation of divergences between hominoid species and then have a look at known examples of particular differences. Other recent reviews are also devoted to various aspects of this problem,[30-33] but this field is very rapidly changing.

# General Characteristics of Interspecies Divergence and Intraspecies Polymorphisms in Hominoids

Two very general characteristics of a species genome are its intraspecies polymorphism and divergence from the genomes of other, related species. These characteristics depend on many factors such as population history and dynamics, particularly on the mutation rate which is the primary cause for the polymorphism and divergence.

## *Intraspecies Variability (Polymorphisms)*

Understanding the evolutionary context of human variation is indispensable for explaining current phenotypic variation, which is a consequence of our genetic history. Each extant species polymorphism has been caused by an interplay of mutation and internal and external factors including developmental constraints, natural selection, demographic and historical processes, leading eventually to nonrandom distribution of the polymorphisms along the genome both within an individual and among various human populations. A characteristic feature of this evolution is extinction—only a tiny fraction of the variations ever created in a species lineage still persists nowadays. It is especially true for the human population that, as it will be discussed below, experienced an explosion-like demographic expansion from a rather small ancestor population so that only a small remaining fraction currently impacts on human phenotypes.[34]

### The Average Diversity of the Human Genome Is Relatively Low, Probably Due to Bottleneck and Subsequent Explosion-Like Expansion of Our Ancestors Population

The human molecular polymorphism was first recognized in 1919 since the study of ABO blood-group frequencies (reviewed in ref. 35). But the real scale of the human genome polymorphism was fully understood only in the course of the Human Genome Project. The publication of a draft genomic sequence[36,37] was accompanied by the appearance of data on single nucleotide polymorphism (SNP) at more than 1.4 million sites mapped along the draft sequence, reported by an SNP working group.[38] These identified SNPs are scattered over the human genome at a density of one SNP per 1.9 kilobases. The human DNA variation is a subject of intense research summarized in several seminal reviews.[4,31,32,34,39,40] It was realized that the genetic history of a separate locus could be different from that of other loci in the genome, and, consequently, its extant population characteristics such as polymorphism, haplotype conservation etc. can substantially differ from the corresponding characteristics of even closely located loci. Therefore, to correctly construct phylogenetic trees of the species (and not a separate locus) one should rely on many loci and, even better, on the whole genome comparisons.

The variation can be measured[b] as so called nucleotide diversity $\pi$, which is roughly the average number of differences between two genomes, divided by the number of base pairs compared.[4,41] There are various estimates of $\pi$ value (reviewed in ref. 34), the order of the values can be exemplified by a most recent report.[42] The authors presented a compilation of the data available at that time and showed that autosomal regions had, on the average, the highest $\pi$ value (0.091%), while X-linked regions had a somewhat lower value (0.079%) and Y-linked regions the least value (0.008%). The variation between individual nonsex chromosomes is also small and lies within the range of 0.0519% (for chromosome 21) to 0.0879% (for chromosome 15).[38] Characteristic values of variation for sex chromosomes are explained by their patterns of mutation and recombination that differ both from each other and from nonsex chromosomes. Moreover, fewer ancestors have contributed to sex chromosomes that made them less variable than nonsex chromosomes.

The human genome variability is not restricted to SNP. A polymorphism of mini- and microsatellites and insertion /deletion polymorphisms of retroelements, mainly Alu (see for example ref. 43) and L1, are used by scientists working in the field of human population genetics and evolution. It is significant that for all kinds of polymorphisms the frequencies of particular alleles are strongly population dependent. An analysis of the peculiarities of the allele frequencies distribution among various human populations is one of the tools for reconstruction of the evolutionary history of human populations. Levels of genetic variation at equilibrium are used to estimate so called effective population size, $N_e$, of the ancient population of species (see footnote b) representing the size of an "ideal" population, which acts the same as the real population in question.[44] The calculated $N_e$ value depends on many simplifying assumptions concerning the population, such as no selection, random mating, random chance of each offspring having a particular parent etc. It certainly strongly depends on correct evaluation of the mutation rate $\mu$ value, considered above. Generally, $N_e$ is less, sometimes even much less than the actual number of individuals in the population. Despite this variety of uncertainties, different methods using different loci for diversity determination give surprisingly similar $N_e$ values clustering around an approximate long term average effective size of roughly 10,000 for ancestor human population.[39,42,45,46] The variability studies give also a basis for assumption of rapid explosion-like population growth started in late Pleistocene within the past 100,000 years.[45] Together with the material discussed in the above cited reviews, the data on diversity of the human genome provide a strong basis for an, however still debatable,[42] out-of-Africa hypothesis of modern human expansion within the past 150,000 years or so. These data on humankind history are indispensable for those who are interested in the genetic background of the human specific traits.

### Diversity among Chimpanzee and Our Other Nearest Relatives Is Higher Than among Humans

To better understand the evolution of our own it is necessary to find out whether the peculiarities of the gene diversity in *Homo sapiens* are also shared by our closest relatives: chimpanzee, gorilla and orangutan. The most recent reviews[4,31,32] consider this problem in detail, although the information on nonhuman primate genomes is rather scanty as compared to that for the human genome. However, current efforts in chimp genome sequencing (see for example http://sayer.lab.nig.ac.jp/~silver/) and RNA expression patterns determination give us a hope to have more comprehensive databases of DNA sequences already in the near future.

The first research on mtDNA[47] RFLP typing of great apes demonstrated a much higher diversity of this DNA than in human. A higher diversity for apes has been also demonstrated for the first hypervariable region of mtDNA (HVR1).[32,48] However, a study of some proteins (reviewed in ref. 49) unexpectedly revealed that in these cases humans were the most variable primates, even more variable than most other vertebrates. Also, microsatellite variations[49] have indicated that the diversities at these loci are approximately the same in human and chimpanzee. These discrepancies indicate the need of more comprehensive studies of the diversity in nonhuman hominoids.

One of the efforts towards getting more detail on the divergence has been made by Pääbo group (reviewed in ref. 32) which sequenced about 10,000 bp at Xq13.3 in chimpanzee (*Pan troglodytes*), bonobo (*P. paniscus*), gorilla (*Gorilla gorilla*), and orangutan (*Pongo pygmaeus*).

---

[b] Genetic diversity can be measured as nucleotide diversity ($\pi$) based on the average nucleotide differences per site between two sequences randomly selected from a sample. Under equilibrium conditions with respect to mutation and drift, $\pi$ measures the neutral parameter $4N_e\mu$ for autosomal loci, $3N_e\mu$ for X-linked loci and $2N_e\mu$ for Y-chromosome and mitochondrial DNA (mtDNA) loci, where $N_e$ is the effective population size and $\mu$ is the mutation rate.[4]

Intraspecific variation was evaluated for 30 chimpanzees of different *P. troglodytes* subspecies, as well as for five bonobos. The diversity in chimpanzees appeared to be approximately three times as high as in humans. The authors give some evidence that selection is unlikely to be the primary cause for the low genetic variation seen at Xq13.3 in humans. Therefore, the diversity of the chimpanzee genome seems to be generally greater than that of the human genome. The diversity in this locus was also studied in subspecies of gorilla and orangutan. It was shown that orangutans are characterized by approximately 3.5 times and gorillas twice as high Xq13.3 sequence diversity as humans.

Higher average diversity in chimpanzee as compared to human is now demonstrated with most other loci studied so far, including nuclear autosomal, X-linked, and mitochondrial loci, among them the NRY region having substantially higher levels of variation.[4,49,50]

The greater average diversity in chimpanzee suggests a larger effective population size $N_e$, more ancient origins and a high degree of subdivision of chimpanzees. Significantly larger $N_e$ (52,000 to 96,000, see also Chapter 3) values for common ancestors of humans and apes have also been estimated from interspecies polymorphisms at more than 50 loci.[51] Other estimates (see for example refs. 4, 52) also give higher values for the effective population size both for chimpanzees and for the common ancestor of the two species. It was proposed that the human lineage had experienced a large reduction in effective population size after its separation from the chimpanzee lineage. The explosion like expansion of the population discussed above has happened after this reduction.

To conclude the discussion of the diversity of human and its closest relatives, it should be reminded that the diversity in itself is an experimentally determined value, whereas the conclusions drawn using this value for calculations of effective population size, however brilliant and tempting they were, are just speculations based on many assumptions and thus leading to idealized patterns of events. It therefore leaves much room for doubts in main points of the evolutionary scenario based on these theories. In particular, the debates around the origin of modern humans are far from being finished. One of the two main models suggests a recent (about 100,000-150,000 years ago) out-of-Africa origin of modern humans who then dispersed throughout the Old World, replacing preexisting archaic hominids with little or no admixture (review in refs. 32, 45). This model is quite compatible with the existing genomic data. But, as correctly indicated by Relethford,[45] compatible does not mean proved. A competing multiregional evolution model is still not unequivocally rejected. This model views all hominid evolution since the origin of *Homo erectus* as taking place within a single evolutionary lineage. According to this hypothesis, archaic human ancestors (*H. erectus*) evolved into modern humans in Africa, Asia, and Europe with some migration (gene flow) in between these areas.[42,45] Moreover, sober voices are being raised saying that both hypotheses are too simple to explain the evolution of modern humans.[42,53]

## Interspecies Divergence and Human's Place in Primate Phylogeny As a Basis for Understanding the Genetic Processes Operated in Primate Genome Evolution

Considerable evidence on primate phylogeny has been accumulated over last years based on a great many DNA sequence comparisons. It is internally congruent and compatible with fossil osteological evidence.[52,54] This fossil evidence allows one to convert the DNA sequence divergences between orthologous sequences in different species into the rate of evolutional genomic changes ("molecular clock") and thus to evaluate the time of species divergence. Long debates (for review see refs. 2, 32, 55) have finally led to a current consensus concerning the succession of the Hominoid species branching: first, the human ancestral lineage branched off from gibbons, then from orangutans, gorillas, and, finally, chimpanzees (Fig. 1). Hot debates, however, boil down to the rate of molecular clock, its constancy (for most recent reviews see refs. 56, 57), and therefore to the times of branching off. The discussion of this complicated problem being beyond this review's scope, we will therefore describe here experimental data on genomic divergences among different representatives of the Hominoidea. Combined with similar data for other primates, they form a basis for plenty of (often contradictory) conclusions regarding the rate of evolution and timing of the branching points (reviewed in refs. 2, 51, 58).

The most recent data on the divergence among Hominoid species are based on the average values obtained[51] by comparing 53 autosomal intergenic nonrepetitive DNA segments from the human genome and the genomes of chimpanzee, gorilla, and orangutan. The average sequence divergence appeared to be 1.24 ± 0.07% for the human-chimpanzee pair (this figure is very close to that obtained later by other authors[59]), 1.62 ± 0.08% for the human-gorilla pair, and 1.63 ± 0.08% for the chimpanzee-gorilla pair. According to these authors, the orangutan genome differs by about 3% from each of the genomes of human, chimpanzee and gorilla. All these figures are markedly lower than previously reported. For example, a figure for the human-chimpanzee genome divergence obtained from data on DNA hybridization[60] or on the DNA sequence comparison for the η-globin pseudogene region[61] was 1.6%. The hybridization data might be insufficiently exact for evaluation of divergences,[62] though they were probably good enough for evaluation of the branching orders.[60,63] As to the selected genome region, the figures of divergences are known to considerably vary from region to region (reviewed in ref. 51). In addition, the regions with interspersed repeats, especially GC reach like Alu, can have increased diversity due to enhanced mutation rate of CpG dinucleotides in the genomes.[59] Therefore, the data of Chen and Li based on the average divergence of the 53 intergenic loci provide a better approximation to the real average genomic divergences among hominoids, the more so that the observed frequency of CpG is much lower in the 53 (noncoding) segments (0.69%) than in the gene sequences (2.77%). Chen and Li[51] also analyzed a large collection of chimpanzee and gorilla sequences available in GenBank. They found the largest (about 2%) differences in Alu sequences, followed by pseudogenes and then by intergenic regions, synonymous sites, introns and nonsynonymous sites. A coding region analysis identified amino acid sequence divergence between human and chimpanzee (1.34%), and between human and gorilla (1.58%) proteins (for review see ref. 31).

In a recent work by Svante Pääbo group[59] the sequences of totally 3 Mb from > 10,000 regions in chimpanzee genome were compared to the human genome. Two thirds of them could be unambiguously aligned to the human DNA sequence having shown that unequal content of differences is characteristic of different chromosomes: from 1.0% of differences for X-chromosome to 1.9% for Y. Autosomes also contained variable number of differences, from a minimum of 1.1% for human chromosome 11 and the corresponding chimpanzees chromosome to a maximum of 1.5% for the pair corresponding to human chromosome 21.

A human-chimpanzee comparative clone map[64] was constructed by means of paired alignment of 77,461 chimpanzee bacterial artificial chromosome end sequences (BES) with the human genome sequences available. In total, 48.6% of the whole human genome was covered by the chimpanzee bacterial artificial chromosomes (BACs). The authors detected some positions, among them two clusters on human chromosome 21, that suggest the existence of large, nonrandom regions of difference between the two genomes, in accord with previous observations that different regions of the genome have different mutation rates.

Many of the chimpanzee BACs could not be mapped on the human genome. It is premature to discuss the reason(s) for this variance because at least some of them might correspond to unsequenced human genome regions, however it is also possible that the BACs represent sequences present in chimpanzee, but absent from the human genome. In this regard it should be mentioned that, according to Ebersberger et al,[59] 7% of the chimpanzee genome sequence had no similarity with the human genome. Whether these sequences correspond to nonsequenced part of the human genome or were acquired in evolution after the species branching remains to be clarified.

A large number (2,407) of deletion/insertion differences almost equally distributed between the two genomes was detected, suggesting equal rates of these mutations in the both species. The presence of deletion/insertion differences between the genomes of human and chimpanzees was confirmed also by subtractive hybridization[65-68] (see Chapter 2) which allows genome-wide comparisons of the genomes. Some of these stretches can be involved in the regulatory networks determining phenotypic differences between the Hominoid species.

## Divergence Dates, Mutation Rates and Molecular Clock

As soon as the paleontological dating for some of the branching points is available, the divergences of the genomes of the extant species, that are descendants of the common ancestors existing before these branching points, can be recalculated into mutation rates. In turn, the mutation rates can be used for timing of other branching points, providing either overall constancy of the molecular clock or at least approximate constancy e.g., for a given clade. The molecular clock problem was for a long time a subject of hot debates as well as divergence dates between human and other hominoids (e.g., refs. 52, 60, 69). It was noted that some organisms seemingly have a higher rate of substitutions than the others. In particular, the rate is higher for warm-blooded than for cold-blooded vertebrates, and much higher for rodents as compared to primates. As more and more comparative molecular data was collected, ample evidence against global constant molecular clock has been accumulated. It is now widely accepted that the molecular clock rate is not constant but different both within and between various genomes. Within scopes of the neutral evolution and molecular clock concepts, subsidiary hypotheses were proposed to explain the reasons for the rate heterogeneity, though with certain constraints (review in ref. 70). Recently, it was observed[57] that (i) the rates of evolution change erratically between and within lineages, and that (ii) the patterns of the change for different genes are discordant. These observations are inconsistent with the molecular clock and any of the subsidiary hypotheses. Probably, local rate variations are determined by peculiarities of the genetic history of a given locus, which is not directly connected with the species history. The authors[57] indicate that their results "amount to a denial of there being a molecular clock, although there would be an overall correlation between amount of change and time elapsed".

The scale of errors in the evaluation of species divergence dating can be dramatic due to these "vagaries" in rates of various genes evolution. Thus, using the average rate of *Drosophila* evolution over the last 60 Myr for estimating the time of other organisms' divergence, the divergence of the three multicellular kingdoms is dated at 7,045 Myr by *GPDH*, and at 451 and 398 Myr by *SOD* and *XDH*, respectively. Here the former value is clearly a gross overestimate, while the latter values are blatant underestimates.[57]

Therefore, the extension of mutation rate values obtained for one or few loci to the general mutation rate of taxons should be done with great caution, which is also true for primate evolution. This caution should be extended to the presently popular hominoid slowdown hypothesis (review in refs. 51, 54, 71) which postulates that the rate of molecular evolution has become slower in the hominoid (apes and humans) lineage since its separation from the Old World monkey lineage and slowed down even more in the human and chimpanzee lineages as compared to the rate in the gorilla lineage.[71]

However, some regions in the gorilla lineage were shown to evolve significantly slower than their counterparts in the human and chimpanzee lineages.[51] An analysis of 53 intergenic loci led the authors[51] to a conclusion that there is no strong trend toward slowdown in the human or chimpanzee lineages, and the molecular clock hypothesis generally holds well for intergenic regions and synonymous sites in the human, chimpanzee, and gorilla lineages. This conclusion looks convincing even though some of the evolutionary implications of the hominoid slow-down hypothesis are very attractive: for example that life-history strategies that favor intelligence and longer life spans could also select for decreases in de novo mutation rates

**Table 1. Some recent estimates of the branching points timing for the hominoids (Mya). Modified from ref. 2 with additions.**

| | | Branching Time for Hominoid Species | | | |
|---|---|---|---|---|---|
| Technique | Authors | Gibbon | Orangutan | Gorilla | Chimpanzee |
| NuclearDNA | Hasegava et al (1987)** | | 9 – 15 | 4.1 – 7 - 8 | 3.1 – 7.0 |
| sequence | Eastel et al (1995)** | | 7.3 – 9.8 | About 5 | 3.2 – 4.5 |
| | Holmes et al (1988)** | | 14.5* | 16.2 – 9.2 | 4.7 – 5.1 |
| | Page and Goodman[71] | 18 | 14 | 7 | 5 – 6 |
| | Takahata and Satta[75] | | | 8 | 4.5 |
| | Stauffer et al[58] | 14.9+/- 2.0 | 11.3+/- 1.3 | 6.4+/- 1.5 | 5.4 +/- 1.1 |
| | Chen and Li[51] | | 12 – 16* | 6.2 – 8.4 | 4.6 – 6.2 |
| mtDNA | Adachi and | | 13 – 16* | 7 – 9 | 4 – 5 |
| sequence | Hosigava (1995)** | | | | |
| | Horai et al[76] | | 13* | 6.3 – 6.9 | 4.7 – 5.1 |
| DNA | Sibley et al[63] | 16.4 – 23 | 12.2 – 17* | 7.7 – 11 | 5.5 –7.7 |
| hybridization | Ruvolo (1995)** | | | 7.4 – 8.9 | 5 – 6 |
| Paleontology | | | 13 – 16 | | >4.4 |
| data** | | | | | |

*Based on paleontology data; **References from ref. 2.

perhaps via improved mechanisms for DNA repair and for detoxifying free radicals.[54,71,72] The sequencing of about 10,000 bp at Xq13.3 in chimpanzee, bonobo, gorilla and orangutan also demonstrated the absence of significant differences in the mutation rates of this region.[32,55]

It should be also noted that not all agree that the global molecular clock does not exist.[73] The coming sequencing of the apes genomes[33,74] will make possible a truly genome wide-comparison of divergences among hominoids and other animals thus contributing to more comprehensive understanding of average and local rates of evolution.

At this point, keeping in mind uncertainties in paleontological standards taken by different authors for calculations of mutation rates and times of branching, different loci used for analysis and different concepts of the constancy of the molecular clocks, a considerable spread in divergence times obtained by different authors does not seem surprising.

In view of extreme importance of the divergence times for various evolutionary implications, we compiled the data for Hominoid branching times reported by different authors in Table 1.

## Cytogenetic Differences

Chromosome sets are actively rearranged in evolution due to inversions, translocations, Robertsonian rearrangements, and other types of chromosomal mutations. The more remote the organisms are from each other, the more pronounced the distinctions between their chromosome sets. Human and higher primate chromosomes are structurally similar.[77] However, the human karyotype differs from that of higher primates in one pair of chromosomes: 2n=46 in human against 2n=48 in higher primates. Human chromosome 2 is the result of telomeric fusion of two ancestor chromosomes equivalent to the present-day chimpanzee chromosomes 12 and 13 and gorilla chromosomes 13 and 14.[7,78,79] The formation of human chromosome 2 is obviously an example of Robertsonian fusion. However, human chromosome 2 is not a precise replica of the two higher primate chromosomes.

Other structural differences between human and higher primate chromosomes were also described. They include pericentromeric inversions of human chromosomes 1 and 18,[12] inversions in the gorilla chromosomes homologous to human chromosomes 5 and 17,[80] different distribution of alpha-satellites along human and higher primate chromosomes,[81,82] pericentromeric inversions in the chimpanzee chromosomes homologous to human chromosomes 4, 9, and 12. In the latter case inversions in the chromosomes homologous to the human chromosome were identified for chimpanzee and gorilla.[8]

It was proposed that chromosome rearrangements occur with the involvement of repeated elements. There is an example of an extended for more than 62 kb interchromosomal exchange at the periphery of alpha-satellite DNA that includes nonalpha-satellite genomic regions with the formation of paralogous loci on several human and higher primate chromosomes. The spreading of these sequences in the genome happened both before the divergence of the human and higher primate ancestor lines and after it.[83] It was also shown that the 32-bp AT-rich subterminal satellites located near telomeres in chimpanzee and gorilla are absent from man.[84] Another example of human-specific structural alterations of chromosomal material is the duplication of a longer than 15 kb fragment of human chromosome 6 which contains the minisatellite lambda MS29 sequences, with the formation of a paralogous locus on chromosome 16.[85] A Hominidae family-specific X-Y transposition followed by an inversion in Yp by means of LINE-LINE recombination was described.[13]

Chromosomal rearrangements can potentially lead to gene duplications, formation of new coding sequences due to the emergence of new combinations of exons, and formation of new gene clusters and families such as clusters of olfactory receptor genes.[86] However, no examples of human-specific chromosomal rearrangements leading to the emergence of new functional sequences are known so far. Nevertheless, a possible evolutionary role of some above-described distinctions in the interspecies reproductive insulation and speciation can be suggested.

## Differences between Particular Elements of the Hominoid Genomes

General differences allow to understand the place of humans in the phylogenetic hierarchy of Hominoids and general trends in the genome evolution. However, to understand why humans are so peculiar as a species, one will have to search particular genomic differences distinguishing humans from other hominoids. Unfortunately, at this point only highly scattered data on such comparisons are available, and little evolutionary information relevant to species peculiarities can be drawn from the data available. However, a brief outline of what is presently known could be of help in future systematic genome-wide analyses performed in parallel with the sequencing of hominoid genomes.

### Mitochondrial Genome

Mitochondria are organelles that are present practically in almost all eukaryotic cell types. They represent a unique type of animal cell organelles and have their own genetic apparatus: a circular DNA molecule of about 16.5 kb in length encoding 12S and 16S rRNA, 22 tRNA, and 13 polypeptides, a majority of which are elements of the mitochondrial chain of electron transport.[87] The main function of mitochondria is energy supply to the cells. Energy provision is a central problem for all living entities, and it is not surprising that defects in mitochondrial function often lead to physiological disorders. The number of diseases presumably caused by mitochondrial mutations is increasing. Mutations in mtDNA were suggested to be accumulated with age reducing the efficiency of energy provision. The 'leakage' of electrons which occurs during $O_2$ metabolism results in the formation of small amounts of damaging and mutagenic free radicals that contribute to the rate of mutation. Mitochondria may also play a key role in such vital cell functions as apoptosis.[88] In this regard it would be not surprising if changes in mitochondria itself or in networks involving mitochondria were responsible for some important events in divergence of human and chimpanzee lineages. These networks include, in particular, nuclear genes such as those encoding subunits of cytochrome c oxidase (*COX*) which form a multisubunit complex on the mitochondrial inner membrane, catalyzing the terminal step in electron transport.

A great increase in body and brain size of the earliest *H. sapiens* ancestors[89] required increased metabolic resources. In particular, the human brain, which is 2% of the body weight, uses some 20–25% of the human body metabolic energy. Therefore, the system responsible for energy supply might well be involved in human evolution. In this regard it should be mentioned that there is some evidence in favor of positive selection of the *COX* genes during primate evolution[90] (to be discussed in more detail below).

In general, human and higher primate mitochondrial genomes are structurally organized in similar way.[91] At the same time the average level of divergence between the human and chimpanzee mitochondrial genomes is about 8.9%, markedly, approximately tenfold, exceeding that for nuclear genomes of these species.[51,91,92] The level of divergence of coding parts of the human and gorilla mitochondrial genomes is even higher and amounts up to around 10.5%.[93] The divergence of the chimpanzee and gorilla mitochondrial sequences is about 10%, and by this parameter man and chimpanzee are evolutionary closer to each other than each of them to gorilla. The mitochondrial D-loop (a segment of about 1100 bp responsible for replication of mitochondrial genome) is the most variable region of the mitochondrial genome, and the level of its sequence divergence between man and chimpanzee makes up 13.9%.[91] It has been shown that the sequence polymorphism level in gorilla is higher than in chimpanzee, while the latter is higher than in human.[93,94]

The higher the general divergence of human and higher primate mitochondrial genomes may arise due to peculiarities of the mitochondrial genomic DNA replication, the higher susceptibility to mutations because of free radicals,[95] as well as to the peculiarity of distribution in cells and inheritance of mitochondrial genetic material.[96]

Differences in the mitochondrial sequence polymorphism in human and higher primate species can be determined by various factors, such as distinctions in migration principles of male and female individuals of these species, population subdivisions, a later effect of a founder in human population, differences in population effective sizes, and different selective pressure.[31,32] However, we are still very far from conclusive understanding what was the role, if any, of all these changes in the human phenotypic evolution.

## Repetitive Elements

The primate genomes are abundant with various repetitive elements starting from microsatellites and finishing with different transposon-like sequences. In this section we will not consider potentially evolutionary important differences between humans and nonhuman primates in content and positions of transposon-like elements like Alu, LINE, and endogenous retroviruses. This topic will be discussed elsewhere in this book (see Chapters 6 and 8).

As to other repeats like various satellites, numerous examples of differences between human and chimpanzee were reported (see for example ref. 84). It is not surprising in view of enhanced rate of changes and general instability of such simple repeats. Differences for the repeats located near telomeres[84] or centromeres[97,98] of human and other Hominoid chromosomes, and rapid evolution of microsatellites[99,100] were reported. Moreover, comparisons of allele length distributions for 42 human microsatellites with their homologues in related primates revealed a significant trend for the loci to be longer in humans, showing that microsatellites can evolve directionally and at different rates in closely related species.[99,100]

## Differences in Gene Structures and Expression Levels

### Gene Inactivation Sometimes Occurs in One Species but Not in Others: Is Less Sometimes More?

Recently a 'less-is-more' hypothesis, accentuating the importance of loss-of-function mutations on a recently evolved novel lineage such as the human, has been formulated. It develops the idea that human is a 'degenerate ape' in many biological respects.[101] In a very recent review[33] it is put in the following way: "To some degree, genetic loss in the human lineage seems almost certain to have caused some of the differences that exist between chimpanzees and humans: examples such as delayed postnatal development and loss of muscle strength and hair in humans all seem 'degenerative'. However, it is unclear how far it is safe to extrapolate from these conspicuously retrograde phenotypes to broader aspects of human and chimpanzee biology". Here can be added that a well known example of the neoteny, when infant chimpanzees look very similar to adult humans could be considered in the same way. Indeed, it seems rather straightforward to suggest that if a particular phenotypic trait is lost, then it is most probably due to a loss of a particular gene(s) function. But actually a degenerative phenotype can be potentially caused by any kind of mutations causing loss of gene function, changing the function, causing heterochronic shifts, or gain of function. In the latter case the gain can lead to an appearance of a repressor that either inactivates the gene expression or inhibits its product at the protein level etc.

Nevertheless, a few examples of human-specific gene inactivation are already known (for review see refs. 30, 31). They include the inactivation of the *V10* cellular receptor gene by a point mutation in a donor splice-site,[102] the inactivation of a great number of human olfactory receptor genes as compared to those in chimpanzee and gorilla,[18] and probably a most interesting case of inactivation of the CMP-sialic acid hydroxylase gene due to an Alu-dependent recombinational deletion of this gene exon in man.[19,103,104] So far it is a unique example of genetic distinctions leading to global alterations at the biochemical level. The inactivation of this gene resulted in a complete disappearance of N-glycolyl-neuraminic acid (Neu5Gc) from the human cell surface and in an increase in the amount of N-acetyl-neuraminic acid (Neu5Ac). A possible biological significance of the Neu5Gc absence is changes in intercellular interactions, and a higher resistance of human, as compared to higher primates, to certain bacterial and viral pathogens which use neuraminic acid derivatives as ligands for attachment and penetration into host cells. Possible effects of the Neu5Gc absence on the development and functional activity of human brain are discussed.[30]

There is also a couple of other similar examples, but however interesting all they are, a paucity of data available (which the authors of the hypothesis clearly admit) and a great misbalance between what we know about the human and chimpanzee genomes make us refrain from considering human specific loss of functions as a major engine of human evolution. Moreover, simple "yes-no" logic may be not dominant in evolution. Simple quantitative changes in gene expression could play a vital role in combination with other modifications, as discussed above.

### Gene Duplications Could Proceed Differently in Different Primate Species after Their Divergence

One of the most attractive candidates on the role of speciation pace makers is continually going on segmental genome duplications that could provide fresh material for functional novelties. Interspersed duplications having appeared over the past 35 million years of evolution now occupy approximately 5% of the human genome.[36,37] They include both segmental duplications between nonhomologous chromosomes (transchromosomal duplications) and duplications mainly restricted to a particular chromosome (chromosome-specific duplications).

Many of them span large genomic distances (1-100 kb) and exhibit a high degree of sequence identity. The role of duplications in evolution is discussed in more detail in Chapter 1, and here we shall limit the discussion to some examples of the hominoid specific duplications with an emphasis on human specific ones.

The evidence that duplications of genome segments in the primate lineage actively proceeded during the past 35 million years is now abundant (reviewed in refs. 105,106). Some of them were shown to be a subject of very fast changes and possibly positive selection. The most exiting example of such a supply of material for evolutionary change is provided by a 20 kb segment of chromosome 16, termed LCR16a (reviewed in ref. 106). Its recent (12–5 Mya) duplication(s) has resulted in 15–30 new copies dispersed throughout 15 Mb of the short arm of human and chimpanzee chromosome 16. In some of these human duplicates a novel hominoid gene family (termed *morpheus*) was discovered. An unusual feature of this family was an increased rate of mutations within the exons as compared to the introns. Most of the mutations (>95%) led to amino acid substitutions, so that the average coding sequence divergence (15–20%) between human and chimpanzee is considerably higher than the divergence of the intronic sequences (1–2%). The enhanced amino acid replacement started basically after the separation of human, chimpanzee and gorilla lineages from orangutans. The data suggest that this duplication evolved under strong positive selection pressure. It remains to be understood what this selection means and whether it was important for the adaptation of primate organisms to their environment.

Another widely discussed example of a human specific duplication is the Yp11.2 block that arose due to a translocation of Xq21.3 with a subsequent paracentric inversion after the separation of the chimpanzee and hominid lineages.[107] The Yp11.2/Xq21.3 human specific homology block constitutes the largest shared region among the sex chromosomes, spanning some 3.5 Mb. Only three transcribed sequences have been mapped to this segment: the protocadherin genes *PCDHX/Y*, the X-linked poly(A)-binding protein *PABPC5* gene whose Y-homolog has been lost during human evolution and a homeodomain-containing gene, *TGIFLX/Y* (ref. 108 and references therein). A very interesting hypothesis was suggested that this region and in particular protocadherin XY is somehow involved in handedness, language lateralization, brain asymmetry and schizophrenia.[107,109,110] But however tempting this hypothesis may appear, it should be kept in mind that all of the above mentioned traits are complex and can not be assigned to one certain gene. A further discussion of this problem can be found in Chapter 1.

The investigations on the human specific duplications, their functional significance and possible roles in evolution are just in the very beginning, but it is safe to predict that the research in this field will bring a lot of unexpected and exiting results.

### Differences in Known Active Gene Structures: Was Darwinian Positive Selection a Driving Force Involved in Making Us Humans?

Multiple differences in structures of active genes among humans and other Hominoids were reported. They were shown to include single nucleotide synonymous and nonsynonymous substitutions, deletions/insertions in the coding and noncoding regions of the genes, up to deletions of entire exons, changes in splicing signals etc. (see for examples refs. 111-115).

There are also rather numerous examples[31] in favor of possible involvement of positive selection in evolution of this or that particular gene (nonsynonymous/synonymous (N/S) substitution rate is expected to be increased for a positively selected protein during the period of evolution when the positive selection was in action[44]). However, these examples are insufficient to even roughly reconstruct the networks, in which these proteins are presently involved, or identify their roles before the selective forces started to act. The situation is even worse with regulatory cis-acting elements because in this case there are no criteria for detecting positive selection of protein encoding genes. The same is with positive selection of nonprotein molecules such as noncoding RNAs, a seemingly important component of the cellular regulatory machinery.[116]

One of the most systematic researches devoted to the role of positive selection in primate evolution was carried out by Goodman's group on positively selected adaptive changes in the biochemical machinery for aerobic energy metabolism.[72,90,117-119] These changes were quite probably among important molecular events in Hominoid evolution. The authors suggested that a likely evolutionary pressure for adaptive changes in energy metabolism in the anthropoid primates had been the emergence of a larger neocortex, one of the most aerobic and energy consuming tissues. Such a pressure may have been in part responsible for the peculiarities of evolution of cytochrome c oxidase (COX) subunits in the anthropoid primates. COX plays a vital role in aerobic energy metabolism catalyzing the final step of electron transfer through the respiratory chain. COX is a multisubunit enzyme complex and the authors presented evidence of higher nonsynonymous (N) than synonymous (S) mutation rates in COX subunit IV in the lineage encompassing catarrhine and hominid stems. This suggested a positive selection for adaptive amino acid replacements in this subunit. A marked deceleration of N rates with much lower N than S rates in the terminal lineages to gorilla, human, and chimpanzee genes indicated that the positively selected changes were then preserved. Indications of positively selected changes in our ancestors were reported also for other COX subunits and some of other proteins active in the respiratory chain.[90,118-120]

The authors proposed the analysis of rate speedup/slowdown pattern of nucleotide substitutions similar to that used in the case of COX IV subunit as a paradigm for detection of evolutionary significant changes among the oceans of other human and nonhuman primate genomes.

A few other examples of N/S increase as an evidence of positive selection were reviewed in reference 31.

It was reported that 5-7 of 70 genes analyzed were under the effect of positive selection. Extrapolating these values to 30,000 genes in the human genome suggests that about 3,000 genes have undergone positive selection since the time of the most recent common ancestor of human and chimpanzee.[121] If confirmed, this enormous number of positively selected genes demonstrates how complex was the process of the evolutionary divergence of human and chimpanzee and how difficult it would be to answer the question of what genes make us human.

Goodman et al[72] believe that the similar rate speedup—rate slowdown paradigm could also work for detection of evolutionary important changes in cis-regulatory elements. Sequence changes within cis-elements are expected to occur at high rate in those species which experience selective pressure. As soon as the adaptation is achieved the descendants of the species will be subjects of purifying selection resulting in the conservation of the properly changed cis-element sequences. At the same time, nonfunctional flanking sequences will continue to rapidly mutate. Thus, a functional element starting from a certain evolutionary point will become a conservative island surrounded by continually changing genomic areas. In this regard, it would be interesting to compare the divergences among orthologous retroelements and their flanking sequences in the primate genomes, since these elements are real candidates for being involved in evolutionary significant changes. Such an approach, though correct in principle, is however extremely difficult to use in practice, because it would require many genome-wide interspecies sequence comparisons which are highly time and cost consuming.

Some examples of possible positive selection not connected with N/S changes were reported, and different arguments were put forward in favor of positive selection for this or that allele of loci under investigation. One of interesting though highly speculative examples is the highly variable human dopamine receptor D4 (DRD4) gene.[122] Most part of the gene diversity is due to the length of a 48-bp tandem repeat (VNTR) in the third exon and single-nucleotide polymorphism within the repeat. In a world-wide population sample, the most commonly occurring allele (65.1%) contains four repeats (4R), a less frequent form (7R) contains 7 repeats (19.2%), and more rare forms contain from 2 to 11 repeats (2R-11R). The authors provide evidence that the 7R allele has arisen from the 4R allele due to recombination and mutation and is probably 5–10-fold younger than the 4R allele. This allowed the authors to suggest that the high frequency of the 7R allele is the result of positive selection. This example is interesting because it may demonstrate the selection of one of the causative factors determining personality traits specific only for humans. The 7R allele in modern human population is associated with attention deficit/hyperactivity disorder (ADHD) that manifests itself in affected children as inattentiveness, impulsivity and hyperactivity. The authors speculated that the benefit provided by the 7R polymorphism that emerged 40,000-50,000 years ago, could be in risk taking, drive and impulsiveness. Again, this example shows how far we are from real understanding of the relations between genomic characters, even shown to be positively selected, and the phenotypes underlain by these genomic features.

### Differences in Expression Levels of Genes in Various Tissues

According to a hypothesis of King and Wilson[5] that changes in gene regulation is a key factor in speciation, one can expect differences in gene expression levels in homologous cells of different species even if their gene contents are similar. The differences will manifest themselves both in different rates at which messenger RNA and proteins are made from a gene in the same type of cells and in shifts in the gene expression to distinct cell types in one but not the other species.

Recently, Pääbo's group[123] reported experimental evidence in favor of the regulatory hypothesis for human and chimpanzee species. Moreover, their data suggest that most essential changes occurred in gene expression in the human brain. Using microchips and membrane-based DNA arrays containing fragments of about 18,000 genes, Enard et al[123] compared mRNA contents in brain, liver, and blood samples from humans, chimps, macaques, and orangutans died of natural causes. Protein contents were also compared by two-dimensional gel electrophoresis. In blood leukocytes and liver the human expression patterns appeared to be more similar to those of chimpanzee than to those of macaques reflecting the evolutionary relationship of the species. In contrast, the expression pattern in the chimpanzee brain cortex was more similar to that of macaque than human. This result suggests that the rate of evolutionary changes of the gene expression levels in the brain was accelerated in the human evolutionary lineage relative to chimpanzee, whereas no such acceleration was evident in liver or blood. The authors compared also the gene expression patterns in different strains of mice and concluded that many quantitative changes in gene expression could be detected between closely related mammals, but the differences had been particularly pronounced in recent evolution of the human brain. They also indicated that "The underlying reasons for such expression differences are likely to be manifold, for example, duplications and deletions of genes, promoter changes, changes in levels of transcription factors, and changes in cellular composition of tissues. A challenge for the future is to investigate the relative contributions of these factors to the expression differences observed. A further challenge is to clarify how many of the differences have functional consequences". It also remains to guess when the changes observed happened during hominid evolution. They could be accumulated gradually since the human-chimp lineages divergence, or start e.g., 2-3 Mya, when a dramatic increase in the brain volume accompanying the appearance of *H. erectus* was first observed, or begin in the latest period, a couple of hundred thousand years ago when modern human was formed.

There are quite a few reports dealing with various particular gene expression differences. A good example of how apparently insignificant changes in promoter structures could cause considerable differences between human and chimpanzee is the expression of the atherothrombogenic lipoprotein (a) gene. In the blood plasma of chimpanzee its level is much higher than the mean value in man. Although the promoter regions of human and chimpanzee *apo(a)* genes were shown to be 98% identical, "minor" nucleotide differences in the 5' region of the gene resulted in five-fold enhancement of the chimpanzee promoter as compared to that of human.[124]

A comparative analysis of proteins from the human and higher primate blood plasma revealed distinctions between man and higher primates in the amount of haptoglobin isoforms (see above) and in transthyretin concentration.[125] Transthyretin is expressed in liver and brain and is one of three proteins - carriers of thyroid hormones in the blood plasma and blood-brain barrier. The transthyretin concentration in the chimpanzee blood plasma and spinal fluid is approximately twice as high as in human.[125] An independent analysis of transthyretin expression level in the human and chimpanzee brain showed similar results.[126] The level of free thyroid hormones in the chimpanzee blood plasma is higher than in human.[125] These results seem to be interesting because of essential role of thyroid hormones in the processes of brain development and functioning. The findings deserve intent attention and require further investigation. Other examples can be found in a recent review.[31]

Of course, it should be stressed that, as correctly put by Olson and Varki:[33] "the potential for artifacts in such studie—given their reliance on autopsy material from both species and on the anatomic complexity of primate brains—is sobering. For example, it is possible that there are age or species-specific differences in the rate at which particular brain tissues deteriorate after death—or simply differences in the handling of chimpanzee and human tissue that are difficult to eliminate".

The data on differences in gene expression demonstrate them to be mostly quantitative. The question is how significant these quantitative differences can be regarding their importance in evolution. Recent examples show that as low as two fold differences in the level of a transcription factor can dramatically change the developmental program. A study by Niwa et al[127] (see also comment by Stewart[128]) shows that changes in the transcription factor Oct-4 levels define the differentiation of embryonic stem (ES) cells along three different pathways: a "normal" level is compatible with blastocyst inner cell mass phenotype, repression of Oct-3/4 induces loss of pluripotency and dedifferentiation to trophectoderm, while a less than twofold increase in expression causes differentiation into primitive endoderm and mesoderm. Thus a critical amount of Oct-4 is required to sustain stem-cell self-renewal, and up- or downregulation induce divergent developmental programs.

Clearly, Oct-4 does not operate alone, and it would seem that it is the ratios between the levels of Oct-4 and the other factors that ultimately determine cell fate. An extrapolation of the data on Oct-4 indicates that development may in general strongly depend upon quantities of transcription factors which might be therefore a possible driving force of evolution.

## Conclusion: A Supercomplex Disease—To Be a Human

About 5 million years ago something happened with the population of the common ancestors of human and chimpanzee, and the two lineages started their independent evolution. First representatives of the hominid lineage right after the split between chimpanzee and the hominids such as *Ardipithecus ramidus*,[129] were better adapted for climbing but a later evolved group of species, the australopithecines, though clearly belonged to the great apes, already started to walk upright. This new adaptation was most probably caused by certain genetic changes—the first Genetic Revolution had occurred (Fig. 3).

Somewhere between 2.5 and 1.8 Mya and ~3 Myr after the branching of human and chimpanzee lineages, human evolution has been marked by an appearance of a species with a never seen before combination of features. Its fossils showed a significant change in cranial size corresponding to an increase in brain size.[3,89] The change in brain size was accompanied by changing dental function with a greater emphasis on grinding and less on crunching, a strong increase in body size and numerous changes in body proportions. The change in brain size was

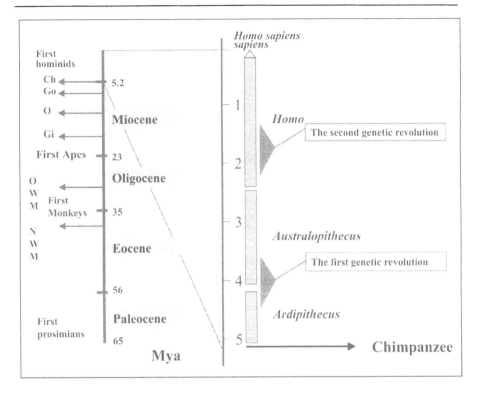

Figure 3. Two genetic revolutions in the evolution of modern humans. The first genetic revolution has led to upright walking. The second genetic revolution has caused a significant change in the cranial size, reflecting an increase in the brain size, a strong increase in body size and numerous changes in proportions. The change in the brain size was larger than could be explained by the body size alone. These differences seem to happen all together, at the time of the species origin.[3,89]

greater than could be explained by body size alone. These changes seem to happen all at once, at the time of the species origin. The anatomy of already the very first representative of the genus Homo (*H. erectus*) was characterized by a quite new unique combination of features never known before.[3] Very similar features are also characteristic of modern humans—*H. sapiens*. Unique characteristics of Homo among primates also include long lifespan, delayed reproduction and prolonged childhood.

These anatomic and physiological peculiarities correlate in modern humans with distinct changes in behavior which is substantially more complex and flexible than that of any other member of the animal kingdom. They correlate with a human sexual psychology, favoring pair-bonding over promiscuity, with complex social relations, with highly enhanced cognitive function and the most striking human characteristic that has no equivalent in other primates—language.[130] Though many of these features, such as relative size of particular brain sectors[131-134] which were long thought to be different between humans and great apes are actually not so different and were probably developed even before the divergence of humans and great apes, it is evident that a large class of adaptations had appeared during the transition from australopithecines to *H. erectus*. The numerous adaptations might be caused by equally numerous gene changes, as proposed by Wildman et al[121] (see above).

A great increase in body size and even greater one in size of brain seem to be major factors in the behavioral changes of the earliest *H. sapiens* ancestors[89] because larger body size requires increased metabolic resources. Being only 2% of the body weight, the human brain uses some 20–25% of the organism's metabolic energy. The increased energy requirements have probably been satisfied due to changes in the diet, in particular, due to increased meat consumption. From molecular and cellular point of view they were possibly satisfied by means of changes in mitochondrial networks (see above).

The behavioral changes are far more massive and sudden than any earlier changes known for hominids. In combination with the anatomic evidence they suggest major genetic novelties that have come into operation with the appearance of *H. erectus*. This was the second Genetic Revolution,[89] probably much more radical than the first one.

The populations of human ancestors have survived dramatic bottlenecks and expansions, some branches disappeared while the others flourished for long evolutional periods before the only extant Hominid species remained.

There are many hypotheses on the causes and consequences of adaptations successively accumulated in the course of the hominid evolution. However, we are still very far from understanding of the nature of the genetic changes that caused the two genetic revolutions. The fossil record is too scarce to thoroughly track down the transitions on the way to modern humans. Moreover, the genetic changes will hardly be traced in their continuity, and we can only try to recover them from variations in the structure of the extant primate genomes which reflect only fragments of the genetic history.

A great progress in the human genome sequencing coupled with nearing almost complete sequence of the chimpanzee genome and future sequencing of the genomes of other primates will allow us to move to systematic whole genome analyses and to identify both anciently conserved features common to all living primates and phylogenetically more restricted features that arose later in evolution including the human-specific ones. This will in particular allow to separate polymorphisms, evidently not important for being human, from really essential interspecies differences. Using microchip technologies, it will be possible to reveal differences in tissue specific gene expression etc. Eventually, we will most probably find ourselves in the ocean of differences still unable to separate those of them that are relevant to the "making us humans". The situation will be much more complicated than that we have now with the analyses of the genetic background of complex diseases: despite great efforts, using both genotyping and expression approaches we are still far from deciphering the networks of the genes and their regulations involved in the disease etiology. These complex traits are extremely simple as compared to multitudes of interdependent adaptations, each of which represents a complex product of functioning of a genetic network and nonlinearly contributes to phenotypic peculiarities of a species. The adaptations essential for being humans will be inevitably masked by a great many of by-products, which consistently differ humans from great apes but are present just because they are linked to traits that were positively selected. Finally, we should keep in mind that many of the adaptations that have really played an adaptive role in certain periods of evolution are presently involved in activities unrelated to their original functions. Thus, in a certain sense, to be a human is a supercomplex "disease", and deciphering of its genetic basis is a much greater challenge than have ever been met by science.

In the absence of systematic approaches to singling out essential adaptive interspecific genomic differences, the progress in understanding of the genetic basis of humanity may depend upon generating and refuting hypotheses, exactly like we do it with candidate genes when trying to reveal the genes involved in a complex disease.

Any of these hypothesis will have to answer at least one of numerous questions, part of which is listed below:

1. Why human organismal evolution proceeded much faster than that of any great ape, whereas the rate of the molecular evolution was approximately equal in all these species? This question was put forward by King and Wilson.[5]

2. Which kind of changes is connected with upright walking?
3. Which kind of changes caused the changes in the body and brain size?
4. What type of genetic changes is causally related to changes in the brain functioning?
5. What type of changes may be responsible for language?
6. Why a multitude of complex adaptations seemingly occurred simultaneously?
7. How these changes influenced developmental processes?
8. How the novelties were incorporated into preevolved and well harmonized networks to be in accord with them, not to be rejected by purifying selection and to add new capacities to already existing adaptations without damaging them?

To get the answers we should be prepared "to move away from the cataloging of individual components of the system, and beyond the simplistic notions of "this binds to that, which then docks on this, and then the complex moves there. ..." to the exciting area of network perturbations, nonlinear responses and thresholds...".[37] We should expect that "... in organisms with complex nervous systems, neither gene number, neuron number, nor number of cell types correlates in any meaningful manner with even simplistic measures of structural or behavioral complexity. Nor would they be expected to; this is the realm of nonlinearities and epigenesis".[37]

Taking all this in mind, we will still face a problem of how we can approach to the decryption of the paramount complexity of the phenotypic consequences of the genetic differences between us and chimpanzees. Unfortunately, no special technique to investigate complex system functioning "at one blow" was invented so far, despite ample discussions on this problem (see for example refs. 135-138). Moreover, the data we know are restricted to genetic changes, and an enormous task will be to decipher the epigenetic differences among species. The best reported so far approximation to a systemic analysis of a regulatory network, realized for a particular case of a gene regulatory network that controls the specification of endoderm and mesoderm in the sea urchin embryo,[139] still used just traditional, though intensified, functional genomic analysis coupled with extensive computational methodologies.

There is probably only little hope for principally new techniques of analyzing complex processes. But it is possible to further improve, intensify and authomatize the techniques and technological principles already in use in functional genomics and proteomics, and to employ them for monitoring cellular processes at whole genome, transcriptome and proteome scales. Integration of the molecular information with that at cellular and organismal levels combined with computational methods will facilitate the progress towards understanding of the transformation of the genetic information into phenotypic traits. Hopefully, the 21st century will be a century of phenome deciphering that will be a brilliant extension of the achievements of the 20th century in the genome research. And probably this century will give us a hint why we are humans and they are chimps.

## Acknowledgments

The authors thank Boris O. Glotov and Nina N. Belyaeva for help in manuscript preparation. This work was partially supported by Physico-Chemical Biological Program of the Russian Academy of Sciences and the Russian Foundation for Basic Research 02-04-48712 and 2006.20034 grants.

## References

1. Moffat AS. Primate origins meeting. New fossils and a glimpse of evolution. Science 2002; 295:613-615.
2. Begun D. Hominid family values: Morphological and molecular data on relations among the great apes and humans. In: Parker ST, Mitchell RW, Miles HL, eds. The Mentalities of Gorillas and Orangutans: Comparative Perspectives. New York: Cambridge University Press, 1999:419.

3. Wood B, Collard M. The human genus. Science 1999; 284:65-71.
4. Gagneux P. The genus Pan: Population genetics of an endangered outgroup. Trends Genet 2002; 18:327-330.
5. King MC, Wilson AC. Evolution at two levels in humans and chimpanzees. Science 1975; 188:107-116.
6. O'Brien S, Seuanez HN, Womack JE. On the evolution of genome organization in mammals. In: MacIntyre RJ, ed. Molecular Evolutionary Genetics. New York: Plenum Press, 1985:xxi-610.
7. IJdo JW, Baldini A, Ward DC et al. Origin of human chromosome 2: An ancestral telomere-telomere fusion. Proc Natl Acad Sci USA 1991; 88:9051-9055.
8. Nickerson E, Nelson DL. Molecular definition of pericentric inversion breakpoints occurring during the evolution of humans and chimpanzees. Genomics 1998; 50:368-372.
9. Conte RA, Samonte RV, Verma RS. Evolutionary divergence of the oncogenes GLI, HST and INT2. Heredity 1998; 81(Pt 1):10-13.
10. Tanabe H, Ishida T, Ueda S et al. Molecular anatomy of human chromosome 9: Comparative mapping of the immunoglobulin processed pseudogene C epsilon 3 (IGHEP2) in primates. Cytogenet Cell Genet 1996; 73:92-96.
11. Verma RS, Luke S. Evolutionary divergence of human chromosome 9 as revealed by the position of the ABL protooncogene in higher primates. Mol Gen Genet 1994; 243:369-373.
12. McConkey EH. The origin of human chromosome 18 from a human/ape ancestor. Cytogenet Cell Genet 1997; 76:189-191.
13. Schwartz A, Chan DC, Brown LG et al. Reconstructing hominid Y evolution: X-homologous block, created by X-Y transposition, was disrupted by Yp inversion through LINE-LINE recombination. Hum Mol Genet 1998; 7:1-11.
14. Assum G, Gartmann C, Schempp W et al. Evolution of the chAB4 multisequence family in primates. Genomics 1994; 21:34-41.
15. Monfouilloux S, Avet-Loiseau H, Amarger V et al. Recent human-specific spreading of a subtelomeric domain. Genomics 1998; 51:165-176.
16. McEvoy SM, Maeda N. Complex events in the evolution of the haptoglobin gene cluster in primates. J Biol Chem 1988; 263:15740-15747.
17. Eichler EE, Lu F, Shen Y et al. Duplication of a gene-rich cluster between 16p11.1 and Xq28: A novel pericentromeric-directed mechanism for paralogous genome evolution. Hum Mol Genet 1996; 5:899-912.
18. Rouquier S, Taviaux S, Trask BJ et al. Distribution of olfactory receptor genes in the human genome. Nature Genet 1998; 18:243-250.
19. Chou HH, Takematsu H, Diaz S et al. A mutation in human CMP-sialic acid hydroxylase occurred after the Homo-Pan divergence. Proc Natl Acad Sci USA 1998; 95:11751-11756.
20. Normile D. Comparative genomics. Gene expression differs in human and chimp brains. Science 2001; 292:44-45.
21. Yoder JA, Walsh CP, Bestor TH. Cytosine methylation and the ecology of intragenomic parasites. Trends Genet 1997; 13:335-340.
22. Mighell AJ, Markham AF, Robinson PA. Alu sequences. FEBS Lett 1997; 417:1-5.
23. Minghetti PP, Dugaiczyk A. The emergence of new DNA repeats and the divergence of primates. Proc Natl Acad Sci USA 1993; 90:1872-1876.
24. Hwu HR, Roberts JW, Davidson EH et al. Insertion and/or deletion of many repeated DNA sequences in human and higher ape evolution. Proc Natl Acad Sci USA 1986; 83:3875-3879.
25. Batzer MA, Rubin CM, Hellmann-Blumberg U et al. Dispersion and insertion polymorphism in two small subfamilies of recently amplified human Alu repeats. J Mol Biol 1995; 247:418-427.
26. Hardison RC. Conserved noncoding sequences are reliable guides to regulatory elements. Trends Genet 2000; 16:369-372.
27. Kelley RL, Kuroda MI. Noncoding RNA genes in dosage compensation and imprinting. Cell 2000; 103:9-12.
28. Sharp PA. RNA interference-2001. Genes Dev 2001; 15:485-490.
29. Wassarman KM, Repoila F, Rosenow C et al. Identification of novel small RNAs using comparative genomics and microarrays. Genes Dev 2001; 15:1637-1651.
30. Gagneux P, Varki A. Genetic differences between humans and great apes. Mol Phylogenet Evol 2001; 18:2-13.

31. Hacia JG. Genome of the apes. Trends Genet 2001; 17:637-645.
32. Kaessmann H, Paabo S. The genetical history of humans and the great apes. J Intern Med 2002; 251:1-18.
33. Olson MV, Varki A. Sequencing the chimpanzee genome: Insights into human evolution and disease. Nature Rev Genet 2003; 4:20-28.
34. Chakravarti A. Population genetics-making sense out of sequence. Nature Genet 1999; 21(1 Suppl):56-60.
35. Stoneking M. Single nucleotide polymorphisms. From the evolutionary past. Nature 2001; 409:821-822.
36. Lander ES, Linton LM, Birren B et al. Initial sequencing and analysis of the human genome. Nature 2001; 409:860-921.
37. Venter JC, Adams MD, Myers EW et al. The sequence of the human genome. Science 2001; 291:1304-1351.
38. Sachidanandam R, Weissman D, Schmidt SC et al. A map of human genome sequence variation containing 1.42 million single nucleotide polymorphisms. Nature 2001; 409:928-933.
39. Harpending HC, Batzer MA, Gurven M et al. Genetic traces of ancient demography. Proc Natl Acad Sci USA 1998; 95:1961-1967.
40. Kaessmann H, Wiebe V, Weiss G et al. Great ape DNA sequences reveal a reduced diversity and an expansion in humans. Nature Genet 2001; 27(2):155-156.
41. Chakravarti A. To a future of genetic medicine. Nature 2001; 409:822-823.
42. Yu N, Zhao Z, Fu YX et al. Global patterns of human DNA sequence variation in a 10-kb region on chromosome 1. Mol Biol Evol 2001; 18:214-222.
43. Sherry ST, Harpending HC, Batzer MA et al. Alu evolution in human populations: Using the coalescent to estimate effective population size. Genetics 1997; 147:1977-1982.
44. Kimura M. The Neutral Theory of Molecular Evolution. Cambridge, New York: Cambridge University Press, 1983.
45. Relethford JH. Genetics of modern human origins and diversity. Annu Rev Anthropol 1998; 27:1-23.
46. Przeworski M, Hudson RR, Di Rienzo A. Adjusting the focus on human variation. Trends Genet 2000; 16:296-302.
47. Ferris SD, Brown WM, Davidson WS et al. Extensive polymorphism in the mitochondrial DNA of apes. Proc Natl Acad Sci USA 1981; 78:6319-6323.
48. Morin PA, Moore JJ, Chakraborty R et al. Kin selection, social structure, gene flow, and the evolution of chimpanzees. Science 1994; 265:1193-1201.
49. Satta Y. Comparison of DNA and protein polymorphisms between humans and chimpanzees. Genes Genet Syst 2001; 76:159-168.
50. Deinard A, Kidd K. Evolution of a HOXB6 intergenic region within the great apes and humans. J Hum Evol 1999; 36:687-703.
51. Chen FC, Li WH. Genomic divergences between humans and other hominoids and the effective population size of the common ancestor of humans and chimpanzees. Am J Hum Genet 2001; 68:444-456.
52. Ruvolo M. Molecular phylogeny of the hominoids: Inferences from multiple independent DNA sequence data sets. Mol Biol Evol 1997; 14:248-265.
53. Harris EE, Hey J. X chromosome evidence for ancient human histories. Proc Natl Acad Sci USA 1999; 96:3320-3324.
54. Goodman M, Porter CA, Czelusniak J et al. Toward a phylogenetic classification of primates based on DNA evidence complemented by fossil evidence. Mol Phylogenet Evol 1998; 9:585-598.
55. Kaessmann H, Wiebe V, Paabo S. Extensive nuclear DNA sequence diversity among chimpanzees. Science 1999; 286:1159-1162.
56. Lio P, Goldman N. Models of molecular evolution and phylogeny. Genome Res 1998; 8:1233-1244.
57. Rodriguez-Trelles F, Tarrio R, Ayala FJ. Erratic overdispersion of three molecular clocks: GPDH, SOD, and XDH. Proc Natl Acad Sci USA 2001; 98:11405-11410.
58. Stauffer RL, Walker A, Ryder OA et al. Human and ape molecular clocks and constraints on paleontological hypotheses. J Hered 2001; 92:469-474.
59. Ebersberger I, Metzler D, Schwarz C et al. Genomewide comparison of DNA sequences between humans and chimpanzees. Am J Hum Genet 2002; 70:1490-1497.

60. Sibley CG, Ahlquist JE. DNA hybridization evidence of hominoid phylogeny: Results from an expanded data set. J Mol Evol 1987; 26:99-121.
61. Bailey WJ, Fitch DH, Tagle DA et al. Molecular evolution of the psi eta-globin gene locus: Gibbon phylogeny and the hominoid slowdown. Mol Biol Evol 1991; 8:155-184.
62. Schmid CW, Marks J. DNA hybridization as a guide to phylogeny: Chemical and physical limits. J Mol Evol 1990; 30:237-246.
63. Sibley CG, Comstock JA, Ahlquist JE. DNA hybridization evidence of hominoid phylogeny: A reanalysis of the data. J Mol Evol 1990; 30:202-236.
64. Fujiyama A, Watanabe H, Toyoda A et al. Construction and analysis of a human-chimpanzee comparative clone map. Science 2002; 295:131-134.
65. Lisitsyn NA, Launer GA, Wagner LL et al. Isolation of rapidly evolving genomic sequences: Construction of a differential library and identification of a human DNA fragment that does not hybridize to chimpanzee DNA. Biomed Sci 1990; 1:513-516.
66. Toder R, Grutzner F, Haaf T et al. Species-specific evolution of repeated DNA sequences in great apes. Chromosome Res 2001; 9:431-435.
67. Toder R, Xia Y, Bausch E. Interspecies comparative genome hybridization and interspecies representational difference analysis reveal gross DNA differences between humans and great apes. Chromosome Res 1998; 6:487-494.
68. Ueda S, Washio K, Kurosaki K. Human-specific sequences: Isolation of species-specific DNA regions by genome subtraction. Genomics 1990; 8:7-12.
69. Takahata N, Satta Y, Klein J. Divergence time and population size in the lineage leading to modern humans. Theor Popul Biol 1995; 48:198-221.
70. Ayala FJ. Molecular clock mirages. Bioessays 1999; 21:71-75.
71. Page SL, Goodman M. Catarrhine phylogeny: Noncoding DNA evidence for a diphyletic origin of the mangabeys and for a human-chimpanzee clade. Mol Phylogenet Evol 2001; 18:14-25.
72. Goodman M, Grossman LI, Schmidt TR et al. Primate genomics: The search for genic changes that shaped being human. http://www.uchicago.edu/aff/mwc-amacad/biocomplexity/MGHGEP.html.
73. Easteal S, Herbert G. Molecular evidence from the nuclear genome for the time frame of human evolution. J Mol Evol 1997; 44(Suppl 1):S121-132.
74. Normile D. Genomics. Chimp sequencing crawls forward. Science 2001; 291:2297.
75. Takahata N, Satta Y. Evolution of the primate lineage leading to modern humans: Phylogenetic and demographic inferences from DNA sequences. Proc Natl Acad Sci USA 1997; 94:4811-4815.
76. Horai S, Hayasaka K, Kondo R et al. Recent African origin of modern humans revealed by complete sequences of hominoid mitochondrial DNAs. Proc Natl Acad Sci USA 1995; 92:532-536.
77. Yunis JJ, Sawyer JR, Dunham K. The striking resemblance of high-resolution G-banded chromosomes of man and chimpanzee. Science 1980; 208:1145-1148.
78. Yunis JJ, Prakash O. The origin of man: A chromosomal pictorial legacy. Science 1982; 215:1525-1530.
79. Wienberg J, Jauch A, Ludecke HJ et al. The origin of human chromosome 2 analyzed by comparative chromosome mapping with a DNA microlibrary. Chromosome Res 1994; 2:405-410.
80. Stanyon R, Wienberg J, Romagno D et al. Molecular and classical cytogenetic analyses demonstrate an apomorphic reciprocal chromosomal translocation in Gorilla gorilla. Am J Phys Anthropol 1992; 88:245-250.
81. Samonte RV, Ramesh KH, Verma RS. Comparative mapping of human alphoid satellite DNA repeat sequences in the great apes. Genetica 1997; 101:97-104.
82. Archidiacono N, Antonacci R, Marzella R et al. Comparative mapping of human alphoid sequences in great apes using fluorescence in situ hybridization. Genomics 1995; 25:477-484.
83. Horvath JE, Viggiano L, Loftus BJ et al. Molecular structure and evolution of an alpha satellite/nonalpha satellite junction at 16p11. Hum Mol Genet 2000; 9:113-123.
84. Royle NJ, Baird DM, Jeffreys AJ. A subterminal satellite located adjacent to telomeres in chimpanzees is absent from the human genome. Nature Genet 1994; 6:52-56.
85. Wong Z, Royle NJ, Jeffreys AJ. A novel human DNA polymorphism resulting from transfer of DNA from chromosome 6 to chromosome 16. Genomics 1990; 7:222-234.
86. Trask BJ, Massa H, Brand-Arpon V et al. Large multi-chromosomal duplications encompass many members of the olfactory receptor gene family in the human genome. Hum Mol Genet 1998; 7:2007-2020.

87. Taanman JW. The mitochondrial genome: Structure, transcription, translation and replication. Biochim Biophys Acta 1999; 1410:103-123.

88. Rich P. Chemiosmotic coupling: The cost of living. Nature 2003; 421:583.

89. Hawks J, Hunley K, Lee SH et al. Population bottlenecks and pleistocene human evolution. Mol Biol Evol 2000; 17:2-22.

90. Grossman LI, Schmidt TR, Wildman DE et al. Molecular evolution of aerobic energy metabolism in primates. Mol Phylogenet Evol 2001; 18:26-36.

91. Arnason U, Xu X, Gullberg A. Comparison between the complete mitochondrial DNA sequences of Homo and the common chimpanzee based on nonchimeric sequences. J Mol Evol 1996; 42:145-152.

92. Brown WM, George Jr M, Wilson AC. Rapid evolution of animal mitochondrial DNA. Proc Natl Acad Sci USA 1979; 76:1967-1971.

93. Xu X, Arnason U. A complete sequence of the mitochondrial genome of the western lowland gorilla. Mol Biol Evol 1996; 13:691-698.

94. Wise CA, Sraml M, Rubinsztein DC et al. Comparative nuclear and mitochondrial genome diversity in humans and chimpanzees. Mol Biol Evol 1997; 14:707-716.

95. Wei YH, Kao SH, Lee HC. Simultaneous increase of mitochondrial DNA deletions and lipid peroxidation in human aging. Ann N Y Acad Sci 1996; 786:24-43.

96. Giles RE, Blanc H, Cann HM et al. Maternal inheritance of human mitochondrial DNA. Proc Natl Acad Sci USA 1980; 77:6715-6719.

97. Haaf T, Willard HF. Chromosome-specific alpha-satellite DNA from the centromere of chimpanzee chromosome 4. Chromosoma 1997; 106:226-232.

98. Rocchi M, Archidiacono N, Antonacci R et al. Cloning and comparative mapping of recently evolved human chromosome 22- specific alpha satellite DNA. Somat Cell Mol Genet 1994; 20:443-448.

99. Rubinsztein DC, Amos W, Leggo J et al. Microsatellite evolution-evidence for directionality and variation in rate between species. Nature Genet 1995; 10:337-343.

100. Amos W, Rubinstzein DC. Microsatellites are subject to directional evolution. Nature Genet 1996; 12:13-14.

101. Olson MV. When less is more: Gene loss as an engine of evolutionary change. Am J Hum Genet 1999; 64:18-23.

102. Zhang XM, Cathala G, Soua Z et al. The human T-cell receptor gamma variable pseudogene V10 is a distinctive marker of human speciation. Immunogenetics 1996; 43:196-203.

103. Muchmore EA, Diaz S, Varki A. A structural difference between the cell surfaces of humans and the great apes. Am J Phys Anthropol 1998; 107(2):187-198.

104. Hayakawa T, Satta Y, Gagneux P et al. Alu-mediated inactivation of the human CMP-N-acetylneuraminic acid hydroxylase gene. Proc Natl Acad Sci USA 2001; 98:11399-11404.

105. Eichler EE. Recent duplication, domain accretion and the dynamic mutation of the human genome. Trends Genet 2001; 17:661-669.

106. Samonte RV, Eichler EE. Segmental duplications and the evolution of the primate genome. Nature Rev Genet 2002; 3:65-72.

107. Crow TJ. Schizophrenia as the price that Homo sapiens pays for language: A resolution of the central paradox in the origin of the species. Brain Res Brain Res Rev 2000; 31:118-129.

108. Blanco-Arias P, Sargent CA, Affara NA. The human-specific Yp11.2/Xq21.3 homology block encodes a potentially functional testis-specific TGIF-like retroposon. Mamm Genome 2002; 13:463-468.

109. Crow TJ. Handedness, language lateralisation and anatomical asymmetry: Relevance of protocadherin XY to hominid speciation and the aetiology of psychosis. Point of view. Br J Psychiatry 2002; 181:295-297.

110. Nicholson TR, Yang J, DeLisi LE et al. Allele sharing for schizophrenia and schizo-affective disorder within a region of Homo sapiens specific XY homology. Am J Med Genet 2002; 114:637-640.

111. Dufour C, Casane D, Denton D et al. Human-chimpanzee DNA sequence variation in the four major genes of the renin angiotensin system. Genomics 2000; 69:14-26.

112. Hacia JG, Fan JB, Ryder O et al. Determination of ancestral alleles for human single-nucleotide polymorphisms using high-density oligonucleotide arrays. Nature Genet 1999; 22:164-167.

113. Madeyski K, Lidberg U, Bjursell G et al. Characterization of the gorilla carboxyl ester lipase locus, and the appearance of the carboxyl ester lipase pseudogene during primate evolution. Gene 1999; 239:273-282.

114. Ichinose H, Ohye T, Fujita K et al. Increased heterogeneity of tyrosine hydroxylase in humans. Biochem Biophys Res Commun 1993; 195:158-165.

115. Szabo Z, Levi-Minzi SA, Christiano AM et al. Sequential loss of two neighboring exons of the tropoelastin gene during primate evolution. J Mol Evol 1999; 49:664-671.

116. Eddy SR. Noncoding RNA genes and the modern RNA world. Nature Rev Genet 2001; 2:919-929.

117. Wu W, Goodman M, Lomax MI et al. Molecular evolution of cytochrome c oxidase subunit IV: Evidence for positive selection in simian primates. J Mol Evol 1997; 44:477-491.

118. Wu W, Schmidt TR, Goodman M et al. Molecular evolution of cytochrome c oxidase subunit I in primates: Is there coevolution between mitochondrial and nuclear genomes? Mol Phylogenet Evol 2000; 17:294-304.

119. Schmidt TR, Goodman M, Grossman LI. Molecular evolution of the COX7A gene family in primates. Mol Biol Evol 1999; 16:619-626.

120. Adkins RM, Honeycutt RL, Disotell TR. Evolution of eutherian cytochrome c oxidase subunit II: Heterogeneous rates of protein evolution and altered interaction with cytochrome c. Mol Biol Evol 1996; 13:1393-1404.

121. Wildman DE, Grossman LI, Goodman M. Human and chimpanzee functional DNA shows they are more similar to each other then either is to other apes. http://www.uchicago.edu/aff/mwc-amacad/biocomplexity/conference_papers/papers.html.

122. Ding YC, Chi HC, Grady DL et al. Evidence of positive selection acting at the human dopamine receptor D4 gene locus. Proc Natl Acad Sci USA 2002; 99:309-314.

123. Enard W, Khaitovich P, Klose J et al. Intra- and interspecific variation in primate gene expression patterns. Science 2002; 296:340-343.

124. Huby T, Dachet C, Lawn RM et al. Functional analysis of the chimpanzee and human apo(a) promoter sequences. Identification of sequence variations responsible for elevated transcriptional activity in chimpanzee. J Biol Chem 2001; 276:22209-22214.

125. Gagneux P, Amess B, Diaz S et al. Proteomic comparison of human and great ape blood plasma reveals conserved glycosylation and differences in thyroid hormone metabolism. Am J Phys Anthropol 2001; 115:99-109.

126. Nadezhdin EV, Vinogradova TV, Sverdlov ED. Interspecies subtractive hybridization of cDNA from human and chimpanzee brains. Dokl Biochem Biophys 2001; 381:415-418.

127. Niwa H, Miyazaki J, Smith AG. Quantitative expression of Oct-3/4 defines differentiation, dedifferentiation or self-renewal of ES cells. Nature Genet 2000; 24:372-376.

128. Stewart CL. Oct-4, scene 1: The drama of mouse development. Nature Genet 2000; 24:328-330.

129. Haile-Selassie Y. Late Miocene hominids from the Middle Awash, Ethiopia. Nature 2001; 412:178-181.

130. Bishop DV. Putting language genes in perspective. Trends Genet 2002; 18:57-59.

131. Semendeferi K, Damasio H. The brain and its main anatomical subdivisions in living hominoids using magnetic resonance imaging. J Hum Evol 2000; 38:317-332.

132. Semendeferi K, Lu A, Schenker N et al. Humans and great apes share a large frontal cortex. Nature Neurosci 2002; 5:272-276.

133. Hopkins WD, Marino L, Rilling JK et al. Planum temporale asymmetries in great apes as revealed by magnetic resonance imaging (MRI). Neuroreport 1998; 9:2913-2918.

134. Gannon PJ, Holloway RL, Broadfield DC et al. Asymmetry of chimpanzee planum temporale: Humanlike pattern of Wernicke's brain language area homolog. Science 1998; 279:220-222.

135. Hartwell LH, Hopfield JJ, Leibler S et al. From molecular to modular cell biology. Nature 1999; 402(Suppl):C47-52.

136. Kitano H. Systems biology: A brief overview. Science 2002; 295:1662-1664.

137. Koch C, Laurent G. Complexity and the nervous system. Science 1999; 284:96-98.

138. Weng G, Bhalla US, Iyengar R. Complexity in biological signaling systems. Science 1999; 284:92-96.

139. Davidson EH, Rast JP, Oliveri P et al. A genomic regulatory network for development. Science 2002; 295:1669-1678.

140. White T. Paleoanthropology. Early hominids-diversity or distortion? Science 2003; 299:1994-1997.

# CHAPTER 5

# Retroviruses, Their Domesticated Relatives and Other Retroinvaders:
## Potential Genetic and Epigenetic Mediators of Phenotypic Variation

Eugene D. Sverdlov

## Abstract

This chapter is a very brief introduction to exogenous and endogenous retroviruses and other genomic mobile elements whose mobility is conditioned by transient passage through an RNA stage. This passage is mediated by transcription of the element with host cell RNA polymerase, followed by reverse transcription of the resulting RNA and integration of the cDNA copy into a new position within the genome. Due to this reverse transcription stage, all of such mobile elements are termed retroelements (REs). The chapter is not a comprehensive review, but rather a textbook-style narrative aimed at making the reading of the following chapters easier.

### Introduction: Continuum of The Retroworld

Due to their fascinating life cycle, retroviruses occupy a special place among other viruses: like a genie from Arab myths they change their genome from RNA to DNA inserted into the genome of a host cell, and than back to RNA. In addition, they are probably the only RNA viruses capable of becoming heritable components of the host DNA and being fixed in evolution. It has been hypothesized (for discussion see ref. 1) that the earliest infectious exogenous retroviruses could originate from endogenous retroviral-like (ERV) elements (retrotransposons) that became horizontally transmissible by acquisition of a cellular *envelope* gene (*env*). In turn, by infecting vertebrate germ cells they gave birth to a variety of current endogenous retroviral elements. Indeed, it is not difficult to notice a striking similarity of exogenous infectious retroviruses to endogenous retroviral-like genome elements. Thus, there is a continuum of exogenous and endogenous retroviruses, the border between them being rather conditional. For example, a recent report that porcine endogenous retrovirus (PERV) can infect both porcine and human cells (see for example ref. 2) supports its productive infection capability. In the aspect of genome evolution, each hereditable acquisition of a retrovirus is a small but still "genomic revolution" able to affect genomic functions in the region of the retrovirus integration. Apart from retroviruses, there are many other genomic revolutionists jumping from place to place in the genome, each time with a risk of regulatory intervention. These revolutionists are very similar to retroviruses in their way of movement—they start from a genomic DNA

---

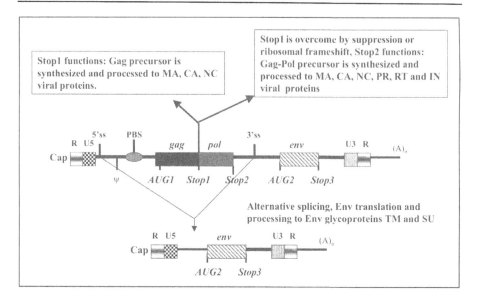

Figure 1. Structure of the retroviral genomic RNA and scheme of its functioning in the cell as mRNA.

copy which is then converted into an RNA copy and again into DNA followed by its integration into the genome, but in a different place. All such movements can be important factors in evolution. The problem of a possible evolutionary role of such "convertible" elements termed retroelements (REs) is central for this book.

This chapter is a very brief introduction to the problem, written to facilitate the reading of the following chapters. Therefore it is not a comprehensive review but rather a textbook-style narrative.

## Retroviral Particles

Retroviruses are RNA viruses whose virions (virus or viral particles) contain two copies of a single stranded 7000–11000 nucleotide long RNA molecule. Each RNA has a structure schematized in Figure 1. It contains a 5'-CAP site and a 3'-poly(A) tail thus looking exactly as most mammalian mRNAs. The viral RNA strand is called (+)-strand since it seems to be able to serve as mRNA for translation (however, this is not the case after the virus has penetrated the cell, see below). The RNA encodes three viral genes—*gag, pol* and *env* characterized by a genetic architecture principally similar for all known nondefective retroviruses. The functional role of the genes will be considered below. In more complicated retroviruses this general structure can be extended by addition of some extra genes, like in lentiviruses. Discussion of the complex retroviruses containing additional regulatory genes like lentiviruses is beyond the scope of the chapter.

In the viral particle (Fig. 2) the RNAs are directly embedded into an inner protein "bag", consisting of a viral capsid (CA) protein surrounding the RNAs bound to nucleocapsid (NC) proteins. Within this CA shell there are also tRNA molecules (usually Trp, Pro or Lys) which serve as primers during first and second strand DNA synthesis, as well as enzymes—reverse transcriptase (RT) and integrase (IN). The RNA genome together with NC, RT and IN proteins forms a so called viral core.

The internal "bag" is, in its turn, wrapped into an envelope consisting of the host cell membrane lipids acquired when the virus bursts out of the cell (Fig. 3). The envelope

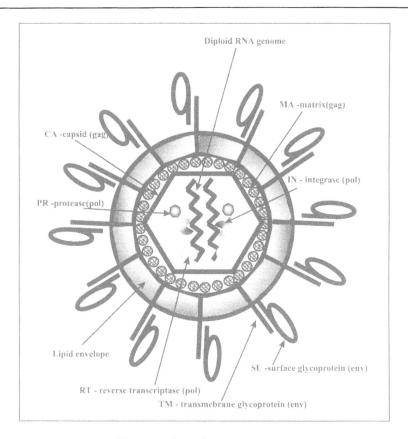

Figure 2. The general scheme of the retroviral particle organization.

accommodates a viral transmembrane glycoprotein (TM) and a surface glycoprotein (SU) linked to TM and thus anchored to the viral particle. These virus-encoded glycoproteins form spikes in the membrane.

The space between the envelope and inner CA shell is filled with a matrix protein (MA).

The abbreviations in brackets used above represent a universal nomenclature of viral proteins. For reader's convenience the nomenclature is given in Table 1.

## Retroviral RNA Genome

The RNA included in the retroviral particles represents the retroviral genome in its RNA form. A unique property of the retroviruses is that their RNA genome with a few exceptions (e.g., lentiviruses) is produced entirely by cellular transcriptional machinery. Retroviruses are also the only (+)sense RNA viruses whose genome does not function as mRNA immediately after infection and requires a specific cellular RNA (tRNA) for replication. The principle of the organization of retroviral RNA genomes is presented in Figure 1.

As indicated above, the genome is capped at 5'- and polyadenylated at 3'- end. The cap sequence is of type 1: m7G5'ppp5'GmpNp. Both the cap and poly(A) tail are attached to identical short (18-250 nt) sequences (R) which form direct repeats at both ends of the genome.

The R repeat at the 5'-end of the genome is adjacent to a unique sequence called U5. It is a noncoding region of 75-250 nt which is the first part of the genome to be reverse transcribed,

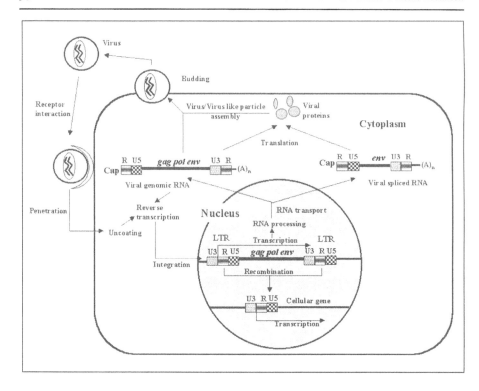

Figure 3. Schematic presentation of the retroviral life cycle and endogenization with the formation of solitary LTRs.

## Table 1. Proteins of retroviral particles

| Protein Name | Encoding Gene | Function |
|---|---|---|
| TM, Transmembrane glycoprotein | *env* | The inner component of the mature envelope glycoprotein |
| SU, Surface glycoprotein | *env* | The outer envelope glycoprotein, major virus antigen |
| MA, Matrix | *gag* | lines envelope |
| CA, Capsid | *gag* | most abundant protein of the virus, protects the core |
| NC, Nucleocapsid | *gag* | Bound to RNA, protects the genome |
| PR, Protease | *pol* | Essential for gag and env polyprotein cleavage during maturation |
| RT Reverse transcriptase | *pol* | Reverse transcribes the RNA genome; also has RNAseH activity |
| IN Integrase | *pol* | Involved in integration of the provirus DNA |

forming the 3'-end of the DNA provirus genome (see below). U5 is followed by Primer Binding Site (PBS) which is a 18 nt stretch complementary to the 3'-end of a specific tRNA primer used by the viral RT to initiate reverse transcription.

The R repeat at the opposite side of the RNA genome is flanked by a unique noncoding U3 region of 200-1200 nt, forming the 5' end of the provirus DNA after reverse transcription. Just upstream of U3 there is a short (~10 nucleotides) stretch of purine nucleotides used for initiating (+)strand synthesis during reverse transcription.

Between these border elements there are three common genes—*gag* (an acronym of *Group AntiGene*), *pol* (polymerase) and *env* (envelope). These genes encode all viral proteins listed in Table 1.

## Retroviral Life Cycle

The most important stages of the retroviral life cycle are very schematically illustrated in Figure 3.

The infection starts with the binding of the SU envelope glycoprotein to a specific receptor on the surface of the host cell. The receptor binding finally results in the fusion of the virus envelope with the cell membrane. Penetration and uncoating are not completely understood, but uncoating is only partial, resulting eventually in a core (nucleocapsid) particle within the cytoplasm. Reverse transcription occurs inside this particle. The cDNA synthesis is primed with a specific cellular tRNA bound to the virion genomic RNA. This is a rather complicated process, and its description is beyond the scope of this review. It is important that it leads to the formation of long terminal repeats (LTRs) flanking the viral double-stranded cDNA. The genetic content of the cDNA remains identical to that of the viral genomic RNA. But the LTRs formed by joining together U3, R and U5 sequences of the original RNA during the cDNA synthesis are a remarkable invention of retroviruses. These LTRs are carriers of transcriptional regulatory information with promoters, enhancers, various transcription factors' binding sites and polyadenylation signals, all concentrated within an about 1000 bp long DNA stretch.

The cDNA copy then integrates into the host cell genome. The process of integration is an important part of the retroviral replication cycle. It involves the IN enzyme probably acting in cooperation with cellular proteins.[3] As a result, the retroviral cDNA becomes forever integrated into the host genome DNA as a newly acquired element. The choice of the integration site is determined by a multitude of factors and at first glance looks almost accidental. However, the situation is probably not so simple (for further review see ref. 4). The integration process also involves the removal of 1–2 nucleotides from the ends of each LTR. The termini of the integrated LTRs always have the same sequence (5' - TG...CA - 3'), and 4-6 bp of the host cell DNA flanking the integrated provirus are duplicated. The integrated DNA is called a provirus. The provirus can not be exactly excised from the host cell DNA due to the lack of mechanisms for its excision.

The integrated provirus can be transcribed provided that its integration site is located in a transcriptionally competent genome region. As already mentioned above, retroviruses use the cellular transcriptional machinery for their expression (although a few of them, like HTLV or HIV, encode additional transcriptional and post-transcriptional regulatory factors). The U3 region contains the promoter elements responsible for transcription of the provirus. R and U5 LTR regions also contribute to the regulation of expression.[5,6] The resulting transcript is a copy of the viral genomic RNA.

Retroviruses have to synthesize a variety of proteins (see Table 1) using only one LTR for transcription regulation. Therefore they use posttranscriptional processing of the primary RNA transcript, regulation of translation due to either ribosomal frameshifts or stop codon suppression, as well as posttranslational processing of proteins. The simplest scheme of these processes,

**Table 2.   The taxonomy of vertebrate retroviruses (The International Committee on Taxonomy of Viruses http://www.ncbi.nlm.nih.gov/ICTV/)**

**RNA Reverse Transcribing Viruses**

| Family | Genus | Representative | HERV Class |
|---|---|---|---|
| Retroviridae | Alpharetrovirus | ALV | II |
| | Betaretrovirus | MMTV | II |
| | Gammaretrovirus | MLV | I |
| | Deltaretrovirus | BLV | |
| | Epsilonretrovirus | Walley dermal sarcoma virus | |
| | Lentivirus | HIV | |
| | Spumavirus | Human foamy virus | III |

utilized by some of retroviruses, is shown in Figure 1. Complex viruses, such as lentiviruses, use much more complicated alternative splicing schemes.

Finally, the viral proteins are assembled with the genomic RNAs ($\psi$ signal shown in Fig. 1 is essential for this process) to give mature virus particles, which then release from the cell acquiring the lipid membrane and embedding the Env protein in it.

Certainly, life cycles of various types of retroviruses do not exactly follow the general pattern above, however the deviations look more like variations of a common motif.

## Taxonomy of Retroviruses

Historically, retroviruses were divided into groups based on their morphology in negatively stained electron microscopy (ICTVdB Virus Descriptions http://life.bio2.edu/ICTVdB/). These groups are listed below.

A-type: Also known as 'intracisternal particles'. Nonenveloped, immature particles only seen inside cells, believed to result from endogenous retrovirus-like genetic elements

B-type: Enveloped, extracellular particles with a condensed, acentric core and prominent envelope spikes, e.g., Mouse mammary tumor virus (MMTV).

C-type: The same as B-type, but with a central core and barely visible spikes, e.g., most mammalian and avian retroviruses:Murine leukemia virus (MLV), Avian leukosis virus (ALV), Human T-cell leukaemia viruses (HTLV), Human immunodeficiency virus (HIV).

D-type: Usually slightly larger (to 120 nm) with less prominent spikes, e.g., Mason-pfizer monkey virus (MPMV).

The modern taxonomy (see Chapters 10 and 11), which is used in some reviews here, is shown in Table 2.

## Endogenization of Exogenous Retroviruses

Although retroviruses mostly infect somatic cells, sometimes germ cells can also be infected. In the latter case, the inserted viruses are hereditarily transmitted from one generation to another as stable Mendelian genes. With the lapse of time, these newly acquired genes lost their capacity to form mature retroviral infectious particles and accumulated a multitude of mutations deleterious for their genes expression and viral protein synthesis. The noninfectious proviruses are said to be endogenized—they become endogenous retroviruses (ERVs). According to a human genome sequence,[7] ERV related sequences occupy up to 8% of the human genome. It is not known whether ERVs retained their transpositional capacity in the human genome, but the detection of human-specific integrations of ERV-related elements (see Chapter 6 and ref. 8) suggests that some of them are still active. There are many types of ERVs in the human genome, their phylogeny and relation to exogenous retroviruses are discussed in Chapters 10 and 11. Many of the proviruses recombine between their flanking LTRs with the removal of all viral genes, so that only solitary LTRs remain in place of former retroviruses (see Fig. 3). The nomenclature of HERVs is rather confusing (see for further discussion Chapters 10 and 11). Human endogenous retroviruses (HERVs) have been divided into three broad classes according to their genomic sequence similarities to mammalian retroviruses: class I members are related to type C, class II - to type B/D sequences, and class III - to foamy viruses (see Table 2). It should be noted that there are many problems in such assignments of the ERVs probably due to fast evolution of exogenous retroviral genomic sequences. These problems are discussed in chapter 10. Each of the HERV classes includes individual families that were grouped according to PBS structures in broad families such as HERV-K (PBS structure suggests the priming by Lys tRNA) or HERV-H (PBS corresponds to His tRNA) and further on the base of sequence similarity of their members which is greater within each family than among the constituents of different families. For example, class II HERVs include most characterized HERV-K family and those of class I - HERV-H, ERV-9, HERV-W and HERV-E families. The subdivision can be extended even further according to more restricted sequence similarities. When the PBS affinity of a HERV cannot be determined, a variety of other designations are used, such as a nearby gene (e.g. HERV-ADP), an amino acid motif present within the sequence, (e.g. HERV-FRD) or the cosmid number of the prototype member. However, it is now obvious that many distantly related families share the same tRNA primer, for example class I and II HERVs both being primed by $tRNA^{Lys}$ and class I and III HERVs being primed by $tRNA^{Leu}$. This leads to rather ambigous differentiation between the families based for example on an alphabetical character, as in the case of three different families that are all primed by arginine and are referred to as HERV-R type (a), (b) and (c). We will try to keep with "HERV-X(narrow group)" nomenclature where **X** is the tRNA specificity of the PBS, and **group** represents a cluster name based on sequence similarity (see Chapter 11). However, it is not always possible in view of a lot of uncertainties in the assignment of a particular HERV to a certain family (see Chapter 10 for discussion), In this case we used the author's nomenclature. A more rational nomenclature aimed at consolidation of the nomenclatures of all human genome elements (see for example www.gene.ucl.ac.uk/nomenclature/guidelines.html) is still not in wide use among ERV researchers.

## Army of Retroelements in the Human Genome

It is now a common knowledge that a good half of the human genome consists of retroelements (REs) or their derivatives.[7,9] The mouse genome is of the same constitution.[10] A common feature of all these retroelements is their origin by retrotranspositions in germ line which proceed through:

- transcription of DNA templates by host cell RNA polymerase,
- reverse transcription into DNA by a reverse transcriptase,
- insertion into new genomic locations increasing the number of genomic copies of the retroelement.
- Two broad classes of retrotransposons include LTR and nonLTR retrotransposons (see Chapter 6 and ref. 11). LTR retrotransposons were already discussed above. NonLTR retrotransposons comprise a significant fraction (about 30%) of the human genome.
- Transposable elements can also be grouped according to their capacity to transpose independently from other retrotransposons (though with the participation of host cell factors).

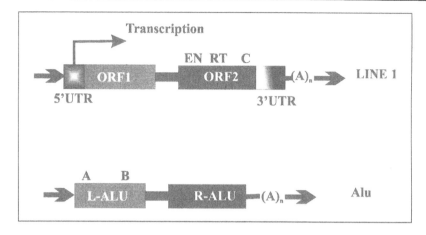

Figure 4. Structures of LINE-1 and Alu retrotransposable elements. A full-length (~6.1 kilobase) L1 element includes a 5'-untranslated region (5'-UTR), two open reading frames (ORF1 and ORF2), separated by an intergenic spacer, a 3'-UTR and a poly(A) tail. ORF1 encodes a 40 kDa protein with a capacity of nucleic acids binding; ORF2 encodes reverse transcriptase (RT), endonuclease (EN) and a cysteine-rich region (C). Genomic L1 insertions are often flanked by 7–20-nt target-site duplications (TSDs) denoted by arrows. 5'-UTR possesses RNA Pol II promoter activity. Alu elements in the human genome are 300-nt long noncoding sequences comprised of two homologous G+C-rich monomers, the left (L-Alu) and the right (R-Alu) connected by an A-rich linker. An Alu element contains a poly(A) tail at the 3' terminus and is flanked by short direct repeats depicted as arrows, which are formed by duplications of the host genomic DNA at the insertion sites. The element is transcribed from an internal RNA polymerase III promoter. (Modified from ref. 11).

Autonomous transposable elements include both LTR and nonLTR retrotransposons, e.g., long interspersed nuclear elements (LINE-1 or L1 retrotransposons). The transposition of nonautonomous transposable elements, such as short interspersed nuclear elements (SINEs), is dependent upon other transposable elements. For instance, the L1 transposition machinery is probably used for Alu retrotranspositions. At the same time both L1 and Alu are dependent on cellular RNA-polymerases at the stage of transcription. Two most abundant and successful families of retrotransposons are schematized in Figure 4. As discussed in Chapter 6, there are also other retroelements in the mammalian genomes. It should be emphasized that the distinction between autonomous and nonautonomous elements is often just conditional.

## Some Traditionally Discussed Functional Potentials of Retroelements

In most cases, the newly inserted elements cause deleterious effects, including hereditary diseases induced by insertion mutations. The host cells, however, sometimes exploit the ability of REs to generate variations for their own benefit. Three main properties of the retrotransposons make them attractive candidates for being evolutionary important factors capable of creating functional novelties for subsequent natural selection:

mobilization to new genomic positions in the vicinity of various genes due to retropositions

possession of a powerful transcription regulatory machinery able to affect the regulation of neighboring genes, e.g., to confer on them new tissue specificity of expression, which may cause spatial or temporal distortions of the organism ontogenic trajectory. This is the most probable way of creating morphological novelties for selection.

capacity of causing transcriptional interference, a powerful mechanism of impacting the expression of possibly any mammalian gene. The interference means changes in the expression of one gene due to some activity of another gene. Mechanisms of the interference can be different. When speaking of REs, it can be caused by initiating the synthesis of antisense transcripts that leads to the RNA interference effect.[12] There can be also other ways of interference, like competition for enhancer effects or read-through transcription.[13] These phenomena are widespread in animals.

Moreover, REs can affect stability of the genome by introducing recombination hot spots or enrich the genome with new genetic information, such as genes of reverse transcriptase or viral resistance. Insertional mutagenesis, creation of pseudogenes and exon shuffling are also the tools of genome reshaping used by REs. The number of described cases where retroelement sequences have been shown to impart useful traits to the host is constantly growing.

Detailed discussions of these and other REs features can be found in this book or in other recent reviews [e.g., ref. 11]. To avoid repetition of this material, I will restrict myself to a newly emerging concept of the role of retroelements as an important source of genetic noise with unpredictable consequences for phenotypes of mammalian individuals.

## Stochastic Drivers of Organismal Epigenetic Mosaicism. All of Us Are Probably Complex Epigenetic Mosaics, and What?

Nearly half of mammalian genome consists of REs. Hundreds of thousands of them retain the capacity to regulate transcription of nearby and remote genes.[13] One can only imagine what chaos they could introduce into the genome regulation if there were no ways to suppress their activity. In mammals methylation is widely believed to be a way for such a suppression (for recent review see ref. 14). Indeed, most of methylated C residues in mammalian somatic cell genome are concentrated in REs, and there is a good evidence that this methylation is important for suppression of the REs transcriptional activity. Retrotransposons can be also silenced by RNA-mediated cosuppression, a phenomenon in which mRNA degradation, gene silencing and methylation are triggered by homologous RNA. These two ways are probably interdependent in mammalian cells. The suppression is, however, not total - expressed sequence tag (EST) databases are full of retroelement-derived ESTs from cDNA libraries made from normal tissues, tumor cells and early embryos.

An extremely challenging example of somatic retrotransposon activity in somatic cells is provided by the IAP insertions upstream of the mouse coat color *agouti* gene.[13,14] Several *agouti viable yellow* alleles ($A^{vy}$ or $A^{IAP}$) were formed due to the insertion of an intracisternal A particle (IAP) 100 kb (!) upstream of the *agouti* hair cycle specific promoters. It appeared that when the IAP was silent and methylated, the agouti protein was expressed specifically in hair follicles at a particular stage of the hair cycle, the fur color being agouti. In contrast, when the IAP was active and not methylated, *A* was transcribed from a promoter in the IAP LTR. In this case the normal program of *A* expression was broken, and the agouti protein was expressed in all cells. This abnormal expression of *A* results in yellow fur along with many deviations from a normal phenotype, such as obesity, diabetes and an increased incidence of tumors. Remarkably, the phenotype was variegated, as indicated by the patches of yellow and agouti fur (mottled fur) in most mice. It means that the IAP was active in a part of somatic cells that widely varied between genetically identical individuals obtained by intensive inbreeding. This case reflects a quite expected general situation when a great many of retrotransposons can be variably active in different somatic cells even of one and the same tissue. This incomplete, stochastic, variegated silencing of transcriptionally competent retrotransposons, if it exists, makes every individual mammal a compound epigenetic mosaic with a unique epigenotype. If, in addition, transcriptional activity of retrotransposons influences the expression of neighboring genes, this unique epigenotype will affect the phenotype in a very complex way.[13] How this influence might contribute to complex traits and evolutionary processes remains to be guessed.

## Concluding Remarks

A Japanese proverb that I came across when reading various literature says: "To be important an entity must be simple and frequently repetitive". Transposons in animal genomes are relatively simple and highly repetitive. Therefore, following this oriental wisdom, they are important, and not only for poetic reasons. Is not half of genome important just from quantitative point of view? Are not the elements to be transcribed important as cellular energy consumers? Is it not too expensive to keep all this armada without any quid pro quo? What can be such a compensation? One example is certainly the participation of some (rare!) REs in gene regulation, discussed in this book as well as in numerous reviews devoted to retrotransposons. But is it sufficient? I think this question is meaningless. The organism is hardly organized on principles of maximum economy. Each of its cells is a stochastic system. Each event in such a system is probabilistic, that is not exactly determined. Fascinating examples of a rather chaotic organization of even such an important process as transcription have been reported recently.[15,16] One should not expect a strict order like "erste Kolonne marshiert, zweite Kolonne marshiert" in a system which operates with small numbers of molecules - single molecules (genomic DNA), tens of molecules (rare mRNAs) or even thousands of molecules – abundant (!) mRNAs. It is far less than Avogadro's number $6 \times 10^{23}$ molecules, an order usual for chemists and even than a $6 \times 10^{11}$ molecules level more familiar to molecular biologists. In the latter cases the statistics works perfectly: in a definite time a definite amount of molecules will be converted into another sort of molecules in a certain chemical reaction. Here statistical averaging enables one to describe the reaction using reaction rate constants and halftimes. The situation is different when individual cells are considered. For example, the transcription start time and the number of transcripts synthesized in a definite time will be different in two cells with the same genomes even when the transcription is initiated at identical promoters in twin cells. In an extreme case, the transcription can be not initiated at all in one cell and proceed normally in the other one. Therefore, even the cells of the same tissue will inevitably be different in their patterns of expression, epigenetic modifications, numbers of RNA and protein molecules etc., as illustrated above by the *agouti* gene expression. Despite of this, the organism is able to maintain its integrity, exact timing of development etc.. Whether evolution came to using the uncertainty (noise) to the benefit of organisms is currently a subject of hot debates (e.g., see refs. 17-19). Whatever mechanism(s) are used by organisms and cells to function reliably as a whole, transposons as part of the cell should not disturb them. Of course, they are noisy and unpredictable but they are part of the intracellular molecular chaotic microcosm underlying seemingly well organized cells and multicellular organisms. Are transposons just useless, neutral junk not adding much to the intrinsic cellular noise, or they are able to decrease cellular fitness? These questions will, hopefully, be answered when we will comprehend the cell life in its integrity. Retroelements are an indispensable attribute of the cells, and they should be studied to understand what life is.

## References

1. Coffin JM, Hughes SH, Varmus HE, eds. Retroviruses. Cold Spring Harbor: Cold Spring Harbor Laboratory Press, 1997.
2. Specke V, Schuurman HJ, Plesker R et al. Virus safety in xenotransplantation: First exploratory in vivo studies in small laboratory animals and non-human primates. Transpl Immunol. 2002; 9:281-288.
3. Coffin JM, Rosenberg N. Closing the joint. Nature 1999; 399:413-416.
4. Sverdlov ED. Retroviruses and primate evolution. BioEssays 2000; 22:161-171.
5. Seiki M, Hikikoshi A, Yoshida M. The U5 sequence is a cis-acting repressive element for genomic RNA expression of human T cell leukemia virus type I. Virology 1990; 176:81-86.

6. Domansky AN, Kopantzev EP, Snezhkov EV et al. Solitary HERV-K LTRs possess bi directional promoter activity and contain a negative regulatory element in the U5 region. FEBS Lett 2000; 472:191-195.

7. Lander ES, Linton LM, Birren B et al. Initial sequencing and analysis of the human genome. Nature 2001; 409:860-921.

8. Buzdin A, Khodosevich K, Mamedov I et al. A technique for genome-wide identification of differences in the interspersed repeats integrations between closely related genomes and its application to detection of human-specific integrations of HERV-K LTRs. Genomics 2002; 79:413-422.

9. Venter JC, Adams MD, Myers EW et al. The sequence of the human genome. Science 2001; 291:1304-1351.

10. Mouse Genome Sequencing Consortium. Initial sequencing and comparative analysis of the mouse genome. Nature 2002; 420:520-562.

11. Prak ET, Kazazian HH, Jr. Mobile elements and the human genome. Nature Rev Genet 2000; 1:134-144.

12 Cerutti H. RNA interference: Traveling in the cell and gaining functions? Trends Genet 2003; 19:39–46.

13. Whitelaw E, Martin DIK. Retrotransposons as epigenetic mediators of phenotypic variation in mammals. Nature Genet 2001; 27:361-365

14. Jaenish R, Bird A. Epigenetic regulation of gene expression: How the genome integrates intrinsic and environmental signals. Nature Genet 2003; 33(Suppl):245-254.

15. Dundr M, Hoffmann-Rohrer U, Hu Q et al. A kinetic framework for a mammalian RNA polymerase in vivo. Science 2002; 298:1623–1626.

16. Couzin J. Chaos reigns in RNA transcription. Science 2002; 298:1538.

17. McAdams HH, Arkin A. It's a noisy business! Genetic regulation at the nanomolar scale. Trends Genet 1999; 15:65-69.

18. Huang S. The practical problems of post-genomic biology. Nature Biotechnol 2000; 18:471-472.

19. Ohlsson R, Paldi A, Graves JAM. Did genomic imprinting and X chromosome inactivation arise from stochastic expression? Trends Genet 2001; 17:136-141.

# CHAPTER 6

# Genomic Distributions of Human Retroelements

**Dixie L. Mager, Louie N. van de Lagemaat and Patrik Medstrand**

## Abstract

Nearly half of the human and primate genome is derived from ancient transposable elements, primarily retroelements. This surprising fact alone suggests that retroelements have played a major role in genome organization and evolution. Here we review studies performed in the last 20 years on the chromosomal arrangements of human retroelements, including the exogenous and endogenous retroviruses. We also discuss biological mechanisms or evolutionary forces that may influence their ultimate distribution patterns.

## Introduction

Since Barbara McClintock discovered that the maize genome contains transposable elements (TEs)[1] it has been well established that such elements are universal. Their movements have been linked to a variety of detrimental mutations but it is also becoming clear that their presence adds beneficial diversity to genomes.[2-4] As is discussed elsewhere in this volume (Chapters 1, 7 and 11), there is increasing evidence that TEs affect their hosts in different ways. While there are examples of both loss and increase of host fitness due to the activity of transposable elements, their population dynamics are far from being understood, and the forces underlying their genomic distributions and maintenance in populations are a matter of debate.[5,6] One common view is that TEs are essentially selfish DNA parasites with little functional relevance for their hosts.[7,8] According to this hypothesis, the interaction of TEs with the host is primarily neutral or detrimental and their abundance is a direct result of their ability to replicate autonomously. Another theory to explain the ubiquitous presence of TEs has recently gained increasing attention. It is now becoming accepted that TEs give rise to selectively advantageous adaptive variability which contributes to the evolution of their hosts.[2-4] However, the mechanisms responsible for maintenance, dispersion, fixation and genomic clearance of TEs are likely a result of dynamic processes that have taken various avenues during evolution. The recent completion of whole genome sequences of various eukaryotes offers unique opportunities to explore the evolutionary history and genomic distribution patterns of transposable elements with a view to increasing our understanding of the forces that shape genomes and their mobile inhabitants. Here we will discuss human retroelements in this context, focusing on the retroviral-like sequences.

## Types of Human Retroelements

The human genome is very repeat-rich and in depth analysis of the draft human genome has revealed that close to 45% of the sequence is derived from various TEs[9] (Table 1). Collectively, retroelements (TEs dispersed via an RNA intermediate) are the major type of TE present in the human genome and make up about 43% of the sequence, whereas ancient DNA

*Retroviruses and Primate Genome Evolution*, edited by Eugene D. Sverdlov.
©2005 Eurekah.com.

**Table 1. Major classes of retroelements in the human genome[a]**

| Class | Subclass | Copy Number (x 1000) | Proportion of Genome (%) | Appearance (Myr)[b] |
|---|---|---|---|---|
| SINEs | MIR | 393 | 2.2 | 150 |
| | Alu | 1090 | 10.6 | 80[c] |
| LINEs | L2 | 315 | 3.2 | 150 |
| | L1 | 516 | 16.9 | 100[c] |
| LTR elements | MaLR (MLT and MST) | 240 | 3.7 | 100 |
| | class III | 83 | 1.4 | 100 |
| | class I,MER4 | 112 | 2.9 | 40-50 |
| | class II | 8 | 0.3 | 40-50[c] |
| Other elements[d] | | 406 | 3.6 | |

[a] Data derived from ref. 9; [b] Approximate age of the subclass; Myr, million years; [c] Some members are polymorphic in humans; [d] Represents MIR3 and LINE3 retroelements and the DNA transposons.

transposons occupy less than 2%.[9] For the purposes of this review, these retroelements will be classified into 3 primary categories (see also Chapter 5) and are listed in Table 1. SINEs or Short Interspersed Nuclear Elements are nonautonomous retroelements derived from small functional RNAs. In primates, the Alu family is the most abundant with more than one million copies in the haploid genome.[9,10] SINE elements have most probably amplified in the genome using the reverse transcriptase machinery provided by LINEs or Long Interspersed Nuclear Elements.[11] LINEs are autonomous retroelements because intact copies encode reverse transcriptase and other proteins necessary for their retrotransposition. The major LINE family in primates is L1. While the vast majority of LINE/L1 elements in the genome are 5' truncated and defective, a series of studies have shown that dozens of coding-competent L1 elements still exist in the genome and retain the ability to retrotranspose in experimental systems.[11,12] Some members of the youngest subfamilies of L1 and Alu elements are polymorphic in the human population[13,14] and have caused disease in individuals by inserting into genes (see Chapter 11).[10,11] Thus, there is a certain level of Alu and L1 retrotranspositional activity in the present day and it has been estimated that a new retrotransposition event occurs in the germ line of every 10 to 100 individuals.[15]

Long Terminal Repeat (LTR) containing retroelements include the endogenous retroviruses (ERVs), which are presumably the result of exogenous retroviral infection of the germ line (see refs. 16, 17 and elsewhere in this volume). Dozens of ERV families have been identified in humans and other primates, either due to directed investigations or as a result of the systematic cataloging of all human repeats by the curators of Repbase, the widely used database of repetitive sequences.[18] Human ERV (HERV) families (see also Chapters 5, 10 and 11) can be divided into 3 major classes (Table 1) and individual families that have been most well characterized include the class II HERV-K family and the class I HERV-H, ERV-9, HERV-W and HERV-E families. All HERV families appear to have first entered the germ line at least 25-30 million years ago since they are found in Old World Monkeys.[19-23] Some, such as HERV-H and the class III HERV-L elements, are also found in New World primates, indicating an even older origin.[24,25] Given these ancient origins, it is not surprising that almost all ERVs in the human genome are defective. However, the relatively young HERV-K family contains coding competent members that have integrated after the divergence of humans and the great apes.[26,27] Furthermore, it has recently been shown that some HERV-K elements integrated in the last 200,000 years and are polymorphic in humans raising the possibility of ongoing HERV-K activity.[28,29]

In addition to ERVs, the diverse group of LTR retroelements includes a multitude of other sequences which are flanked by LTR-like structures but which have limited, if any, similarity to retroviral genes. The largest group of such elements are the MaLRs (Mammalian LTR Retroelements), some of which can be traced back over 100 million years.[30] Most ERVs and other LTR retroelements exist today only as a solitary LTR, the result of recombination between the 5' and 3' LTRs of a full length element. Indeed, many repeats are simply defined in Repbase as "LTR-like" based on structural features even if no associated full length element has been identified.

## Effects of Transposable Elements on Genomes

The study of genomic arrangements of human retroelements is likely to give us major insight into their role in primate genome and gene evolution. The involvement of Alu elements in causing both pathogenic and ancient gene deletions or other rearrangements is well known[10] but other potential effects of these ubiquitous sequences may be revealed with better knowledge of their distributions. As discussed elsewhere in this volume (Chapters 7 and 11), many isolated cases of retroelement involvement in gene expression have been documented. It is therefore likely that understanding distribution patterns of these elements will help to establish the global significance of their effects on genes. It is possible that predictions of retroelement influences could be made if an increased understanding of their dispersal patterns is obtained. For example, an important role for L1 elements in X-chromosome inactivation has been hypothesized based on studies of their distinct distribution on the X chromosome compared to the autosomes.[31] L1 elements are also likely responsible for processed pseudogene formation[32] and may have helped shape the genome by their ability to transduce or "move" cellular sequences.[33] Two recent reports have also shown that genomic rearrangements during primate evolution have been mediated by endogenous retroviruses.[34,35] As discussed below, there is evidence suggesting that the characteristic banding pattterns of stained human chromosomes are due to the uneven densities of Alu and L1 elements. It is also clear that large variations in genome size of related species are often the result of different numbers of mobile elements.[36] Finally, it has been suggested that the primary function of cytosine methylation is to silence TEs and that the role of methylation in gene regulation may be secondary.[37] Indeed, Whiteclaw and Martin[38] have recently proposed that phenotypic variation in mammals could be partly due to incomplete and variable silencing of retrotransposons in somatic cells. These are just a partial list of potential genomic effects of TEs but they serve to illustrate the importance of gaining a comprehensive understanding of how these elements have been acquired and how they have been assimilated into the genome.

## Integration Patterns of Exogenous Retroviruses

When considering genomic distribution patterns of endogenous retroviruses or other retroelements, it is useful to briefly review efforts to examine the integration sites of infectious vertebrate retroviruses. In general, it is thought that local base composition, chromatin structure, regional genomic differences, specificity of the viral integrase and other host and viral proteins may all influence target site selection. In the late 1980's, several studies on chicken and mouse retroviruses reported that they integrate preferentially into transcriptionally active regions, at DNAse I hypersensitive sites or near CpG islands.[39-42] These studies used relatively small sample sizes and could have been biased by the selection of cloned proviruses. Nonetheless, a preference for retroviruses to integrate into "open" or transcriptionally active chromatin domains has been widely accepted. However, some more recent reports have challenged this view. An unbiased study using a PCR-based assay to isolate integration sites for avian leukosis virus (ALV) has found that transcriptionally active DNA is not a favored site for ALV integration.[43] Indeed, transcriptional activity seemed to correlate with a decrease in integration frequency, possibly due to blocking of the integration machinery by transcriptional complexes. Another recent study also failed to find an association between transcriptionally active DNA and HTLV-I integration sites and found that this virus can integrate into centromeric alphoid repeats.[44] On the other hand, there is evidence that HIV is excluded from alphoid repeat regions[45] and, furthermore, that it integrates into sites which are competent for transcription since all proviruses can be transcriptionally activated by viral Tat regulatory protein.[46] Initial studies of flanking sites of HIV proviruses reported a propensity for integration within L1 sequences[47] or near Alu repeats[48] but these findings were not confirmed in a larger study.[45]

Bernardi has taken a different approach to examining integration preferences of retroviruses. Originally using density gradient centrifugation, he has defined genomic domains, 200 kb to 1 Mb in size, termed isochores, which are relatively homogeneous in base composition.[49] By determining the isochore locations of relatively small numbers of integration sites, his group showed that retroviruses tend to integrate within isochores which are close to their own GC content. That is, retroviruses with a low GC content, such as MMTV and HIV, are found primarily in GC-poor DNA, whereas high GC viruses, such as MLV, BLV and HTLV-I, integrate preferentially in regions of higher GC.[50,51] He suggests that proviruses are initially targeted to "open chromatin" regions but that, in the absence of selection, isopycnicity is associated with stability of integration.[50]

## Impact of Genetic Drift and Selection

Understanding integration site preferences of infectious retroviruses may give us some insight into site preferences of endogenous retroviruses. However, it should be remembered that genomic locations of endogenous retroviruses and all other mobile elements have been ultimately determined by random genetic drift and by selective forces such that their current distributions may bear little resemblance to initial patterns of integration. Insertions with a significant negative impact on host fitness, i.e., due to gene disruption, would have been rapidly eliminated from the breeding population. Furthermore, as depicted in Figure 1, even the selectively neutral TE insertions represent only those few which were fixed in the species by chance. The probability of fixation of an insertion, or any mutation, that is effectively neutral is $1/2N$ where N is the population size.[52] To illustrate the effect of the above formula, we performed a series of simulations using the PopGen software (http://cc.oulu.fi/~jaspi/popgen/popgen.htm). We estimated the likelihood that a neutral insertion in one individual of a population of 20 would reach fixation (number of generations: 100; allele frequency: 0.025; 1000 simulations) and observed that in about 90% of the cases the insertion was already lost after 10-20 generations. After 50 generations only a few had reached an allele frequency above 0.5 or had drifted to fixation. In a slightly larger population of 100 (number of generations: 100; insertion allele frequency: 0.005; 1000 simulations) only 6 of 1000 insertions reached a frequency above 0.5 within 100 generations. Thus, even with very small ancestral population sizes, it is clear that only a tiny fraction of retroelement insertions remain in the genome (Fig. 2). When viewing TEs in this context, the finding that nearly half of the human genome is composed of these sequences is even more remarkable. Considerations such as these make distributions of retroelements a fascinating subject.

## Insights from Other Species

Analysis of the human genome draft sequence demonstrates that the majority of the repeats were actively transposing at various stages prior to and during the radiation of mammals and are now deeply fixed in the primate lineage.[9] Essentially only the youngest subtypes of L1[13] and Alu[14] elements and possibly a few ERVs belonging to the HERV-K family[28,29] are still actively retrotransposing in humans. These observations stand in strong contrast to TEs of other eukaryotic genomes, for example the fly and the worm, which have a higher frequency of DNA transposons and more recently transposed elements.[9,53] The TEs of mouse are also generally younger and the spectrum of elements that are active in mouse compared to humans are different: the young retroelement fraction of mouse is dominated by L1 and LTR elements and the "active" fraction is at least 5-6 times more abundant in mouse than in humans.[9] Possible

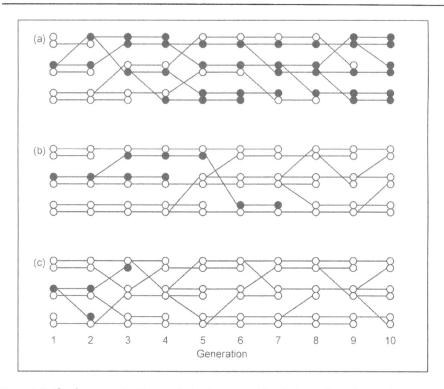

Figure 1. Drift to homozygosity. Parts a-c depict three potential evolutionary fates of a retroelement insertion in a population of three individuals. Circles represent alleles of a particular locus with the filled circle indicating the insert-bearing allele. In every generation, some individuals carrying the insertion will not reproduce whereas others may donate several copies of the insertion by chance. Lines between the circles indicate that the particular allele has been passed to the next generation. When the insertion fails to spread to the next generation, it is lost from the population. In panel (a), the insertion has become fixed (all alleles are filled circles) in generation 10 of the population by random drift. The most likely fate of an insertion is that it is lost in the population.[95]

explanations that may account for differences of active TEs in different organisms include variations in population size, inbreeding and passage of populations through "bottlenecks", all factors that have been suggested as responsible for genomic redistribution, transposition bursts or clearance of TEs.[54]

While this chapter focuses on the distribution of TEs in humans, genome-wide studies of transposable elements and their distributions in organisms such as the fly, worm, yeast and maize are highly informative for a general understanding of the dynamics and interactions between host and TEs.[2] Specifically, chromosomal repetitive DNA profiles demonstrate that TEs are found in most genomic regions investigated and their distribution patterns are clearly nonrandom in the species under study. In Drosophila melanogaster, TEs form clusters in the heterochromatin that are distinct for different TE families.[55,56] For example, HeT-A and TART elements specifically target the telomere regions of the Drosophila chromosomes.[57] However, not all TEs of Drosophila end up in heterochromatin. Some elements are more abundant in the euchromatin[58] and P elements are found in the vicinity of genes.[59] In the Mediterranean fly (Ceratitis capitata), the pattern of mariner elements belonging to different subfamilies also shows variation in the euchromation and heterochromatin distribution.[60] Differences in distributions are also observed for Caenorhabditis elegans TEs on different chromosomes[61] and in DNA regions of various recombination rates.[62] LTR retroelements are the most prevalent class

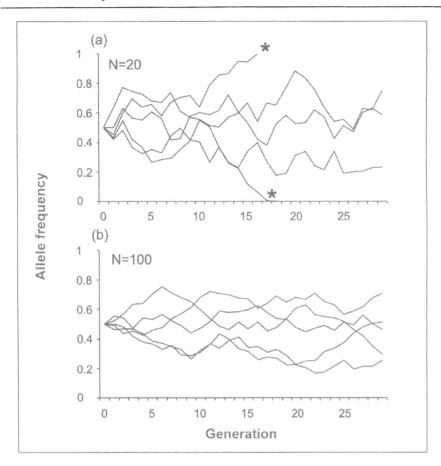

Figure 2. Changes in allele frequencies due to random genetic drift in (a) a population of 10 individuals, and (b) 100 individuals based on five repeat simulations for a two-allele locus having an initial gene frequency of 0.5. Fixation (gene frequency of 1) or loss (gene frequency of 0) is indicated with an asterisk and is reached faster in a smaller population. An allele of a larger population requires a longer time to reach fixation or loss but eventually reaches homozygosity.[52,95] Simulations were performed using the Genetic Drift Model (http://www.utm.edu/~rirwin/Drift.htm).

of mobile elements in maize but are almost absent in hypomethylated DNA regions suggesting that these elements are under-represented in regions upstream of genes[63] whereas the MITE transposons of maize are preferentially found in gene regions.[64] Another interesting example is represented by the Ty 1-4 retroelements of the yeast Saccharomyces cerevisiae which are located specifically upstream of RNA polymerase II transcribed genes.[65] These observations suggest that various elements have a specific integration preference (niche) or, in the case of fixed elements, are preferentially maintained at specific sites in the genome.

## Experimental Determinations of Distributions

Before the availability of large amounts of genomic sequence, distributions of human retroelements were studied using in situ hybridization or cell hybrids or by selective analysis of particular regions. For example, in 1988, in situ hybridization studies showed that Alu elements are found predominantly in gene-rich, GC-rich Giemsa light chromosomal bands while L1 elements predominate in gene-poor, GC-poor Giemsa dark bands.[66] Since Alu elements are

GC-rich and L1 elements GC-poor, it was suggested that their distribution may actually partially account for the banding patterns of human chromosomes. Using density gradient centrifugation, Bernardi's group[50] also showed that Alus are concentrated in high GC isochores and L1 elements in GC-poor isochores and others have made similar observations.[67] As will be discussed further below, these experimental findings are firmly supported by genomic database analyses.

Early in situ hybridization experiments or usage of somatic cell hybrid panels showed that HERV-H, HERV-E and HERV-K elements are widely distributed but an uneven distribution on certain chromosomes was suggested.[68-72] In one of these studies, we found evidence for possible clusters of HERV-H elements on 7q31, 1p36 and 11p15.[68] Another group used Southern analysis of cell hybrids to show that HERV-H LTR sequences occur on all human chromosomes with no marked difference in density.[69] An interesting in situ hybridization study in 1991 reported that most HERV-E elements were found in Alu-rich, G-light bands.[70] In addition, a preference for the Y chromosome for some low copy HERV class I elements has been reported.[71] It was also reported that full length HERV-K elements are found only on some chromosomes.[72]

A few specific genomic regions have been analyzed for LTR retroelement distributions. The group of Sverdlov has conducted extensive analyses of HERV-K LTRs on chromosome 19 and found them to be essentially randomly distributed with respect to LTR age[73] but an association of these LTRs with an array of zinc finger genes on chromosome 19 was reported.[74] Three insertions of low-copy number HERV-I elements into the haptoglobin locus have occurred during primate evolution, suggesting that this locus may be a favored site for HERV-I integration.[75] Comparisons of a 136 kb region spanning the ZFY and ZFX genes revealed that LTR retroelements are five fold more dense in the ZFY region compared to its analogous region on the X chromosome.[76] Interestingly, no significant differences in densities of LINEs, SINEs or DNA transposons were found in this region. Analysis of the MHC class I[77] and class II[78] regions showed that these regions contain more HERVs or solitary LTRs than would be expected by chance. Taken together, these types of studies suggest that LTR elements are nonrandomly distributed. However, it is difficult to assess the significance of such findings without the ability to examine the genome as a whole.

## Large Scale Analysis of Retroelement Distributions

As outlined above, many observations on genomic distributions of TEs were performed prior to availability of large amounts of data generated from the genome projects. The near completion and subsequent initial analysis of the draft human genome sequence provides new opportunities and has revealed some provocative findings regarding the distribution and potential relevance of TEs in humans. The phylogenetic analysis presented with the release of the draft sequence supports previous observations that TEs have existed in mammals for hundreds of millions of years.[9] Some retroelement families (the LINEs and SINEs) seem to have maintained their activity for very long time periods, whereas others (DNA transposons and many ERV families) represent true fossils without any apparent activity over the last 5-10 million years[9,79] (Table 1). It is intriguing that retroelements can persist and maintain activity over very long time periods but it is unknown how the balance between transposition, fixation and TE selection is controlled. Some insight into the forces that shaped distributions of TEs in humans has emerged from the availability of the human genome sequence.

Analysis of the human genome sequence clearly indicates that classes of mobile elements are preferentially located in genomic regions of different base composition. As earlier experimental work had shown, bioinformatic studies have demonstrated that L1 elements are found in the AT-rich regions of the human genome and Alu repeats in the GC-rich fractions[9,79,80] (see Fig. 3a). Interestingly, when Alu and L1 elements are separated into different age categories, it was revealed that their distribution patterns shift with time.[9] This is particularly evident for Alu elements which become more prevalent in GC rich regions with increasing age despite the fact

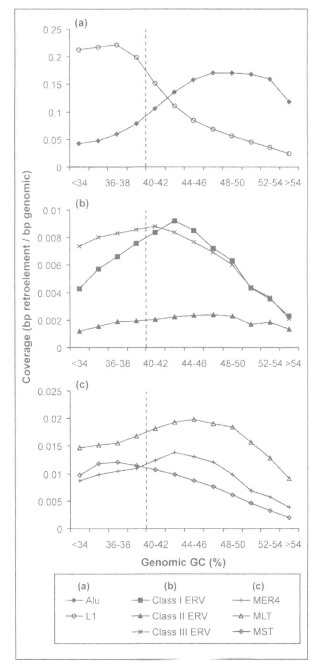

Figure 3. Densities of retroelements in different GC fractions in the human genome, calculated over 20 kb windows across the genome sequence. The vertical axis plots density as "coverage"—defined as the ratio of retroelement base pairs to total genomic base pairs in each GC fraction. The horizontal axis plots GC fraction in increments of 2%. Panel (a) shows coverage by the nonLTR Alu and L1 elements. Panels (b) and (c) show coverage for LTRs from autonomous and nonautonomous retroelements, respectively. The vertical dashed line indicates the average GC content of the total genome (40% GC).[9] Data was derived from the August 6, 2001 draft human genome assembly at http://genome.ucsc.edu/.

that both Alu and L1 elements share a common AT-rich integration preference[9-11] (Fig. 4). When interpreting such findings, it should be remembered that the distribution pattern of the youngest Alu class in Figure 4 likely approximates a true integration site preference, rather than being the result of genetic drift or selection. This is because many of the Alus in this category

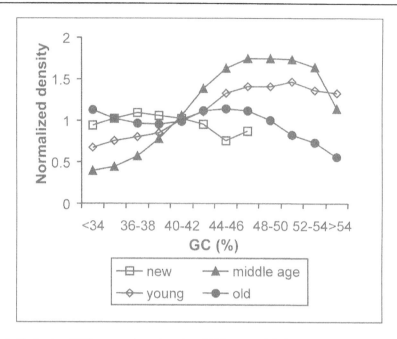

Figure 4. Alu density of different age classes in various GC fractions of the human genome. Age cohorts are defined in terms of percent divergence from their consensus sequence.[9,18,96] New, young, middle age, and old cohorts represent Alus with < 1, 1-5, 10-15, and 20-25 percent divergence from consensus. Assuming that Alus diverge at a rate of 0.2% per million years (Myr),[96] these cohorts correspond roughly to ages of < 5, 5-25, 50-75 and >100 Myr, respectively. Densities in each cohort were normalized by the average density for that cohort over the entire genome. Data points missing in traces are due to GC fractions containing less than 100 elements.

are polymorphic or not yet fixed in humans[81] whereas the other Alu age groups represent truly fixed elements. Thus, in the case of Alu elements, there seems to be a distribution difference between newly integrated and fixed elements. This suggests that evolutionary forces, e.g., selection or genetic drift, have determined the overall distribution of fixed Alu sequences in the human genome. The explanation for this provocative finding is not yet clear. Furthermore, it is even harder to interpret the observed gradual density increase of Alu elements in the GC rich fractions of the genome as Alu elements grow older since natural selection alone cannot clear fixed elements from the genome due to the lack of insert-free alleles.[82,83] Nonetheless, it appears either that Alu elements in GC-poor DNA have been preferentially lost or elements in GC-rich DNA have been preferentially retained during evolution. The latter possibility is in accord with the somewhat controversial hypothesis[9,79,84] that it is beneficial for Alu elements to be located in gene rich or GC-rich regions. It has been suggested that transcription of Alu elements in times of stress plays a useful role in inducing protein translation by inhibiting the double-stranded RNA induced kinase PKR.[84] If true, then positive selection would operate to maintain Alu sequences in GC-rich DNA while allowing Alus in gene poor regions to be lost. However, it is not necessary to invoke a functional role in gene regulation to explain the high density of Alu elements in GC-rich DNA. Bernardi's group proposes that such a distribution maintains genome stability.[85] It is possible that selection operates to maintain the structure of isochores resulting in the selective retention of Alu elements in GC-rich regions because they themselves are GC-rich.[80,85] Similarly, the GC-poor L1 elements may be more stable in GC-poor isochores. A third hypothesis for Alu distribution patterns has been put forward by Brookfield.[83]

He postulates that Alu elements are simply more likely to be retained in gene-rich, GC-rich DNA because their deletion is more likely to simultaneously remove functionally important nearby sequences and hence be detrimental. He argues that an increasing abundance of Alu elements in GC-rich DNA even after they are fixed cannot be the result of selection for positive Alu functions. Although the Bernardi and Brookfield theories are possible explanations, we have shown that the distribution pattern of the oldest Alu elements is not skewed to high GC regions (Fig. 4),[86] suggesting that other mechanisms are involved. Below we suggest that retroelements can be removed after fixation through recombination resulting in recreation of insert-free alleles. This process may provide an explanation for the shifting distributions of retroelements with time.

Compared to Alu and L1 elements, computer-based studies of distribution patterns of other human TEs have been much more limited. Other classes of TEs in humans are less abundant which necessitates the accumulation of large data sets. In addition, the heterogeneous group of LTR elements is actually composed of hundreds of families which adds substantial complexity to any analysis. It has been shown, however, that DNA transposons and LTR elements as a group are more uniformly distributed than Alu or L1 elements.[9,79] Taking this observation as a starting point, we separated the LTR elements into the six major subclasses (Table 1) and have determined their density with respect to GC content across the human genome sequence (Fig. 3b, c).[86] By distinguishing between these classes, an uneven distribution in different GC fractions is also evident. In most cases, the distributions are "broader" than the Alu or L1 patterns, indicating that most classes of LTR elements can occupy a wider range of GC fractions. It is intriguing that, for the class II ERV subfamily, which contains the relatively young HERV-K elements, a significant fraction of elements are located in regions of relatively high GC. This may indicate that, compared to families concentrated in lower GC regions, HERV-K elements or their LTRs may be located closer to genes. Supporting this suggestion is a study which mapped a limited number of HERV-K LTRs relative to genes on chromosome 21 and reported that LTR density roughly correlates with gene density.[87]

## Retroelement Distributions Relative to Genes

To determine if retroelement densities on each chromosome agree with overall densities shown in Figure 3, we plotted densities of Alu, L1 and the six classes of LTR retroelements against estimated gene or GC content of each chromosome (Fig. 5).[86] As expected, the two distribution profiles are almost identical because of a strong correlation between GC content and gene density.[9] Unlike the density of Alu elements, which increases as a strict function of increasing gene content (Fig. 5a), there is generally a negative or no correlation between the density of L1 or LTR elements and gene density (Fig. 5b-h). The Class II ERVs and the MLT elements show little, if any, bias for gene-poor chromosomes, while the L1, Class I, III and MST groups are over represented on these chromosomes. Class I-II elements are dramatically over represented on chromosome Y, as noted before,[9,71,79] and to a lesser extent on 19. Because chromosome 19 is much more gene-dense than the other chromosomes, one possible explanation for the over-representation of the same ERVs on this autosome is that these elements had an initial integration preference for regions near genes or gene-related features such as CpG islands. Similar trends are observed for nonautonomous MER4 distributions and their presumed autonomous class I counterparts (over representation on Y and 19), and for the nonautonomous MaLR (MLT and MST) elements and their apparent autonomous class III ERVs (over representation on 21). Even though some of the LTR classes show a stronger negative correlation than others, the distribution profiles demonstrate that various retroelement families cluster preferentially in different genomic landscapes and are in agreement with the general genomic trends shown in Figure 3.

Given these results, and the fact that gene content and GC content are highly correlated, we looked in more detail at the distribution of retroelements in the draft human genome by locating all elements relative to the nearest annotated gene. We asked the question—can retroelement

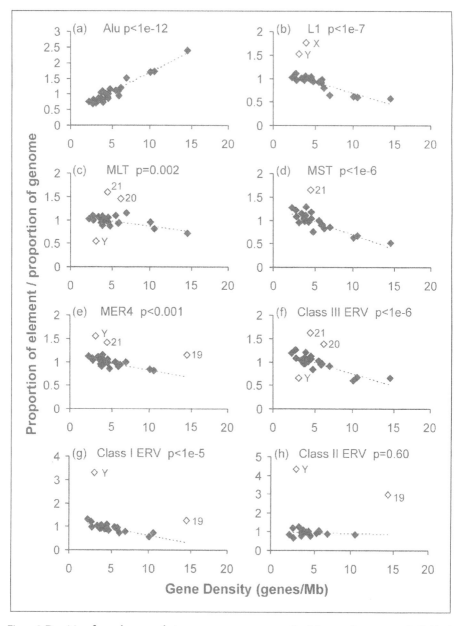

Figure 5. Densities of retroelements relative to average gene content of each human chromosome. Individual chromosomal densities are normalized by the family's average density across the genome. The line connecting solid diamonds indicates the general correlation trend between retroelement and gene density of individual chromosomes. The level of significance (p-values) of the correlation for each data set is indicated. Open diamonds were excluded from the correlation analysis and indicate over or under representation of retroelement density on a particular chromosome. Chromosomes 20, 21 and 22 were excluded from the Class II ERV graph (panel h) due to having less than 100 supporting elements. The number of known genes on each chromosome was derived from alignments of human Reference Sequence database genes with the individual chromosomal sequences.

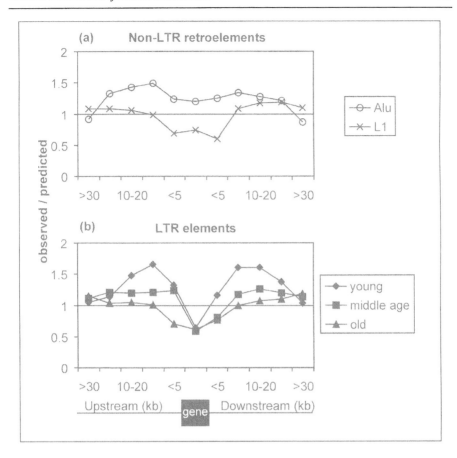

Figure 6. Ratios of observed to predicted retroelement densities with respect to known genes in the human genome. Predictions of base pair contributions in each sequence segment are based on the GC content of the segment and the observed whole genome GC dependence trends in Figure 3. Young, middle age, and old LTR retroelements represent LTRs of the Class II; Class I and MER4; and Class III, MLT, and MST retroelement families, respectively. The points above "gene" and "0-5" of each graph indicate the density in gene regions, and in the first 5 kb either 5' or 3' of genes. The other segments are 5-10, 10-20, 20-30, and >30 kb either upstream or downstream of genes.

densities with respect to genes be accurately predicted based on surrounding GC content or do genes exert independent effects on the density distributions? The density of each class of elements within genes and in various segments of intergenic sequence was determined and then divided by the density predicted from local GC content of the segments. A value of one indicates that density of the retroelement was as expected based on the GC content of the region and the patterns shown in Figure 3. A number of intriguing findings emerged from this analysis (Fig. 6).[86] First, the density of Alu elements is higher than predicted upstream and, to a lesser extent, downstream of genes and is lowest in regions most distant from genes (Fig. 6a). Within genes, the density of Alu elements is slightly higher than that predicted based on GC content. Therefore, not only do Alus have a bias for domains of high gene density, they are most abundant in regions 5' of genes. This pattern supports the hypothesis[9,84] that positive selection may have acted to maintain Alu elements near genes. In contrast to Alu elements, L1s (Fig. 6a) and all LTR elements (Fig. 6b) are markedly underrepresented within the borders of a gene. In addition, although not shown in Figure 6, there is a higher tendency for LTR elements

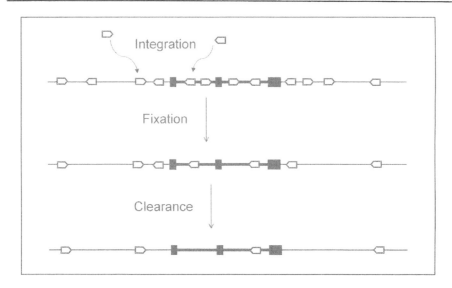

Figure 7. A model for evolution of LTR distribution patterns. A region of genomic DNA containing a gene with three exons (black boxes) is shown with integrated LTR retroelements depicted as arrows. Distributions of the relatively young Class II ERVs suggest that LTR retroelements have an integration preference for sites near (and possibly within) genes. Only those that have no strong negative effect have a chance of becoming fixed by genetic drift. (Note that the fraction which becomes fixed is much smaller than that shown.) Density patterns of the oldest LTR elements (Class III ERVs and MaLRs) further suggest that these sequences are gradually removed from genes and nearby regions after fixation.

within genes to be oriented in the opposite transcriptional orientation[79,86] which likely reflects less fixation because of interference by LTR transcriptional regulatory elements when genes and LTR are located in the same transcriptional direction. These patterns suggest that LTR elements and L1 sequences are "pushed away" from genes by selection. Interestingly, while the relatively young Class II ERVs are underrepresented within genes, they are significantly over-represented in regions 5-20 kb from genes (Fig. 6b). Such a pattern might reflect an integration site preference for gene-rich regions, as suggested above, but strong selection against having these elements within gene borders. In accord with this idea is a recent study showing that 14 of 24 human-specific (young) HERV-K LTR insertions are located within 20 kb of a gene.[88] Unlike Class II ERVs, the older Class III/MLT/MST elements are also significantly underrepresented in regions within 5 kb of genes. One interpretation of this distribution is that selection to remove LTR elements from genes and nearby regions operates over long periods of evolutionary time. That is, older elements are being gradually lost in these regions throughout evolution. These concepts are illustrated in Figure 7.

## Retroelements and the Y Chromosome

The Y chromosome of mammals has presumably originated from two autosomes with sex determining functions.[89,90] During evolution the Y chromosome has undergone several dele-tions resulting in loss of most of the original Y and expansions of a so called nonrecombining Y (NRY) region leading to an almost complete loss of genes and low level of sequence similarity in comparison to its X homologue. The human Y chromosome contains approximately 95% of NRY sequence and is thereby largely suppressed for recombination between X or other chro-mosomes.

Comparison of an analogous but nonrecombining 136 kb region of chromosome X and Y spanning the ZFY and ZFX genes revealed that LTR elements are five fold more dense in the

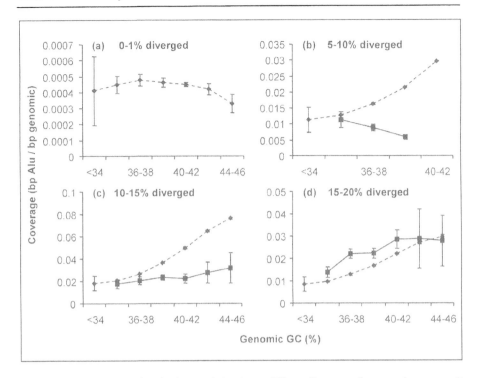

Figure 8. Sequence coverage by Alu elements belonging to different divergence classes on chromosome Y (solid lines) compared to the whole genome (dashed lines). There were insufficient numbers of Alu elements on the Y chromosome in the divergence cohorts < 5 % to be plotted. The coverage by each Alu divergence class is plotted against the local 20-kb genome GC content.

ZFY region compared to its analogous region on the X chromosome,[76] and the ZFY region contains significantly more LTR retroelements than the autosomes. L1 elements are also more abundant on Y than on the autosomes[9] (Fig. 5b) whereas Alu elements are not over-represented on Y[9] (Fig. 5a). The analysis of the human draft sequence revealed that the Y chromosome contains a higher density of young LTR retroelements and L1 elements than any of the autosomes or the X chromosome. Many TEs are present also at high number on Y of Drosophila and the recombinational isolation of the Y chromosome has been implicated in explaining accumulation of TEs of this chromosome.[91] As shown in Figure 5, we also observed an abundance of the youngest LTR elements on chromosome Y whereas the under representation on Y of the old Class III ERVs and MLT retroelements are consistent with major rearrangements and deletions of Y during mammalian evolution.[89,90]

As a way to investigate the change in distribution of younger Alus toward GC-rich regions (Fig. 4), we analyzed Alu density patterns on the Y chromosome and detected a major difference on this chromosome compared to the whole genome (Fig. 8).[86] The density pattern of Alus in the 5-10% divergence class is strikingly opposite to that observed in the whole genome in that they are more prevalent in regions of lower GC content (Fig. 8b). The distribution trends of the next two older Alu classes with respect to GC content gradually become more similar to the patterns seen in the entire genome (Fig. 8c, d). This finding suggests that the density shift of Alus from AT-rich to GC-rich regions during evolution was significantly delayed on the Y chromosome and, therefore, that the ability to recombine with a homologous chromosome greatly facilitated this shift.

## Genomic Clearance of Retroelements

Based on the copy number in humans and other primate genomes, Alu elements seem to be the most successfully transposing element during primate evolution. It is possible that the current Alu distribution pattern was shaped by a quick build up in the AT rich regions followed by extensive loss via recombination. Indeed, it has recently been demonstrated that the efficiency of Alu-Alu recombination in yeast increases as a pair of elements are placed closer together.[92] Alu elements seem quite prone for recombination because two elements up to 20% divergent are still able to recombine efficiently and such closely spaced Alu pairs are found only occasionally in the human genome,[92,93] possibly because of clearance of these elements through the mechanism of inverted repeat (IR)-mediated recombination.[94] Considering the high number of genomic Alu elements and the fact that they preferentially target AT-rich regions, these domains must have suffered a massive build up of Alu integrations, some of which became fixed in the population. Such accumulation may have resulted in increased recombination as the occurrence of closely spaced Alus increased which could have led to loss of both newly integrated and fixed Alu elements in the AT fraction of the genome. We propose that a possible cause of Alu element build up in the GC rich fraction is that these sequences never reach a critical density in GC-rich DNA (because it is not the preferred integration target for Alu elements) which is high enough to allow IR-mediated recombination to operate efficiently.

The selfish DNA hypothesis[7,8] predicts that, due to their neutral or deleterious effects, retroelements will be better tolerated in the AT rich and gene poor part of the genome where they would interfere less frequently with genes due to transcriptional interference or recombination. Most retroelement seem to follow this rule. Older LTR elements accumulate in AT-rich DNA[86] and all types of LTR retroelements are significantly underrepresented in genes (Fig. 6). LTRs contain regulatory signals very similar to those in cellular genes so it seems reasonable that insertion of an LTR close to or within a gene would frequently be disadvantageous. Such insertions with a marked negative impact will be selected against with no chance to spread to fixation. However, mutations with a selective disadvantage can still be fixed through genetic drift, especially if the effective population size is small.[52] Even though LTR retroelements became fixed in the genome, this pattern suggests that the presence of such elements close to genes generally has a slight detrimental effect and are being gradually eliminated with time (Fig. 7). Supporting this concept is our recent observation that densities of most LTR retroelements gradually shift with time, such that older elements tend to be found in regions of lower GC compared to their younger counterparts.[86] Once fixed, it would not be possible for an insertion to be eliminated due to the absence of insert-free alleles. However, intrachromosomal deletions, recombination between homologous chromosome regions or IR-mediated recombination could recreate such alleles and again provide an opportunity for the original insertion to be lost from the population through natural selection or drift (Fig. 9). Indeed, the Alu distribution pattern on the Y chromosome supports the concept that recombination plays a role in the elimination of fixed elements, and that IR-mediated elimination is particularly efficient for removing Alu elements.

## Concluding Remarks

It remains to be determined how natural selection and genetic drift have shaped the distributions of ERVs and other human retroelements throughout evolution. However, the draft human genome sequence has already added greatly to our understanding of the mechanisms that are operating to reshape the distribution pattern of these mobile elements. Completion of the human sequence and further in depth analysis will undoubtedly lead to deeper insights. Furthermore, when the sequences of other mammal genomes, particularly those of primates, become available, comparative studies will provide additional opportunities to explore changing retroelement distributions and their importance in gene and genome evolution.

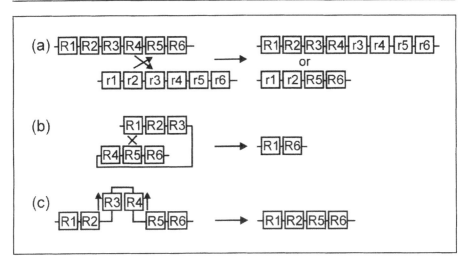

Figure 9. Examples of mechanisms that may regenerate insertion-free alleles: (a) unequal recombination, (b) intrachromosomal recombination and (c) inverted repeat (IR) mediated recombination. The designation R1 to R6 indicates related retroelements on the same chromosome and r1 to r6 indicate the homologous allelic copies of the second chromosome.

## Acknowledgements

We thank Christine Kelly for assistance with manuscript preparation and members of our laboratory for helpful comments. Work in D. Mager's laboratory was supported by a grant from the Canadian Institutes of Health Research. P.M. was supported by the Knut and Alice Wallenberg Foundation and the Ake Wibergs Foundation, Sweden.

## References

1. McClintock B. Controlling elements and the gene. Cold Spring Harbor Symp Quant Biol 1956; 21:197-216.
2. Kidwell MG, Lisch DR. Perspective: Transposable elements, parasitic DNA, and genome evolution. Int J Org Evolution 2001; 55:1-24.
3. Brosius J. Genomes were forged by massive bombardments with retroelements and retrosequences. Genetica 1999; 107:209-238.
4. Hurst GD, Werren JH. The role of selfish genetic elements in eukaryotic evolution. Nature Rev Genet 2001; 2:597-606.
5. Biemont C, Tsitrone A, Vieira C et al. Transposable element distribution in Drosophila. Genetics 1997; 147:1997-1999.
6. Charlesworth B, Langley CH, Sniegowski PD. Transposable element distributions in Drosophila. Genetics 1997; 147:1993-1995.
7. Doolittle WF, Sapienza C. Selfish genes, the phenotype paradigm and genome evolution. Nature 1980; 284:601-603.
8. Orgel LE, Crick FH. Selfish DNA: The ultimate parasite. Nature 1980; 284:604-607.
9. International human genome sequencing consortium. Initial sequencing and analysis of the human genome. Nature 2001; 409:860-921.
10. Deininger PL, Batzer MA. Alu repeats and human disease. Mol Genet Metab 1999; 67:183-193.
11. Ostertag EM, Kazazian Jr HH. Biology of Mammalian l1 retrotransposons. Annu Rev Genet 2001; 35:501-538.
12. Moran JV, Holmes SE, Naas TP et al. High frequency retrotransposition in cultured mammalian cells. Cell 1996; 87:917-927.
13. Boissinot S, Chevret P, Furano AV. L1 (LINE-1) retrotransposon evolution and amplification in recent human history. Mol Biol Evol 2000; 17:915-928.
14. Stoneking M, Fontius JJ, Clifford SL et al. Alu insertion polymorphisms and human evolution: Evidence for a larger population size in Africa. Genome Res 1997; 7:1061-1071.

15. Kazazian Jr HH. An estimated frequency of endogenous insertional mutations in humans. Nature Genet 1999; 22:130.
16. Wilkinson DA, Mager DL, Leong JC. Endogenous human retroviruses. In: Levy J, ed. The Retroviridae. New York: Plenum Press, 1994:465-535.
17. Bock M, Stoye JP. Endogenous retroviruses and the human germline. Curr Opin Genet Dev 2000; 10:651-655.
18. Jurka J. Repbase update: A database and an electronic journal of repetitive elements. Trends Genet 2000; 16:418-420.
19. Mariani-Costantini R, Horn TM, Callahan R. Ancestry of a human endogenous retrovirus family. J Virol 1989; 63:4982-4985.
20. Reus K, Mayer J, Sauter M et al. HERV-K(OLD): Ancestor sequences of the human endogenous retrovirus family HERV-K(HML-2). J Virol 2001; 75:8917-8926.
21. Goodchild NL, Wilkinson DA, Mager DL. Recent evolutionary expansion of a subfamily of RTVL-H human endogenous retrovirus-like elements. Virology 1993; 196:778-788.
22. Voisset C, Blancher A, Perron H et al. Phylogeny of a novel family of human endogenous retrovirus sequences, HERV-W, in humans and other primates. AIDS Res Hum Retroviruses 1999; 15:1529-1533.
23. Costas J, Naveira H. Evolutionary history of the human endogenous retrovirus family ERV9. Mol Biol Evol 2000; 17:320-330.
24. Mager DL, Freeman JD. HERV-H endogenous retroviruses: Presence in the New World branch but amplification in the Old World primate lineage. Virology 1995; 213:395-404.
25. Benit L, Lallemand JB, Casella JF et al. ERV-L elements: A family of endogenous retrovirus-like elements active throughout the evolution of mammals. J Virol 1999; 73:3301-3308.
26. Barbulescu M, Turner G, Seaman MI et al. Many human endogenous retrovirus K (HERV-K) proviruses are unique to humans. Curr Biol 1999; 9:861-868.
27. Medstrand P, Mager DL. Human-specific integrations of the HERV-K endogenous retrovirus family. J Virol 1998; 72:9782-9787.
28. Turner G, Barbulescu M, Su M et al. Insertional polymorphisms of full-length endogenous retroviruses in humans. Curr Biol 2001; 11:1531-1535.
29. Stoye JP. Endogenous retroviruses: Still active after all these years? Curr Biol 2001; 11:R914-916.
30. Smit AF. Identification of a new, abundant superfamily of mammalian LTR-transposons. Nucleic Acids Res 1993; 21:1863-1872.
31. Bailey JA, Carrel L, Chakravarti A et al. Molecular evidence for a relationship between LINE-1 elements and X chromosome inactivation: The Lyon repeat hypothesis. Proc Natl Acad Sci USA 2000; 97:6634-6639.
32. Esnault C, Maestre J, Heidmann T. Human LINE retrotransposons generate processed pseudogenes. Nature Genet 2000; 24:363-367.
33. Pickeral OK, Makalowski W, Boguski MS et al. Frequent human genomic DNA transduction driven by LINE-1 retrotransposition. Genome Res 2000; 10:411-415.
34. Hughes JF, Coffin JM. Evidence for genomic rearrangements mediated by human endogenous retroviruses during primate evolution. Nature Genet 2001; 29:487-489.
35. Jamain S, Girondot M, Leroy P et al. Transduction of the human gene FAM8A1 by endogenous retrovirus during primate evolution. Genomics 2001; 78:38-45.
36. Petrov DA. Evolution of genome size: New approaches to an old problem. Trends Genet 2001; 17:23-28.
37. Yoder JA, Walsh CP, Bestor TH. Cytosine methylation and the ecology of intragenomic parasites. Trends Genet 1997; 13:335-340.
38. Whitelaw E, Martin DI. Retrotransposons as epigenetic mediators of phenotypic variation in mammals. Nature Genet 2001; 27:361-365.
39. Vijaya S, Steffen DL, Robinson HL. Acceptor sites for retroviral integrations map near DNase I-hypersensitive sites in chromatin. J Virol 1986; 60:683-692.
40. Rohdewohld H, Weiher H, Reik W et al. Retrovirus integration and chromatin structure: Moloney murine leukemia proviral integration sites map near DNase I-hypersensitive sites. J Virol 1987; 61:336-343.
41. Mooslehner K, Karls U, Harbers K. Retroviral integration sites in transgenic Mov mice frequently map in the vicinity of transcribed DNA regions. J Virol 1990; 64:3056-3058.
42. Scherdin U, Rhodes K, Breindl M. Transcriptionally active genome regions are preferred targets for retrovirus integration. J Virol 1990; 64:907-912.
43. Weidhaas JB, Angelichio EL, Fenner S et al. Relationship between retroviral DNA integration and gene expression. J Virol 2000; 74:8382-8389.

44. Leclercq I, Mortreux F, Cavrois M et al. Host sequences flanking the human T-cell leukemia virus type 1 provirus in vivo. J Virol 2000; 74:2305-2312.
45. Carteau S, Hoffmann C, Bushman F. Chromosome structure and human immunodeficiency virus type 1 cDNA integration: Centromeric alphoid repeats are a disfavored target. J Virol 1998; 72:4005-4014.
46. Jordan A, Defechereux P, Verdin E. The site of HIV-1 integration in the human genome determines basal transcriptional activity and response to Tat transactivation. EMBO J 2001; 20:1726-1738.
47. Stevens SW, Griffith JD. Human immunodeficiency virus type 1 may preferentially integrate into chromatin occupied by L1Hs repetitive elements. Proc Natl Acad Sci USA 1994; 91:5557-5561.
48. Stevens SW, Griffith JD. Sequence analysis of the human DNA flanking sites of human immunodeficiency virus type 1 integration. J Virol 1996; 70:6459-6462.
49. Bernardi G. The human genome: Organization and evolutionary history. Annu Rev Genet 1995; 29:445-476.
50. Rynditch AV, Zoubak S, Tsyba L et al. The regional integration of retroviral sequences into the mosaic genomes of mammals. Gene 1998; 222:1-16.
51. Glukhova LA, Zoubak SV, Rynditch AV et al. Localization of HTLV-1 and HIV-1 proviral sequences in chromosomes of persistently infected cells. Chromosome Res 1999; 7:177-183.
52. Li WH, Graur D. Fundamentals of molecular evolution. Sutherland MA: Sinauer Associates, 1991.
53. Bowen NJ, McDonald JF. Drosophila euchromatic LTR retrotransposons are much younger than the host species in which they reside. Genome Res 2001; 11:1527-1540.
54. Brookfield JF, Badge RM. Population genetics models of transposable elements. Genetica 1997; 100:281-294.
55. Biemont C, Cizeron G. Distribution of transposable elements in Drosophila species. Genetica 1999; 105:43-62.
56. Maside X, Bartolome C, Assimacopoulos S et al. Rates of movement and distribution of transposable elements in Drosophila melanogaster: In situ hybridization vs Southern blotting data. Genet Res 2001; 78:121-136.
57. Pardue ML, Danilevskaya ON, Lowenhaupt K et al. Drosophila telomeres: New views on chromosome evolution. Trends Genet 1996; 12:48-52.
58. Di Franco C, Terrinoni A, Dimitri P et al. Intragenomic distribution and stability of transposable elements in euchromatin and heterochromatin of Drosophila melanogaster: Elements with inverted repeats Bari 1, hobo, and pogo. J Mol Evol 1997; 45:247-52.
59. Liao GC, Rehm EJ, Rubin GM. Insertion site preferences of the P transposable element in Drosophila melanogaster. Proc Natl Acad Sci USA 2000; 97:3347-3351.
60. Torti C, Gomulski LM, Moralli D et al. Evolution of different subfamilies of mariner elements within the medfly genome inferred from abundance and chromosomal distribution. Chromosoma 2000; 108:523-532.
61. Surzycki SA, Belknap WR. Repetitive-DNA elements are similarly distributed on Caenorhabditis elegans autosomes. Proc Natl Acad Sci USA 2000; 97:245-249.
62. Duret L, Marais G, Biemont C. Transposons but not retrotransposons are located preferentially in regions of high recombination rate in Caenorhabditis elegans. Genetics 2000; 156:1661-1669.
63. Meyers BC, Tingey SV, Morgante M. Abundance, distribution, and transcriptional activity of repetitive elements in the maize genome. Genome Res 2001; 11:1660-1676.
64. Zhang Q, Arbuckle J, Wessler SR. Recent, extensive, and preferential insertion of members of the miniature inverted-repeat transposable element family Heartbreaker into genic regions of maize. Proc Natl Acad Sci USA 2000; 97:1160-1165.
65. Kim JM, Vanguri S, Boeke JD et al. Transposable elements and genome organization: A comprehensive survey of retrotransposons revealed by the complete Saccharomyces cerevisiae genome sequence. Genome Res 1998; 8:464-478.
66. Korenberg JR, Rykowski MC. Human genome organization: Alu, lines, and the molecular structure of metaphase chromosome bands. Cell 1988; 53:391-400.
67. Chen TL, Manuelidis L. SINEs and LINEs cluster in distinct DNA fragments of Giemsa band size. Chromosoma 1989; 98:309-316.
68. Fraser C, Humphries RK, Magcr DL. Chromosomal distribution of the RTVL-H family of human endogenous retrovirus-like sequences. Genomics 1988; 2:280-287.
69. Sugino H, Oshimura M, Mastubara K. Distribution of human endogenous retroviral RTVL-H2 LTR sequences among human chromosomes. Gene 1997; 198:83-87.
70. Taruscio D, Manuelidis L. Integration site preferences of endogenous retroviruses. Chromosoma 1991; 101:141-156.

71. Kjellman C, Sjogren HO, Widegren B. The Y chromosome: A graveyard for endogenous retroviruses. Gene 1995; 161:163-170.

72. Meese E, Gottert E, Zang KD et al. Human endogenous retroviral element k10 (HERV-K10): Chromosomal localization by somatic hybrid mapping and fluorescence in situ hybridization. Cytogenet Cell Genet 1996; 72:40-42.

73. Lavrentieva I, Khil P, Vinogradova T et al. Subfamilies and nearest-neighbour dendrogram for the LTRs of human endogenous retroviruses HERV-K mapped on human chromosome 19: Physical neighbourhood does not correlate with identity level. Hum Genet 1998; 102:107-116.

74. Vinogradova T, Volik S, Lebedev Yu et al. Positioning of 72 potentially full size LTRs of human endogenous retroviruses HERV-K on the human chromosome 19 map. Occurrences of the LTRs in human gene sites. Gene 1997; 199:255-264.

75. Maeda N, Kim HS. Three independent insertions of retrovirus-like sequences in the haptoglobin gene cluster of primates. Genomics 1990; 8:671-683.

76. Erlandsson R, Wilson JF, Paabo S. Sex chromosomal transposable element accumulation and male-driven substitutional evolution in humans. Mol Biol Evol 2000; 17:804-812.

77. Kulski JK, Gaudieri S, Inoko H et al. Comparison between two human endogenous retrovirus (HERV)-rich regions within the major histocompatibility complex. J Mol Evol 1999; 48:675-683.

78. Andersson G, Svensson AC, Setterblad N et al. Retroelements in the human MHC class II region. Trends Genet 1998; 14:109-114.

79. Smit AF. Interspersed repeats and other mementos of transposable elements in mammalian genomes. Curr Opin Genet Dev 1999; 9:657-663.

80. Gu Z, Wang H, Nekrutenko A et al. Densities, length proportions, and other distributional features of repetitive sequences in the human genome estimated from 430 megabases of genomic sequence. Gene 2000; 259:81-88.

81. Carroll ML, Roy-Engel AM, Nguyen SV et al. Large-scale analysis of the Alu Ya5 and Yb8 subfamilies and their contribution to human genomic diversity. J Mol Biol 2001; 311:17-40.

82. Nuzhdin SV. Sure facts, speculations, and open questions about the evolution of transposable element copy number. Genetica 1999; 107:129-137.

83. Brookfield JF. Selection on Alu sequences? Curr Biol 2001; 11:R900-901.

84. Schmid CW. Does SINE evolution preclude Alu function? Nucleic Acids Res 1998; 26:4541-4550.

85. Pavlicek A, Jabbari K, Paces J et al. Similar integration but different stability of Alus and LINEs in the human genome. Gene 2001; 276:39-45.

86. Medstrand P, van de Lagemaat L, Mager DL. Retroelement distributions in the human genome: Variations associated with age and proximity to genes. Genome Res. 2002; 12:1483-1495.

87. Kurdyukov SG, Lebedev YB, Artamonova II et al. Full-sized HERV-K (HML-2) human endogenous retroviral LTR sequences on human chromosome 21: Map locations and evolutionary history. Gene 2001; 273:51-61.

88. Buzdin A, Khodosevich K, Mamedov I et al. A technique for genome-wide identification of differences in the interspersed repeats integrations between closely related genomes and its application to detection of human-specific integrations of HERV-K LTRs. Genomics 2002; 79:413-422.

89. Lahn BT, Pearson NM, Jegalian K. The human Y chromosome, in the light of evolution. Nature Rev Genet 2001; 2:207-216.

90. Graves JA. The origin and function of the mammalian Y chromosome and Y-borne genes—an evolving understanding. Bioessays 1995; 17:311-320.

91. Junakovic N, Terrinoni A, Di Franco C et al. Accumulation of transposable elements in the heterochromatin and on the Y chromosome of Drosophila simulans and Drosophila melanogaster. J Mol Evol 1998; 46:661-668.

92. Lobachev KS, Stenger JE, Kozyreva OG et al. Inverted Alu repeats unstable in yeast are excluded from the human genome. EMBO J 2000; 19:3822-3830.

93. Stenger JE, Lobachev KS, Gordenin D et al. Biased distribution of inverted and direct Alus in the human genome: Implications for insertion, exclusion, and genome stability. Genome Res 2001; 11:12-27.

94. Leach DR. Long DNA palindromes, cruciform structures, genetic instability and secondary structure repair. Bioessays 1994; 16:893-900.

95. Ridley M. Evolution. Cambridge MA: Blackwell scientific publications, 1993.

96. Chen FC, Li WH. Genomic divergences between humans and other hominoids and the effective population size of the common ancestor of humans and chimpanzees. Am J Hum Genet 2001; 68:444-456.

# CHAPTER 7

# Influence of Human Endogenous Retroviruses on Cellular Gene Expression

**Christine Leib-Mösch, Wolfgang Seifarth and Ulrike Schön**

## Abstract

Endogenous retroviruses (HERVs) and retrotransposons are normal components of the human DNA. During evolution these elements have spread by retrotransposition and thus dispersed their regulatory sequences throughout the genome. Novel insertions can have a variety of consequences for adjacent genes, not only by disrupting gene function, but also by enhancing transcription levels, altering tissue specificity of gene expression, or creating new gene products with modified functions via alternative splicing. Therefore, retrotransposable elements have served as a catalyst for genomic evolution and possibly played an important role in primate speciation and adaptation.

## Introduction

Nearly half of the human genome is composed of sequences that have been generated by retrotransposition including approximately 600,000 to 700,000 interspersed copies of elements with long terminal repeats (LTR).[1,2] These sequences are probably remnants of exogenous retroviruses that infected the germ line and became genetically fixed in the primate genome between 70 and 30 million years ago. Based on their similarities to mammalian retroviruses, human endogenous retroviruses (HERVs) have been divided into three classes: class I is comprised of type C (MLV) related HERVs, class II of type A/B/D sequences mainly related to MMTV, and class III of foamy virus related elements (for reviews see Chapter 5, 10, 11 and refs. 3-6). The large number of retroviral sequences presently found in the human genome indicates that endogenous retroviruses have been amplified in primates during evolution by repeated reintegration of reverse transcribed mRNA. Furthermore, the presence of numerous solitary LTRs suggests that extensive reexcision and recombination events have occurred. Whereas the structural genes of most HERVs have lost their functional capacity due to mutations during the course of evolution, their widely spread LTRs still contain active regulatory elements that may affect the expression of neighboring cellular genes.

## Variations in LTR Structure and Activity of Different HERV Families

HERV expression is controlled by the LTRs which flank the structural genes *gag, pol* and *env*. Proviral LTRs have the structure U3-R-U5 with the terminal consensus sequence TG...CA. Transcription is dependent on specific regulatory sequences located primarily within the 3' untranslated (U3) region.[7] This sequence ranges in length from 150 to 1,200 nucleotides (nt) in different retroviruses and contains all the signals required for initiation of transcription, such as promoters, enhancers and transcription factor binding sites. The terminal redundant sequence (R) consists of about 10 to 250 nt and forms the 5' and 3' end of the retroviral RNA. The R sequence is defined by the transcription start site within the 5' LTR and the location of the polyadenylation site for the processing of the 3' end. The U5 region is a unique sequence of 35 to 450 nt at the 5' end of the RNA genome. Downstream of the 5' LTR is the primer binding site (PBS) for the minus strand, which is usually complementary to the last 8 to 18 nucleotides of a specific cellular tRNA. The PBS has been used for designating different HERV families by means of the single letter amino acid code. For example, HERV-L has a primer binding site complementary to leucine tRNA.

LTRs derived from different HERV families are often very heterogeneous in their transcriptional activity and tissue specificity. Members of a distinct HERV family may have completely different expression preferences or expression times varying with cell type.[8] The large copy number of elements within each HERV family makes it difficult to clearly distinguish between the activities of distinct members of a group and to determine whether expression is a property of a whole group or of an individual provirus. Some HERV elements seem to be expressed ubiquitously in every human tissue, whereas others are active in specific cell types (see ref. 8, Seifarth et al., submitted). Furthermore, transcription activity may be influenced by various environmental factors. The following sections will describe characteristic features of the LTRs from the major HERV families and their implications in differential expression.

### HERV-R

HERV-R, a class I element related to type C retroviruses, was first isolated and described as a single copy HERV element (ERV-3).[9] However, BLAST database searches recently revealed that about 100 homologous sequences and 125 solitary HERV-R LTRs are dispersed throughout the human genome.[1] The LTR sequences are about 593 bp long and contain a typical TATA-box (TATAAAAG) and a polyadenylation site (Fig. 1). Compared to other retroviruses, HERV-R has a very short U5 region (35 bp). The HERV-R LTR also contains four regions displaying partial homology with a consensus sequence present in many viral enhancers, which are dispersed over the 3' half of the LTR sequence and the PBS region (Fig. 1, see ref. 9).

### HERV-E

The HERV-E family also belongs to the class I elements with about 250 full-length proviruses and about 1,000 solitary HERV-E LTR sequences detected.[1] The typical HERV-E LTR, represented by clone 4-1, is about 490 bp in length[10] and contains a putative TATA-box, (CCTTAAAAG) and a polyadenylation signal (AATAAA), but has no inverted repeats at its ends (Fig. 1). In addition, the TATA-box is highly degenerate compared to the conserved consensus sequences found in other LTRs. This may explain why the isolated HERV-E (4-1) LTR shows no activity in any of the cell lines tested so far.[8] Other LTRs of the HERV-E family, however, were found to function as promoters or enhancers in vivo and in vitro. For example, the LTRs of the HERV-E elements inserted into the amylase gene cluster (ERV-A) contain a perfect TATA box, CAAT box, polyadenylation signal and a glucocorticoid responsive element (GRE), and regulate tissue-specific expression of human salivary amylases.[11]

### HERV-I

The HERV-I family comprises about 250 proviral elements[1] and about the same number of full-length solitary LTRs. In addition, the human genome harbors about 1,800 truncated LTR copies that lack nearly the entire 5'-half of the LTR sequence and in most cases also 60 nt

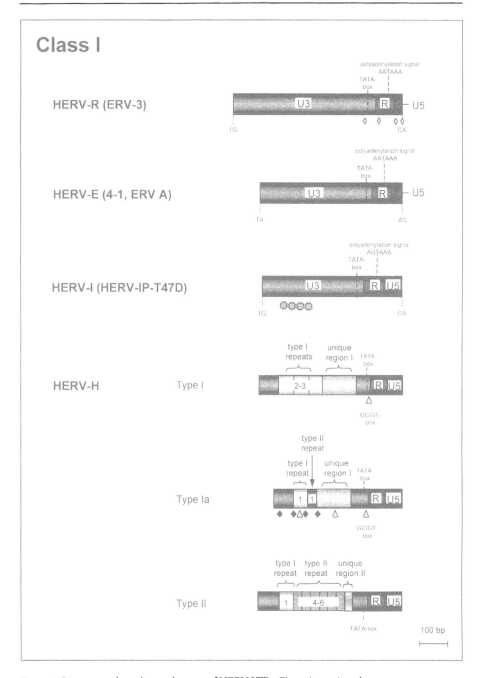

Figure 1. Structure and regulatory elements of HERV LTRs. Figure is continued on next page.

at the 3' end.[12] Located within this core sequence are a putative CAAT and a TATA box (Fig. 1). HERV-IP-T47D, a member of a subgroup of HERV-I elements with a PBS complementary to proline instead of isoleucine tRNA, was originally isolated from particles released by the

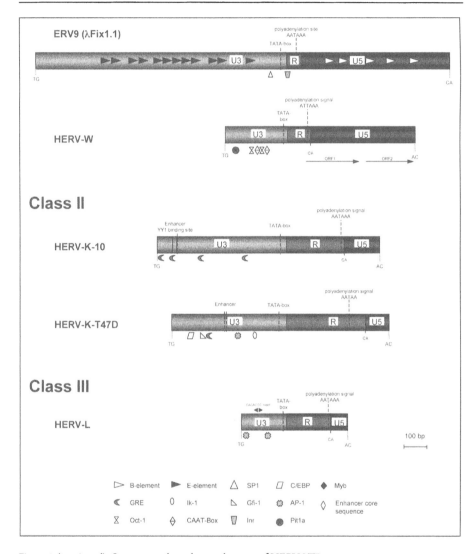

Figure 1 (continued). Structure and regulatory elements of HERV LTRs.

human breast cancer cell line T47D.[13] Computer analysis of its 5' LTR revealed several potential regulatory elements and putative transcription factor binding sites, such as C/EBP, GFI-1 and 4 AP-1 sites arranged in tandem within the U3 region. In transient transfection assays, the HERV-IP LTR was found to be active particularly in liver and kidney cells.[8]

## *HERV-H*

HERV-H sequences comprise a large type C related family of human endogenous retrovirus-like elements, with about 1,000 proviral copies per haploid genome and a similar number of solitary LTRs. Sequence comparison of the HERV-H LTR sequences led to their division into three subtypes: types I, II and Ia (Fig. 1). Type Ia appears to have evolved by recombination between type I and type II LTRs.[14] The LTR sequences of these elements differ mainly in their U3 region, which contains varying copy numbers of tandem repeats found to act as negative regulatory sequences (NREs) when cloned upstream of the human β-globin gene promoter.[15] Type II LTRs containing one type I and 4 to 6 type II repeats are completely inactive. The HERV-H LTR with the strongest promoter function isolated to date is the HERV-H H6 LTR (type Ia), which has lost most of the repressor elements and additionally contains at least two strong synergistically acting SP1 binding sites.[15,16] This LTR displays strong promoter activity in human NTera2D1, 293, Hep2, U373 and LC5 cells when cloned in a reporter plasmid.[17,8] All type I and Ia LTRs isolated so far contain a GC rich region (GC/GT-box) downstream of the TATA box.[16,18] This region has been shown to be important for promoter activity and displays additional binding sites for the transcription factors Sp1 and Sp3, which may be involved in the tissue-specific expression pattern of HERV-H elements.[8,15,18] Interestingly, in these experiments the TATA-box is dispensable since point mutations affecting this sequence do not significantly reduce promoter activity.[18] Several of the HERV-H LTRs tested exhibited bidirectional promoter activity.[17] In addition, they were found to possess enhancer activity when cloned in either orientation upstream of a β-globin- or TK-promoter.[17,18] The enhancer activity seems to be dependent on proteins binding upstream of the TATA and GC/GT box.[18] In the active HERV-H H6 LTR, four potential Myb transcription factor consensus binding sites were identified by DNase I footprinting experiments and mobility shift assays.[19] Moreover, cotransfection assays with a vector expressing Myb revealed a seven-fold increase in promoter activity in human teratocarcinoma cells.

## ERV-9

Another class I family, ERV-9, is represented by about 300 proviral copies and at least 5,000 solitary LTRs.[1] ERV-9 expression is detected mainly in undifferentiated embryonal carcinoma cells (NT2/D1) and is down-regulated when cells are induced to differentiate by retinoic acid.[20] The ERV-9 LTR is the largest retroviral LTR sequence known so far and is composed of a complex arrangement of varying numbers of repetitive elements in the U3 region (41 bp E-element) and in the U5 region (72 to 85 bp B-element) (Fig. 1). In transient expression assays, isolated ERV-9 LTR sequences can drive the expression of a CAT reporter gene in a cell type specific manner.[21] The complete LTR was found to be active only in NT2/D1 cells, whereas the minimal promoter region (-70 to + 4) has activity detectable in NT2/D1 and HeLa cell lines,[22] implying that the minimal promoter does not contain the tissue specific control elements and/or that negative transcriptional control elements probably exist outside the minimal promoter. La Mantia et al.[23] have identified two important regions within the ERV-9 minimal promoter. One region located between -70 to -39 acts as a transcriptional activating sequence and contains an Sp1 binding site important for maximum promoter activity. The second region, from -7 to +6, resembles an initiator element (Inr) site. This site is necessary for correct transcription start site utilization and is recognized by an Inr-binding protein.[23] In cotransfection experiments, Strazzullo et al.[22] demonstrated that the transcription factors Sp1, E1A and VP16 can synergistically induce a high level of transcription when bound to the ERV-9 promoter.

## HERV-W

Members of the HERV-W family (multiple sclerosis associated retroviral element, MSRV) were originally amplified from retroviral particles isolated from patients with multiple sclerosis[24] and were found to be related to ERV-9. They comprise about 40 full-length elements and 1,100 solitary LTRs per human haploid genome.[1] Their LTR sequences are about 800 bp in length and contain, compared to the U3 region of about 247 - 306 bp, an unusually long U5 region of about 450 bp, in which two short ORFs (76 amino acids (aa) and 64 aa) were identified (Fig. 1).[24,25] Whereas HERV-W mRNA expression seems to be restricted in vivo to placenta and brain, isolated HERV-W LTRs are active in a variety of human cell lines including lung fibroblasts, liver cells and pancreatic cell lines.[8] Computer sequence analysis resulted in the identification of putative Pit1a and Oct-1 transcription factor binding sites and two CAAT-boxes in the promoter region. In addition, a TATA box is located in the U3 region and a poly(A) site was found about 75 bp downstream of the TATA box, in the R region. A subgroup of HERV-W LTRs, however, appears to be inactivated by a Sp1 site located between the CAAT box and the TATA box.[8]

## HERV-K(HML-1 to 10)

The HERV-K superfamily was originally identified by cross-hybridization to MMTV *pol*-gene probes and comprises all the type A/B/D-related class II elements. Members of this class have been assigned to at least ten different groups (HML-1 to 10, HML is an abbreviation for human MMTV-like sequence) by sequence analyses of the reverse transcriptase domain.[26,27] The HERV-K(HML-2) prototype provirus is the HERV-K10 clone.[28] About 60 full-length copies are present in the genome and additionally about 2,500 copies of solitary LTRs are scattered throughout the human genome.[1] The typical HERV-K(HML-2) LTR is about 970 bp in length (Fig. 1). In addition to the characteristic structural features such as TATA box, polyadenylation signal and terminal inverted repeats, the HERV-K(HML-2) LTR contains part of a glucocorticoid-responsive element (GRE) at the 5' end of the U3 region followed by a sequence homologue to an enhancer core.[29] The close proximity of the two elements creates a sequence with similarities to the binding site of a progesterone receptor complex.[29] Knossl et al.[30] analyzed one enhancer complex binding sequence that overlaps the putative GRE site and half of the enhancer core. They could localize this site to a 22 bp fragment between nucleotides 62 and 83 (GGAGACTCCATTTTGTTATGTG). Different protein complexes bind to this sequence in the HERV-K expressing cell lines (GH and Tera2). In addition, computer analyses identified three complete GREs within the U3 region, which could also be important for the steroid hormone stimulation of HERV-K expression.[30]

One component of the HERV-K enhancer complex is the zinc finger-type transcription factor YY1. The YY1 core consensus sequence CCATNTT is identical in all HERV-K LTRs. Replacement of this site significantly reduced the HERV-K expression.[30] HERV-K expression was also tested in vivo with transgenic mice that harbor the HERV-K LTR (LTRE3) upstream of a β-galactosidase gene. Expression was limited primarily to the undifferentiated spermato-cytes, to a lesser extend to the brain, and seems to be developmentally regulated.[31] As with the HERV-H family Domansky[32] could show that some solitary HERV-K(HML-2) LTRs can function as promoters in both orientations.

In contrast, the HERV-K-T47D LTR, a representative of the HML-4 subgroup, is active only in forward direction. About 850 HERV-K-T47D related LTRs are dispersed throughout the human genome.[33] The R region and even further downstream regions of HERV-K-T47D probably contain additional regulatory elements that contribute to the promoter activity. Transient transfection assays[34] showed that the HERV-K-T47D LTR is active in the breast cancer cell line T47D and that this LTR is inducible with estradiol and progesterone. Apart from transcripts in steroid-treated T47D cells, high levels of HERV-K-T47D expression were detected only in human placenta. The steroid dependent expression can be explained by the presence of a glucocorticoid responsive element resembling that of members of the HML-2 subgroup and two enhancer-like structures in the U3 region of the 5'LTR. However, compared to HERV-K10, both of these sequence elements were identified at a different location in the U3 region. In addition, a number of putative binding sites for the transcription factors C/EBP, Gfi-1, AP1 and Ik-1 were identified in the HERV-K-T47D LTR by computer analyses.[34] An unusually large space between the polyadenylation signal and the two alternative polyadenylation sites used was found in all HERV-K(HML-4) LTRs analyzed so far.[35] It is possible that formation of a stem-loop structure similar to that in the HTLV-1 Rex responsive element can juxtapose the two signals at a correct distance.

## HERV-L

The HERV-L family, assigned to HERV class III, is most closely related to foamy viruses.[36,37] About 200 full-length copies and 6,000 solitary LTRs of these elements are estimated to exist in the genome.[1] Members of this family have been detected in several mammalian species and are thought to have existed in the mammalian genome for at least 70 million years.[36] The promoter activity of the isolated LTR proved to be strictly cell type specific using a luciferase expression vector.[8] This LTR was found to be highly active in epidermal keratinocytes (HaCaT). The HERV-L LTR is about 462 bp long, is bordered by the usual inverted repeats on both ends (TGT...ACA) and contains a presumptive TATA box and a polyadenylation signal (Fig. 1). Computer analyses of the HERV-L sequence resulted in the identification of two putative Ap1 binding sites.[37] In addition, two CACACCC motifs are located in antisense direction in the HERV-L LTR. These sites are thought to be involved in the keratinocyte-specific expression.[8]

## Endogenous and Exogenous Factors Influencing HERV Expression

Considering the high copy number of HERV sequences in the human genome only a small fraction appears to be transcriptionally active. A variety of mechanisms have apparently developed during evolution to control HERV expression and to prevent retrotransposition mediated changes in the human genome. HERV expression can be strongly repressed by different factors that act in a coordinated way resulting in the complex regulation and expression patterns of endogenous retroviruses found currently. Conversely, a number of factors may reactivate HERV expression or even particle formation.

One way of reducing HERV expression is exemplified by HERV-H LTRs where the insertion of silencers in the U3 region of the type I and II LTRs probably accounts for the repression of most of these LTRs in various human cell lines.[15] Domansky et al. also identified a silencer-like element within the U5 region of some solitary HERV-K LTRs active in human teratocarcinoma cell lines.[32] Another means to restrict expression is the methylation of inserted HERV genomes by the host cell machinery, a mechanism possibly affecting HERV members that are expressed in early developmental stages but are repressed in adult tissues. HERV expression is also controlled by the location of the element in the genome: HERVs integrated in heterochromatic regions or in the close vicinity of a silencer are normally transcriptionally silent. Finally, point mutations accumulated in LTR sequences during evolution may have destroyed regulatory regions such as binding sites for transcription factors.

HERV expression is also tightly down-regulated by agents that induce differentiation, such as retinoic acid. For example, ERV-9 is active in embryonal terratocarcinoma cells but is down-regulated in these cells after induction of differentiation by retinoic acid.[20] The same decrease in promoter activity in response to retinoic acid was found with a member of the HERV-K family and with HERV-H LTRs.[6,32] However, there are some exceptions. The expression of RRHERV-I was found to be activated by retinoic acid in the human teratocarcinoma-derived cell line PA-1.[38] HERV-L which is most active in human keratinocytes is not influenced by retinioc acid in this cell type while in other cell lines such as fibroblasts its transcription is induced by retinoic acid treatment (Schön, unpublished data).

The possible influence of endogenous host factors that may reactivate HERVs under certain conditions has been discussed in detail above. These include the transcription factors Sp1,

YY1 and AP1, which are also proposed to be involved in tissue specific regulation of the HERV promoters. Some LTRs seem to be regulated in a hormone dependent manner, for example HERV-W expression is activated in placenta[25] and HERV-R in placenta and sebaceous glands.[39] The expression of HERV-R may also be regulated by a number of inflammatory cytokines in endothelial and vascular smooth muscle cells.[40] Steroid hormones activate some HERV-K LTRs such as the HERV-K-T47D LTR in the mammary carcinoma cell line T47D, which may be mediated by different glucocorticoid responsive elements in the U3 region. The hormone dependent regulation of an ERV-9 element associated with the zinc finger gene ZNF80 will be discussed later.[41]

Silent HERVs can be reactivated by environmental conditions that induce cellular stress including physical or chemical agents and DNA viruses. For example irradiation with UVB leads to transcription of various endogenous retroviral *pol* sequences in epidermal keratinocytes.[42] Furthermore, in cells derived from patients with multiple sclerosis, the production of retroviral particles containing HERV-W and HERV-H elements is activated by infection with Herpes virus.[43] Many HERVs are also found to be activated in tumors and in transformed cells.[39,44] Enhanced expression is detected in lung, bladder and germ cell cancer, leukemia, seminoma and in teratocarcinoma cell lines.[45] Moreover, several HERV-H LTRs are induced or trans-activated by the SV40 large T antigen.[46]

# Modulation of Cellular Gene Expression

Several chimeric cDNA clones containing HERV LTR sequences have been isolated from human cDNA libraries suggesting HERV elements play a role in the control of adjacent genes. Some transcripts initiate within a 5'LTR and splice from a donor site located upstream of the *gag* region into downstream sequences encoding a cellular gene (Fig. 2A). Others appear to proceed directly from the promoter of a 3' or solitary HERV LTR into flanking cellular sequences (Fig. 2B). In contrast to promoter activity, potential enhancer function of HERVs is hard to detect, because in this case the LTR is not part of the transcript (Fig. 2C). Nevertheless, such effects are probably more common than promoter insertion since they are independent of the distance and orientation of the HERV element with respect to the cellular gene. Furthermore, HERV LTRs may terminate cellular transcripts by providing polyadenylation sites (Fig. 2G). Examples of HERV LTRs involved in the regulation of known human genes or associated with coding sequences related to gene products of other organisms are listed in Table 1.

## *Initiation of Cellular Transcripts From 3' or Solitary LTRs*

A differential hybridization strategy led to isolation of several chimeric clones containing HERV-H R-U5 regions in combination with cellular sequences from a human teratocarcinoma (NTera2D1) cDNA library.[47] One of these clones (CDC4L) represents a read-through transcript starting in an HERV-H LTR and proceeding through an open reading frame with sequence motifs homologous to the yeast cell division cycle (CDC4) protein and to β-transducin, a G protein in photoreceptor rod cells. Expression of the fusion transcript, however, could not be detected in various human cell lines including NTera2D1 cells. Thus, it was suggested that this cDNA may have been derived from a DNA rearrangement or a de novo insertion of an HERV-H element in a subclone of NTera2D1 cells that subsequently expired during passaging of the cell line.

A HERV-K(HML-10) element was found to be involved in tissue-specific expression of the insulin-like growth factor gene *INSL4* in human placenta.[48] In this case the cellular gene is transcribed from the 3'LTR of a truncated provirus that lacks the 3'end of the 5'LTR including the PBS. *INSL4* expression is upregulated more than 10 fold in differentiated syncytiotrophoblasts compared to the corresponding cytotrophoblasts suggesting a possible role in human placental morphogenesis. In contrast to the other insulin-like growth factor

**Table 1. Modification of gene expression by HERV LTRs**

| HERV | Gene Under Influence | Function | Refs. |
|------|---------------------|----------|-------|
| HERV-H | *CDC4L* (cell division cycle protein, β-transducin) | Promoter | (47) |
| HERV-E | *APOCI* (apolipoprotein CI) | Promoter | (49) |
| ERV-9 | *ZNF80* (zinc finger protein) | Promoter | (41, 50, 97, 98) |
| HERV-H | *calbindin* (calcium binding protein) | Promoter | (51) |
| HERV-R (ERV-3) | *plk* (human provirus linked Krüppel gene) | Promoter | (53, 57, 99, 55, 100, 101) |
| HERV-H | *PLA2L* (phospholipase A2-like protein) | Promoter | (71-75) |
| HERV-E | *MID1* (Opitz's syndrome gene) | Promoter | (60,61) |
| HERV-E | *EBR* (endothelin B receptor) | promoter | (49) |
| HERV-K(HML-10) | *INSL4* (early placenta insulin-like peptide) | promoter | (48) |
| HERV-K(HML-6) | *HLA-DRB6* (histocompatibility complex) | 5'UTR, promoter? | (70) |
| HERV-E | *amy1A, 1B, 1C, amy2A* (salivary amylase) | enhancer | (11, 77-81) |
| HERV-E | *PTN* (pleiotrophin) | 5'UTR, enhancer? | (82, 83) |
| HERV-K(HML-8) | *LEP* (leptin) | enhancer | (84) |
| ERV-9 | β-globin gene cluster | promoter/enhancer | (76) |
| HERV-I | *cyt $c_1$* (cytochrome $c_1$) | protein-binding sites | (85) |
| HERV-I | *QP-C* (ubiquinone-binding protein) | protein-binding sites | (85) |
| HERV-H | *PLT* (placental LTR terminated gene) | polyadenylation | (87) |
| HERV-H | *HHLA2* (HERV-H LTR associating gene 2) | polyadenylation | (89) |
| HERV-H | *HHLA3* (HERV-H LTR associating gene 3) | polyadenylation | (89) |
| HERV-H | *HMGI-C* (non histone DNA binding protein) | polyadenylation | (90) |
| HERV-F | *ZNF195* (Krüppel related zink finger protein) | polyadenylation | (91) |
| HERV-K(HML-2) | *LEPR* (leptin receptor) | polyadenylation | (92) |
| HERV-K(HML-4) | *FLT4* (transmembrane tyrosine kinase) | polyadenylation | (35) |
| HERV-E | *PRDM7* (zinc finger protein) | polyadenylation | (61) |

genes which have homologous genes in the mouse, the *INSL4* gene appears to be primate-specific. The HERV element was inserted into the *INSL4* promoter region soon after the emergence of this gene in Old World monkeys.

Screening transcribed sequences of Genbank and EST databases for transcripts containing R-U5 sequences of HERV LTRs[49] revealed a fusion transcript derived from a solitary HERV-E LTR that reads into cellular sequences encoding the apolipoprotein CI (APOCI). The *APOCI* gene is located in a 45 kb gene cluster encoding three proteins involved in lipid metabolism. The solitary LTR was found to be located 300 bp upstream of the native *APOCI* promoter and initiates about 15% of the *APOCI* transcripts expressed in liver. The LTR does not act as an enhancer but contributes to the overall activity of the native promoter as a position-dependent cis-acting element, since its removal from the *APOCI* locus leads to a 40% reduction in expression levels. Integration of this HERV-E LTR into the *APOCI* locus took place after the divergence of hominoids and Old World monkeys about 20 - 30 million years ago. Transient transfection assays showed that the HERV-E LTR increases *APOCI* expression when inserted into the baboon locus. This suggests that upon integration into the primate lineage, the expression level of *APOCI* was immediately altered.

The selective expression of a zinc finger protein (ZNF80) in hematopoietic cell lines is also determined by retroviral regulatory sequences (Fig. 2B). The specific expression of this zinc finger gene is regulated by a solitary LTR element of the ERV-9 family.[41] The 1.7 kb ERV-9 LTR which is inserted in the 5' flanking region of *ZNF80* is able to drive the expression of a CAT reporter gene in Jurkat cells in vitro. However, transcription starts at variable sites, probably due to a mutation in the TATA element (AATAAA to AATGAA). This could also explain why *ZNF80* transcription in vivo starts 16 bp upstream of the predicted start site in the U3 region of ERV-9. Genomic and cDNA sequences are colinear along the entire transcribed region with the exception of a 390 bp intron in the 3' untranslated region (3'UTR). The *ZNF80* gene product comprises 273 aa with a molecular weight of about 31 kDa. This protein contains seven tandemly repeated zinc finger domains. Using FISH (fluorescent in situ hybridization) analyses, *ZNF80* was mapped to the 3q13.3 band, which is a chromosomal region involved in karyotype rearrangements associated with myelocytic disorders. The ERV-9 LTR upstream of the ZNF80 gene was not detected in Old World monkeys but is present in gorilla and chimpanzee indicating that the integration of this element took place between 16 and 9 million years ago.[50]

### Alternative Splicing from 5'LTRs Using Cellular Splice Acceptors

Most HERVs, like exogenous retroviruses, contain a strong splice donor site (SD) located in the leader region downstream of the 5'LTR. The SD is normally used for subgenomic splicing of envelope transcripts. Deletion or mutation of the corresponding viral splice acceptor sequences may lead to alternative splicing to an exon in a downstream gene.

Searching for genes specifically expressed in a cell line derived from a prostate metastasis, Liu and Abraham[51] isolated a chimeric cDNA clone containing 5'LTR sequences and leader region of a HERV-H element fused to the human calbindin gene which codes for a cytosolic calcium-binding protein. A search of the databases suggested that the HERV-H sequence was probably derived from a 5.6 kb proviral element that is located in the sense direction about 7 kb upstream of the calbindin locus and differs by only one point mutation from the retroviral part of the fused transcript (W. Seifarth, unpublished data). In the metastasis cell line, the HERV-H element splices to the second exon of the calbindin gene thereby disrupting the first of six calcium-binding motifs, and introduces a different peptide sequence in its place. It has been suggested that the HERV-H induced deregulation of calbindin expression may contribute to the malignant phenotype of the cell line.[51]

Another example where the insertion of a human endogenous retrovirus upstream of a cellular gene leads to altered splicing is the human provirus-linked Krüppel gene *H-plk*. A HERV-R element, ERV-3, is inserted upstream of *H-plk* (Fig. 2A). Integration of the retroviral element occurred after the divergence of rodent and primate lineages.[52] Three differently spliced transcripts are derived from this HERV-R element.[53,54] One is terminated using the polyadenylation site in the 3' LTR. The other transcripts (9.0 and 7.3 kb) extend through the 3' LTR and are spliced first using the retroviral splice donor and acceptor sites and then a second time approximately 370 nucleotides downstream from the 3' LTR within flanking Alu

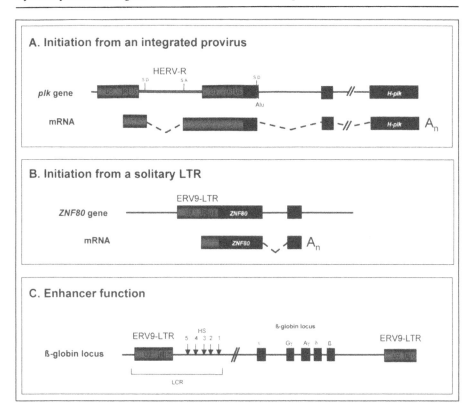

Figure 2. Examples for HERV LTR mediated gene regulation. Figure continued on following pages.

sequences. The splice product contains two ORFs, one for the Env protein and a 1.3 kb ORF which encodes a Krüppel related zinc finger protein with about twelve finger motifs and a molecular weight of about 50 kDa. The resulting transcriptional unit comprising the endogenous retrovirus and the zinc finger protein is probably facilitated by the additional insertion of an Alu element between both sequences. This element contains a cryptic splice acceptor site and was probably a critical feature in the evolution of this unit. The insertion of the ERV-3 element in the 5' end likely leads to the activation of *H-plk* transcription in placenta and was thought to cause down-regulation in choriocarcinoma cells.[55] However, subsequent investigations revealed that ERV-3 is not normally expressed in cytotrophoblasts, from which these tumor cells are derived, but in normal syncythiotrophoblasts.[56] In addition, methylation and demethylation processes have been proposed to affect the regulation of ERV-3.[57]

In a third case of merging HERV regulatory sequences with a zinc finger gene, a fusion transcript of a *KRAB* (Krüppel-associated box) zinc finger gene was detected containing part of a HERV-H *env* gene and a segment of an Alu repeat in the 5' UTR, both in the antisense direction.[58] However, the *KRAB* sequence is probably transcribed from a nonretroviral promoter and the significance of the retroviral sequences within the 5' UTR is not known. Interestingly, a striking association of zinc finger genes with HERV-K LTRs was further detected when mapping human chromosome 19.[59] Chromosome 19 carries a disproportionate number of zinc finger loci including about 80 loci of the *KRAB* subfamily of zinc finger proteins. 10 of these genes were found to be located in close proximity (10 - 40 kb) to HERV-K10 related LTRs.

An element of the HERV-E family fused to the endothelin B receptor gene (EBR) was detected during database searches.[49] The fusion transcript initiates within the 5' LTR of the provirus and utilizes the SD site located in the leader region to splice to an acceptor in the 5' UTR of the *EBR* gene. Whereas the native transcripts of *EBR* are expressed in many tissues, the LTR driven transcripts appear to be restricted to placenta where they make up approximately 25 - 30% of the total *EBR* mRNA. Furthermore, the HERV-E LTR acts as an enhancer of the native promoter, since in a luciferase assay the activity of the *EBR* promoter could be increased significantly when the LTR was inserted in either direction. The HERV element integrated into the *EBR* locus after the split between New and Old World monkeys.

The Opitz syndrome gene *MID1* was also found to be expressed from the 5'LTR of a HERV-E element.[60,61] Transcripts in which the native 5'untranslated regions (UTRs) have been replaced by retroviral LTR sequences alternatively spliced to the second exon of the *MID1* gene were detected preferentially in placenta and embryonal kidney. In these tissues the chimeric transcripts represent about 25 % of the *MID1* mRNAs. Transfection studies supported a role for the HERV-E LTR as a strong tissue-specific promoter. In addition, the LTR was also shown to increase the activity of the nonretroviral *MID1* promoter. The *MID1* gene encodes a microtubule-associated protein that interacts with phosphatase 2A. The HERV-E element was inserted in the primate genome after the separation of Old and New World monkeys.

The sequence variation in the noncoding region of the major histocompatibility complex is mainly conferred by insertions and deletions of retroviral origin and it has been speculated that these sequences may contribute to the high degree of variability of human leukocyte antigens (HLA).[62-69,39] For example, human and chimpanzee *DRB6* genes lack exon 1 which encodes the leader peptide and the first four amino acids of the mature protein in complete *HLA* genes. A retrovirus of the HERV-K(HML-6) subgroup of class II HERVs was inserted into the first intron of the *HLA-DRB6* locus at least about 23 million years ago.[70] The insertion was either accompanied or followed by the deletion of exon 1 and the promoter region of *DRB6*. In the 3' LTR of the HERV element an open reading frame for a new exon arose coding for a sequence with mostly hydrophobic amino acids. This sequence could serve as a leader peptide necessary for the expression of the molecule on the cell surface. The new exon contains a functional SD site and was found to be correctly spliced to exon 2 of the truncated *DRB6* gene.

## Generation of Fusion Transcripts by Intergenic Splicing

The PLA2L transcript that is specifically expressed in teratocarcinoma cells initiates from a HERV-H promoter and splices into downstream cellular sequences (Fig. 2F). The transcript contains an ORF of 689 aa that begins within the HERV-H LTR and has two regions with similarity to the secreted form of phospholipase A.[71] The corresponding genomic locus was mapped to chromosome 8q24.1 – q24.3 and comprises over 20 exons spanning about 50 kb.[72] The HERV-H element must have integrated into this locus about 15 – 20 million years ago since it is present in chimpanzee and gorilla but absent in orangutan and lower primates.

Further analysis of the transcript revealed it to be the product of intergenic splicing between an HERV-H element and two independent downstream genes.[73] Intergenic splicing between adjacent genes is a rare event. In most cases it occurs in human tumors as a consequence of chromosomal rearrangements bringing together normally unlinked genes, such as *bcr* and *abl* in chronic myelogenous leukemias. The PLA2L transcript, however, is derived from two different human genes normally located in close proximity within the same genomic region. The first gene following the HERV-H element, *HHLA1* (HERV-H LTR associating 1), is of unknown function although it is evolutionary conserved among mammals. In the PLA2L clone, the first gene consists of 7 exons and encodes a peptide sequence of 305 amino acids. The original gene appears to have at least one additional 5' exon since the HERV-H element is probably inserted within an intron of *HHLA1*. The second gene located about 10 kb downstream has 14 exons and contains the two PLA-related domains. This gene is the human ortholog of the mouse otoconin-90 (*OC90*) gene, which encodes an inner ear protein essential for sensing gravity.[74]

There is evidence that *HHLA1* and *OC90* are normally expressed independently from different promoters: besides the intergenic spliced PLA2L clone, several HERV-H LTR-driven unspliced *HHLA1* transcripts could be isolated from NTera2D1 cells that are terminated and polyadenylated at variable sites within the terminal exon of the *HHLA1* gene. Furthermore, the intergenic region between *HHLA1* and *OC90* contains sequences homologous to signal structures in the 5' region of mouse otoconin-90. It has been speculated that the high activity of the HERV-H promoter in teratocarcinoma cells may induce the transcriptional fusion of the two genes in this cell type. However, it is still unclear whether the PLA2L transcript plays a functional role as translation of the fusion transcript is suppressed by a potential stem-loop structure formed by HERV-H LTR and adjacent cellular sequences.[75]

## Enhancer Function and Mediation of Tissue Specificity

In spite of the difficulty identifying enhancing effects of HERV LTRs in vivo, the enhancing potential of sequences derived from different HERV LTR regions has been demonstrated for some HERV families in transient transfection assays.[76,49,8] Quite possibly, such sequences have been used frequently during evolution to facilitate the switching of tissue specificity and to alter gene expression levels.

An example of the involvement of a solitary ERV-9 LTR in the regulation of human cellular genes is the β-globin locus (Fig. 2C).[76] Transcription of these genes is regulated by the locus control region (LCR) located far upstream from the gene cluster. Sequencing of this region revealed a solitary ERV-9 LTR near the DNase I hypersensitive site HS5. This LTR contains 23 DNA motifs that can bind to transcription factors expressed abundantly in erythroid cells and to C/EBP and NF-Y expressed in many hematopoietic and nonhematopoietic cells. Transfection assays with recombinant reporter gene constructs and analysis of endogenous LTR-driven transcription showed a higher level of enhancer and promoter activity in erythroid cells than in nonerythroid cells. Therefore, the HS5 LTR may serve several functions relevant to the regulation of transcriptional status of the β-globin LCR.

A very well documented example of enhancer function is the regulation of amylase gene expression in humans by a member of the HERV-E family.[11,77,78] In this case, the insertion of a HERV-E element upstream of an ancestral amylase gene converted the pancreas-specific promoter to a salivary-specific promoter (Fig. 2D). In humans, the amylase gene complex consists of five gene copies with different tissue-specific activities. Two pancreatic genes (*AMY2B*, *AMY2A*) and three salivary genes (*AMY1A*, *AMY1B*, *AMY1C*) are organized in a gene cluster on chromosome 1p21.[79] A complete γ-actin pseudogene is located 200 bp upstream of the first coding exon. In four of the five amylase genes the γ-actin sequence is interrupted by an endogenous retroviral element (Fig. 2D). Each salivary-specific gene contains a full-length copy of this element inserted in an antisense direction compared to the coding sequences. However, the second pancreatic gene only contains a residual solitary LTR. As a result of the retroviral insertion, a cryptic promoter within the γ-actin pseudogene sequence is activated.[80,81] Gene expression studies in transgenic mice demonstrated that the parotid-specific enhancer is located within an 700 bp stretch of the proviral sequence comprising about half of the 5' LTR and part of the 5' untranslated region.[11] The HERV-E insertion is specific for humans and apes and was not detected in the amylase loci of other primates. Thus, an alternative mechanism must regulate salivary amylase expression found in some Old World monkey species.[81]

A HERV element (HERV-E.PTN) inserted between the 5' UTR and the coding region of the human pleiotrophin gene *PTN* was shown to confer trophoblast-specific expression on functional *PTN* gene products (Fig. 2E).[82] Pleiotrophin is a secreted polypeptide growth factor

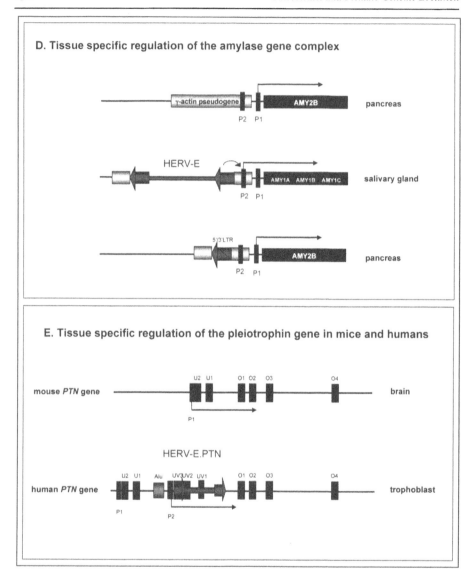

Figure 2 (continued). Examples for HERV LTR mediated gene regulation.

of 18 KDa showing growth-promoting and transforming activity in fibroblasts and epithelial cells, and mitogenic activity in endothelial cells. It further plays an essential role in the growth, angiogenesis and metastasis of human melanomas and the growth of human trophoblast-derived choriocarcinomas. The insertion of the HERV element creates a novel regulatory UTR region that consists of sequences derived from the 5' LTR, *gag* and *pol* of HERV-E.PTN and leads to the activation of an additional promoter in the 5' flanking sequence. Primer extension and luciferase assays demonstrated that this promoter is responsible for the trophoblast-specific expression of the fusion transcript. The HERV element itself has a chimeric structure, probably

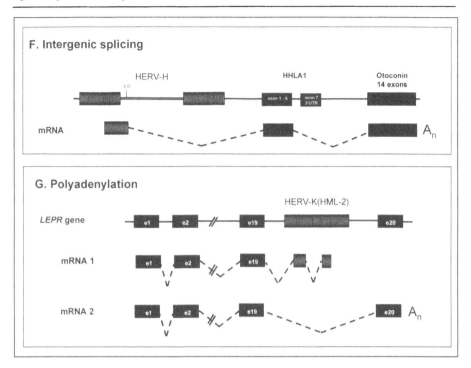

Figure 2 (continued). Examples for HERV LTR mediated gene regulation.

due to a recombination between two unrelated HERV elements. The evidence for such an event is that it contains additional *pol*-env sequences related to HERV-I inserted between the *gag* and *pol* region of a HERV-E element.[83] Recombined HERV-E.PTN homologous sequences were further detected in the *BRCA1* locus, the human hereditary hemochromatosis (hHH) region and the X chromosome. Phylogenetic analysis demonstrates that the insertion into the *PTN* gene occurred about 25 million years ago after the separation of Old World monkeys and hominoids.

A further example is leptin, a hormone chiefly produced in adipose cells. However, in humans it is also expressed in placenta under the control of a MER11A repeat element, a solitary LTR of the HERV-K(HML-8) family.[84] This HERV element acts exclusively as a tissue specific enhancer for the leptin gene in placenta though not in other tissues where leptin is expressed. Within a 60 bp region of the LTR sequence, three independent protein-binding sites were determined to contribute to the enhancer function: an SP1 site and two novel placenta-selective binding sites.

Truncated HERV-I LTR sequences were detected in the 5' flanking regions of the human genes for cytochrome c1 and ubiquinone-binding protein, both nuclear encoded mitochondrial proteins that are subunits of the mitochondrial oxidative phosphorylation complex cytochrome $bc_1$.[85] Within these LTR sequences three common motifs, MT1, MT3 and MT4, that are recognized by at least three binding proteins were identified. One of these sequences, MT1, is very similar to the recognition site of factor GFII involved in the regulation of three cytochrome $bc_1$ subunits in yeast. Therefore, it is thought that these elements may play a role in the coordinate expression of human genes involved in mitochondrial biogenesis.

### Termination and Polyadenylation

Besides regulating transcriptional initiation, retroviral LTRs possess the ability to process the 3' ends of nonviral transcripts. A large number of transcripts polyadenylated by HERV LTRs have already been detected. In most cases the function of these transcripts is unclear since coding sequences could not be identified.[14,86,87,35] These transcripts often contain highly repetitive sequences or tandem arrays of various retrotransposons.[88] One transcript called *PLT* contains an ORF of 223 aa and is apparently generated by alternative splicing at its 3'end.[87] Although *PLT* is differentially expressed in various human cell lines, the LTR chimeric transcript appears to be restricted to placenta.

A further 80 transcripts that fit the criteria of being polyadenylated within HERV-H LTRs were discovered by screening the EST database.[89] Among these, two new human genes of unknown function were identified, HHLA2 and HHLA3. HHLA2 is represented by three independent clones, all polyadenylated within the LTR. The transcripts have an ORF of 414 aa with three immunoglobulin-like domains and are expressed primarily in intestine, kidney and lung. Three types of HHLA3 transcripts were identified that differ in splicing pattern, but all are terminated and polyadenylated within the HERV-H LTR. Interestingly, the ORF of two of these transcripts continues into the LTR. Neither gene uses a nonLTR polyadenylation signal. However, in baboons which lack the HERV-H LTRs, these genes use other polyadenylation signals indicating that the LTR-mediated regulatory functions have developed during human evolution.

The third intron of the gene for HMGI-C, an abundant nonhistone component of chromatin, contains a cluster of almost identical HERV-H elements. Rearrangements in this region are very frequent among human benign tumors of mainly mesenchymal origin. From three such tumors Kazmierczak et al.[90] isolated chimeric cDNAs containing exon 1 – 3 of HMGI-C fused to HERV-H 3'LTRs. The HERV LTRs apparently provide a polyadenylation signal for the fusion transcripts and lead to a truncated HMGI-C coding sequence that retains the DNA binding part of the protein. A member of the HERV-H related HERV-F family, HERV-Fb, was found to be involved in 3'-end processing of a human zinc finger gene *ZNF195*. This is one more example of the intriguing association of zinc finger genes and HERVs. At least some fraction of the *ZNF195* mRNA appears to use the 5' LTR of HERV-Fb as an alternative polyadenylation site adding 636 nt to the transcript.[91]

As in the case of the leptin gene, transcription of the human leptin receptor gene (*LEPR*) is also influenced by a retroviral LTR, albeit in quite a different manner: whereas the HERV-K LTR in the leptin gene acts as an enhancer, the LTR in the receptor gene contributes a new polyadenylation signal (Fig. 2G). Two major alternatively spliced transcripts of the leptin receptor are detected in human cells. The longer form is involved in JAK/STAT signal transduction pathway. The shorter form uses a splice acceptor (SA) site within a HERV-K(HML-2) LTR located in the last intron instead of the SA site of exon 20, and the exon 20 protein sequences are replaced by 67 new amino acids encoded by the LTR.[92] Lacking an intracellular domain, the shorter spliced form is unlikely to be involved in JAK/STAT signaling but may be important for the transfer, clearance and regulation of the concentration of free leptin in the bloodstream. It is speculated that both the expression of leptin and the proportion of the two leptin receptor forms may be regulated by steroids via steroid responsive elements in both LTRs.

An unusual involvement of a HERV-K(HML-4) LTR in the regulation of gene expression was observed for the human transmembrane tyrosine kinase gene *FLT4*.[35] In this case, a partial LTR sequence was found in the antisense orientation with respect to the *FLT4* gene, thus excluding regulation of 3' end processing of *FLT4* by the normal polyadenylation signal located in the R region. Instead, a new polyadenylation signal arose within the U3 region in opposite direction. Another cDNA polyadenylated by an HERV-K(HML-4) LTR contains a sequence with significant homology to an exon of the human tyrosine phosphatase 1 gene (*PTP1*), but also in an antisense orientation. It could be that this transcript acts as an antisense RNA that regulates *PTP1* or a *PTP1*-related gene by RNA:RNA hybridization.[93]

## Concluding Remarks

An important fraction of the human genome is represented by retroviral LTRs and many of these sequences have still retained their functionality. They have been involved in creating new genetic information by reorganizing regulatory and coding sequences. The examples reviewed here demonstrate the wide range of implications that HERV LTRs may have for a genotype: initiating cellular transcripts by promoter insertion, altering tissue specificity of cellular genes by adding new enhancer and transcription factor binding sites, terminating transcripts by providing polyadenylation signals, or creating new UTRs and new exons by alternative splicing leading to modification in RNA stability and to new gene products. Although these are single observations, LTRs and other retroelements bearing regulatory sequences may represent a common evolutionary mechanism for acquiring new gene functions. In the mammalian genome, HERVs can be traced back up to 70 million years. Spreading throughout the genomes at different times,[94] HERVs may have played a key role as evolutionary triggers in speciation and adaptation of primates[95,96] If so, HERVs could be valuable tools to observe how the genome of higher organisms has developed.

Tissue-specific promoters and enhancers are a prerequisite for the construction of targeted retroviral expression vectors and their controlled use in gene therapy. Considering that a number of HERV LTRs have already been recruited during evolution as control elements for gene expression, HERV LTRs might be a valuable source of new tissue-specific regulatory sequences.[8] In contrast to cellular promoters, which often depend on additional signal structures located at some distance up or downstream, the regulatory elements of retroviruses are concentrated in a small and clearly defined region to maintain transcriptional independence regardless of the integration site in the host genome. Since space is limited in cloning vectors, it is often difficult to isolate strong and tissue-specific cellular promoter sequences that are suitable for eukaryotic expression systems. HERV sequences have a number of additional features that make them especially advantageous for the construction of retroviral vectors for use in human gene therapy. They have adapted themselves to their hosts over millions of years and thus most pathogenic sequences have been eliminated during evolution. Recombination of HERV elements with HERV-derived vectors will not create completely new types of retroviruses, as would be the case with vectors based on animal retroviruses or human exogenous retroviruses such as HIV. On the contrary, homologous recombination of HERV sequences with HERV-based vectors might be utilized for targeted gene transfer. Hence, HERV sequences are extremely interesting candidates for the construction of retroviral expression vectors that could be used in humans.

### Acknowledgments

We thank Alex Greenwood and Dixie Mager for critically reading the manuscript.

### References

1. Mager DL, Medstrand P. Retroviral repeat sequences. In: Cooper D, ed. Nature Encyclopedia of the Human Genome. Macmillan Online Publishing: Science, 2001.
2. Li WH, Gu Z, Wang H et al. Evolutionary analyses of the human genome. Nature 2001; 409:847-849.
3. Bock M, Stoye JP. Endogenous retroviruses and the human germline. Curr Opin Genet Dev 2000; 10:651-655.
4. Leib-Mösch C, Seifarth W. Evolution and biological significance of human retroelements. Virus Genes 1995; 11:133-145.
5. Lower R. The pathogenic potential of endogenous retroviruses: Facts and fantasies. Trends Microbiol 1999; 7:350-356.
6. Wilkinson DA, Mager DL, Leong JL. Endogenous human retroviruses, V. 3. New York: Plenum Press, 1994.
7. Majors J. The structure and function of retroviral long terminal repeats. Curr Top Microbiol Immunol 1990; 157:49-92.

8. Schon U, Seifarth W, Baust C et al. Cell type-specific expression and promoter activity of human endogenous. Virology 2001; 279:280-291.

9. O'Connell C, O'Brien S, Nash WG et al. ERV3, a full-length human endogenous provirus: Chromosomal localization and evolutionary relationships. Virology 1984; 138:225-235.

10. Steele PE, Rabson AB, Bryan T et al. Distinctive termini characterize two families of human endogenous retroviral sequences. Science 1984; 225:943-947.

11. Ting CN, Rosenberg MP, Snow CM et al. Endogenous retroviral sequences are required for tissue-specific expression of a human salivary amylase gene. Genes Dev 1992; 6:1457-1465.

12. Seifarth W, Baust C, Schon U et al. HERV-IP-T47D, a novel type C-related human endogenous retroviral sequence derived from T47D particles. AIDS Res Hum Retroviruses 2000; 16:471-480.

13. Seifarth W, Skladny H, Krieg-Schneider F et al. Retrovirus-like particles released from the human breast cancer cell line T47-D display type B- and C-related endogenous retroviral sequences. J Virol 1995; 69:6408-6416.

14. Mager DL. Polyadenylation function and sequence variability of the long terminal repeats of the human endogenous retrovirus-like family RTVL-H. Virology 1989; 173:591-599.

15. Nelson DT, Goodchild NL, Mager DL. Gain of Sp1 sites and loss of repressor sequences associated with a young, transcriptionally active subset of HERV-H endogenous long terminal repeats. Virology 1996; 220:213-218.

16. Anderssen S, Sjottem E, Svineng G et al. Comparative analyses of LTRs of the ERV-H family of primate-specific retrovirus-like elements isolated from marmoset, African green monkey, and man. Virology 1997; 234:14-30.

17. Feuchter A, Mager D. Functional heterogeneity of a large family of human LTR-like promoters and enhancers. Nucleic Acids Res 1990; 18:1261-1270.

18. Sjottem E, Anderssen S, Johansen T. The promoter activity of long terminal repeats of the HERV-H family of human retrovirus-like elements is critically dependent on Sp1 family proteins interacting with a GC/GT box located immediately 3' to the TATA box. J Virol 1996; 70:188-198.

19. de Parseval N, Alkabbani H, Heidmann T. The long terminal repeats of the HERV-H human endogenous retrovirus contain binding sites for transcriptional regulation by the Myb protein. J Gen Virol 1999; 80:841-845.

20. La Mantia G, Maglione D, Pengue G et al. Identification and characterization of novel human endogenous retroviral sequences prefentially expressed in undifferentiated embryonal carcinoma cells. Nucleic Acids Res 1991; 19:1513-1520.

21. Lania L, Di Cristofano A, Strazzullo M et al. Structural and functional organization of the human endogenous retroviral ERV9 sequences. Virology 1992; 191:464-468.

22. Strazzullo M, Majello B, Lania L et al. Mutational analysis of the human endogenous ERV9 proviruses promoter. Virology 1994; 200:686-695.

23. La Mantia G, Majello B, Di Cristofano A et al. Identification of regulatory elements within the minimal promoter region of the human endogenous ERV9 proviruses: Accurate transcription initiation is controlled by an Inr-like element. Nucleic Acids Res1994; 20:4129-4136.

24. Komurian-Pradel F, Paranhos-Baccala G, Bedin F et al. Molecular cloning and characterization of MSRV-related sequences associated with retrovirus-like particles. Virology 1999; 260:1-9.

25. Blond JL, Beseme L, Duret L et al. Molecular characterization and placental expression of HERV-W, a new human endogenous retrovirus family. J Virol 1999; 73:1175-1185.

26. Medstrand P, Blomberg J. Characterization of novel reverse transcriptase encoding human endogenous retroviral sequences similar to type A and type B retroviruses: Differential transcription in normal human tissues. J Virol 1993; 67:6778-6787.

27. Andersson ML, Lindeskog M, Medstrand P et al. Diversity of human endogenous retrovirus class II-like sequences. J Gen Virol 1999; 80:255-260.

28. Ono M, Yasunaga T, Miyata T et al. Nucleotide sequence of human endogenous retrovirus genome related to the mouse mammary tumor virus genome. J Virol 1986; 60:589-598.

29. Ono M, Kawakami M, Ushikubo H. Stimulation of expression of the human endogenous retrovirus genome by female steroid hormones in human breast cancer cell line T47D. J Virol 1987; 61:2059-2062.

30. Knossl M, Lower R, Lower J. Expression of the human endogenous retrovirus HTDV/HERV-K is enhanced by cellular transcription factor YY1. J Virol 1999; 73:1254-1261.

31. Casau AE, Vaughan JE, Lozano G et al. Germ cell expression of an isolated human endogenous retroviral long terminal repeat of the HERV-K/HTDV family in transgenic mice. J Virol 1999; 73:9976-9983.
32. Domansky AN, Kopantzev EP, Snezhkov EV et al. Solitary HERV-K LTRs possess bi-directional promoter activity and contain a negative regulatory element in the U5 region. FEBS Lett 2000; 472:191-195.
33. Baust C, Seifarth W, Schon U et al. Functional activity of HERV-K-T47D-related long terminal repeats. Virology 2001; 283:262-272.
34. Seifarth W, Baust C, Murr A et al. Proviral structure, chromosomal location, and expression of HERV-K- T47D, a novel human endogenous retrovirus derived from T47D particles. J Virol 1998; 72:8384-8391.
35. Baust CW, Seifarth H, Germaier R et al. HERV-K-T47D-Related long terminal repeats mediate polyadenylation of cellular transcripts. Genomics 2000; 66:98-103.
36. Benit L, Lallemand JB, Casella JF et al. ERV-L elements: A family of endogenous retrovirus-like elements active throughout the evolution of mammals. J Virol 1999; 73:3301-3308.
37. Cordonnier A, Casella JF, Heidmann T. Isolation of novel human endogenous retrovirus-like elements with foamy virus-related pol sequence. J Virol 1995; 69:5890-5897.
38. Kannan P, Buettner R, Pratt DR et al. Identification of a retinoic acid-inducible endogenous retroviral transcript in the human teratocarcinoma-derived cell line PA-1. J Virol 1991; 65:6343-6348.
39. Andersson AC, Svensson AC, Rolny C et al. Expression of human endogenous retrovirus ERV3 (HERV-R) mRNA in normal and neoplastic tissues. Int J Oncol 1998; 12:309-313.
40. Katsumata K, Ikeda H, Sato M et al. Tissue-specific high-level expression of human endogenous retrovirus-R in the human adrenal cortex. Pathobiology 1998; 66:209-215.
41. Di Cristofano A, Strazzullo M, Longo L et al. Characterization and genomic mapping of the ZNF80 locus: Expression of this zinc-finger gene is driven by a solitary LTR of ERV9 endogenous retroviral family. Nucleic Acids Res 1995; 23:2823-2830.
42. Hohenadl C, Germaier H, Walchner M et al. Transcriptional activation of endogenous retroviral sequences in human epidermal keratinocytes by UVB irradiation. J Invest Dermatol 1999; 113:587-594.
43. Perron H, Perin JP, Rieger F et al. Particle-associated retroviral RNA and tandem RGH/HERV-W copies on human chromosome 7q: Possible components of a 'chain-reaction' triggered by infectious agents in multiple sclerosis? J Neurovirol 2000; 6:S67-S75.
44. Willer A, Saussele S, Gimbel W et al. Two groups of endogenous MMTV related retroviral env transcripts expressed in human tissues. Virus Genes 1997; 15:123-133.
45. Harris JR. Placental endogenous retrovirus (ERV): Structural, functional, and evolutionary significance. BioEssays 1998; 20:307-316.
46. Feuchter AE, Mager DL. SV40 large T antigen trans-activates the long terminal repeats of a large family of human endogenous retrovirus-like sequences. Virology 1992; 187:242-250.
47. Feuchter AE, Freeman JD, Mager DL. Strategy for detecting cellular transcripts promoted by human endogenous long terminal repeats: Identification of a novel gene (CDC4L) with homology to yeast CDC4. Genomics 1992; 13:1237-1246.
48. Bieche I, Laurent A, Laurendeau I et al. Placenta-specific INSL4 expression is mediated by a human endogenous retrovirus element. Biol Reprod 2003; 68:1422-1429.
49. Medstrand P, Landry JR, Mager DL. Long terminal repeats are used as alternative promoters for the endothelin B receptor and apolipoprotein C-I genes in humans. J Biol Chem 2001; 276:1896-1903.
50. Di Cristofano A, Strazzullo M, Parisi T et al. Mobilization of an ERV9 human endogenous retroviral element during primate evolution. Virology 1995; 213:271-275.
51. Liu AY, Abraham BA. Subtractive cloning of a hybrid human endogenous retrovirus and calbindin gene in the prostate cell line PC3. Cancer Res 1991; 51:4107-4110.
52. Shih A, Coutavas EE, Rush MG. Evolutionary implications of primate endogenous retroviruses. Virology 1991; 182:495-502.
53. Kato N, Pfeifer-Ohlsson S, Kato M et al. Tissue-specific expression of human provirus ERV3 mRNA in human placenta: Two of the three ERV3 mRNAs contain human cellular sequences. J Virol 1987; 61:2182-2191.

54. Kato N, Larsson E, Cohen M. Absence of expression of a human endogenous retrovirus is correlated with. Int J Cancer 1988; 41:380-385.

55. Kato N, Shimotohno K, Van Leeuwen D et al. Human proviral mRNAs down regulated in choriocarcinoma encode a zinc finger protein related to Kruppel. Mol Cell Biol 1990; 10:4401-4405.

56. Boyd MT, Bax CM, Bax CM et al. The human endogenous retrovirus ERV-3 is upregulated in differentiating placental trophoblast cells. Virology 1993; 196:905-909.

57. Abrink M, Larsson E, Hellman L. Demethylation of ERV3, an endogenous retrovirus regulating the Kruppel- related zinc finger gene H-plk, in several human cell lines arrested during early monocyte development. DNA Cell Biol 1998; 17:27-37.

58. Baban S, Freeman JD, Mager DL. Transcripts from a novel human KRAB zinc finger gene contain spliced Alu and endogenous retroviral segments. Genomics 1996; 33:463-472.

59. Vinogradova T, Volik S, Lebedev Y et al. Positioning of 72 potentially full size LTRs of human endogenous retroviruses HERV-K on the human chromosome 19 map. Occurrences of the LTRs in human gene sites. Gene 1997; 199:255-264.

60. Landry JR, Mager DL. Widely spaced alternative promoters, conserved between human and rodent, control expression of the Opitz syndrome gene MID1. Genomics 2002; 80:499-508.

61. Landry JR, Rouhi A, Medstrand P et al. The opitz syndrome gene mid1 is transcribed from a human endogenous retroviral promoter. Mol Biol Evol 2002; 19:1934-1942.

62. Kambhu S, Falldorf P, Lee JS. Endogenous retroviral long terminal repeats within the HLA-DQ locus. Proc Natl Acad Sci USA 1990; 87:4927-4931.

63. Dangel AW, Mendoza AR, Baker BJ et al. The dichotomous size variation of human complement C4 genes is mediated by a novel family of endogenous retroviruses, which also establishes species-specific genomic patterns among old world primates. Immunogenetics 1994; 40:425-436.

64. Svensson AC, Setterblad N, Sigurdardottir S et al. Primate DRB genes from the DR3 and DR8 haplotypes contain ERV9 LTR elements at identical positions. Immunogenetics 1995; 41:74-82.

65. Kulski JK, Gaudieri S, Bellgard M et al. The evolution of MHC diversity by segmental duplication and transposition of retroelements. J Mol Evol 1998; 46:734.

66. Gaudieri S, Kulski JK, Dawkins RL et al. Different evolutionary histories in two subgenomic regions of the major histocompatibility complex. Genome Res 1999; 9:541-549.

67. Donner H, Tonjes RR, Van der Auwera B et al. The presence or absence of a retroviral long terminal repeat influences the genetic risk for type 1 diabetes conferred by human leukocyte antigen DQ haplotypes. Belgian Diabetes Registry. J Clin Endocrinol Metab 1999; 84:1404-1408.

68. Donner H, Tonjes RR, Bontrop RE et al. Intronic sequence motifs of HLA-DQB1 are shared between humans, apes and old world monkeys, but a retroviral LTR element (DQLTR3) is human specific. Tissue Antigens 1999; 53:551-558.

69. Donner H, Tonjes RR, Bontrop RE et al. MHC diversity in Caucasians, investigated using highly heterogeneous noncoding sequence motifs at the DQB1 locus including a retroviral long terminal repeat element, and its comparison to nonhuman primate homologues. Immunogenetics 2000; 51:898-904.

70. Mayer WE, O'HUigin C, Klein J. Resolution of the HLA-DRB6 puzzle: A case of grafting a de novo- generated exon on an existing gene. Proc Natl Acad Sci USA 1993; 90:10720-10724.

71. Feuchter-Murthy AE, Freeman JD, Mager DL. Splicing of a human endogenous retrovirus to a novel phospholipase A2 related gene. Nucleic Acids Res 1993;21:135-143.

72. Kowalski PE, Freeman JD, Nelson DT et al. Genomic structure and evolution of a novel gene (PLA2L) with duplicated phospholipase A2-like domains. Genomics 1997; 39:38-46.

73. Kowalski PE, Freeman JD, Mager DL. Intergenic splicing between a HERV-H endogenous retrovirus and two adjacent human genes. Genomics 1999; 57:371-379.

74. Wang Y, Kowalski PE, Thalmann I et al. Otoconin-90, the mammalian otoconial matrix protein, contains two domains of homology to secretory phospholipase A2. Proc Natl Acad Sci USA 1998; 95:15345-15350.

75. Kowalski PE, Mager DL. A human endogenous retrovirus suppresses translation of an associated fusion transcript, PLA2L. J Virol 1998; 72:6164-6168.

76. Long Q, Bengra C, Li F et al. A long terminal repeat of the human endogenous retrovirus ERV-9 is located in the 5' boundary area of the human beta-globin locus control region. Genomics 1998; 54:542-555.

77. Samuelson LC, Wiebauer K, Gumucio DL et al. Expression of the human amylase genes: Recent origin of a salivary amylase promoter from an actin pseudogene. Nucleic Acids Res 1988; 16:8261-8276.

78. Meisler MH, Ting CN. The remarkable evolutionary history of the human amylase genes. Crit Rev Oral Biol Med 1993; 4:503-509.

79. Emi M, Horii A, Tomita A et al. Overlapping two genes in human DNA: A salivary amylase gene overlaps with a gamma-actin pseudogene that carries an integrated human endogenous retroviral DNA. Gene 1988; 62:229-235.

80. Samuelson LC, Wiebauer K, Snow SK et al. Retroviral and pseudogene insertion sites reveal the lineage of human salivary and pancreatic amylase genes from a single gene during primate evolution. Mol Cell Biol 1990; 10:2513-2520.

81. Samuelson LC, Phillips RS, Swanberg LJ. Amylase gene structures in primates: Retroposon insertions and promoter evolution. Mol Biol Evol 1996; 13:767-779.

82. Schulte AM, Lai S, Kurtz A et al. Human trophoblast and choriocarcinoma expression of the growth factor pleiotrophin attributable to germ-line insertion of an endogenous retrovirus. Proc Natl Acad Sci USA 1996; 93:14759-14764.

83. Schulte AM, Wellstein A. Structure and phylogenetic analysis of an endogenous retrovirus inserted into the human growth factor gene pleiotrophin. J Virol 1998; 72:6065-6072.

84. Bi S, Gavrilova O, Gong DW et al. Identification of a placental enhancer for the human leptin gene. J Biol Chem 1997; 272:30583-30588.

85. Suzuki H, Hosokawa Y, Toda H et al. Common protein-binding sites in the 5'-flanking regions of human genes for cytochrome c1 and ubiquinone-binding protein. J Biol Chem 1990; 265:8159-8163.

86. Tomita N, Horii A, Doi S et al. Transcription of human endogenous retroviral long terminal repeat (LTR) sequence in a lung cancer cell line. Biochem Biophys Res Commun 1990; 166:1-10.

87. Goodchild NL, Wilkinson DA, Mager DL. A human endogenous long terminal repeat provides a polyadenylation signal to a novel, alternatively spliced transcript in normal placenta. Gene 1992; 121:287-294.

88. Kelleher CA, Wilkinson DA, Freeman JD et al. Expression of novel-transposon-containing mRNAs in human T cells. J Gen Virol 1996; 77:1101-1110.

89. Mager DL, Hunter DG, Schertzer M et al. Endogenous retroviruses provide the primary polyadenylation signal for two new human genes (HHLA2 and HHLA3). Genomics 1999; 59:255-263.

90. Kazmierczak B, Pohnke Y, Bullerdiek J. Fusion transcripts between the HMGIC gene and RTVL-H-related sequences in mesenchymal tumors without cytogenetic aberrations. Genomics 1996; 38:223-226.

91. Kjellman C, Sjogren HO, Widegren B. HERV-F, a new group of human endogenous retrovirus sequences. J Gen Virol 1999; 80:2383-2392.

92. Kapitonov VV, Jurka J. The long terminal repeat of an endogenous retrovirus induces alternative splicing and encodes an additional carboxy-terminal sequence in the human leptin receptor. J Mol Evol 1999; 48:248-251.

93. Dolnick BJ. Naturally occurring antisense RNA. Pharmacol Ther 1997; 75:179-184.

94. Medstrand P, Mager DL. Human-specific integrations of the HERV-K endogenous retrovirus family. J Virol 1998; 72:9782-9787.

95. Brosius J. Genomes were forged by massive bombardments with retroelements and retrosequences. Genetica 1999; 107:209-238.

96. Sverdlov ED. Retroviruses and primate evolution. BioEssays 2000; 22:161-171.

97. Costas J, Naveira H. Evolutionary history of the human endogenous retrovirus family ERV9. Mol Biol Evol 2000; 17:320-330.

98. Strazzullo M, Parisi T, Di Cristofano A et al. Characterization and genomic mapping of chimeric ERV9 endogenous retroviruses-host gene transcripts. Gene 1998; 206:77 83.

99. Cohen M, Kato N, Larsson E. ERV3 human endogenous provirus mRNAs are expressed in normal and malignant. J Cell Biochem 1988; 36:121-128.

100. Larsson E. Expression of an endogenous retrovirus, HERV-R in human tissues. J Cancer Res Clin Oncol 1993; 119:S6.

101. Venables PJ, Brookes SM, Griffiths D et al. Abundance of an endogenous retroviral envelope protein in placental trophoblasts suggests a biological function. Virology 1995; 211:589-592.

# Genome-Wide Search for Human Specific Retroelements

Yuri B. Lebedev

## Abstract

Transposable elements, primarily retroelements (REs), were permanently amplified in primate genomes during the last 65 million years suggesting their evolutionary significance. Fixed in the ancestral genome, newly integrated REs are considered as efficient pacemakers in primate evolution. Here we review modern approaches to detection of human specific RE integrations that provide a basis for studies of the hominid lineage evolution. The methods observed include TDGA (Targeted Genomic Difference Analysis), DiffIR (detection of Differences in the integration sites of Interspersed Repeats), and LID (L1 Display). Applications of TDGA and DiffIR techniques to efficient identification of various REs, namely long terminal repeats (LTRs) of endogenous retroviruses (HERVs), and long interspersed repeats (L1) that differ the human and chimpanzee genomes are observed. Peculiarities of the distribution of human specific RE integrations in the human genome are also discussed.

## Introduction: Hypotheses Are Indispensable on the Way towards Understanding the Genetic Basis of Humankind Evolution

It was about 5 million years ago (Mya) when some event(s) within an ancestor highest primate[1] population initiated the divergence of two lineages having led to modern humans and chimpanzees. Later, the hominids passed through several population extensions, 'bottlenecks', the formation and extinction of several species, and finally have evolved into the now dominant species, *Homo sapiens sapiens*.[2] The genetic history of these events is encoded in the genome of modern humans where some DNA sequences can be traced back to very old ancestor organisms whereas the others have arisen more recently, during hominid evolution. Structural analysis of the first draft human genome sequence[3,4] together with the information on the genome sequences of other organisms enables one to identify both common genetic components and those formed due to independent evolution. Thus, a comparison of the mouse and human genome sequences results in the identification of DNA fragments present in the ancestral genome at about 75 Mya. However, for real understanding the genetic reasons for hominoid speciation a comprehensive comparison of their genomes is required. It would allow to reveal the genetic basis of revolutionary phenotypic changes in hominids at relatively conserved ape phenotypes and similar rates of the molecular evolution of human and great apes.[5] An apparent discrepancy in the rates of molecular and phenotypic evolution suggests that phenotypic conversions in hominids could be due to minor genetic changes essential for some regulatory processes.[5]

*Retroviruses and Primate Genome Evolution*, edited by Eugene D. Sverdlov.
©2005 Eurekah.com.

Although the reports on the human and chimpanzee genomes comparison are increasing in number, the available data are insufficient to make definite conclusions on the genetic basis of human phenotype evolution. However, even direct genome-wide comparison of pongids and human will certainly reveal not only causative genetic aberrations but also a much higher amount of random mutations fixed in the human genome either as neutral mutations or due to co-selection with adaptive mutations. As a result, we will face a problem of finding a needle in a hay stack, the informational noise will totally mask the evolutionary important genetic changes. There is no strict way to distinguish between causative genetic differences and genetic noise. Therefore, to single out the genetic changes which led to phenotypic transformations on the evolutionary way to the modern human, it will be necessary to use a well-known and welltried way of creating and verifying hypotheses. In particular, a hypothesis about a significant role of retrotransposons in human evolution is being verified during last years in our laboratory. These mobile elements might induce differences in the regulatory systems of primate genomes.

## Retroelement Families and Subfamilies in the Homo Sapiens Genome

The draft sequence of the human genome[3,4] is in accord with the belief that transposable elements and related sequences occupy almost a half (over 45%) of the total genome sequences, whereas coding sequences comprise only 5%. Three groups of these elements transposing via RNA intermediates and therefore called retroelements (REs) - long interspersed nuclear elements (LINEs), short interspersed nuclear elements (SINEs), and LTR (long terminal repeat) retrotransposons – occupy 13, 20 and 8% of the genome, respectively. Although all of these three types of REs were found also in other vertebrates, there are families and subsets of related sequences distinguishing primate genomes from other mammalian genomes.[6-8] This chapter is devoted to primates, and especially human specific RE integrations. An additional information on RE families can be found in Chapters 6, 7, 10 and 11.

Alu repeats is a classical example of primate-specific SINEs.[9,10] Among LINE elements a subset of L1 family was shown to be specific to primates.[11,12] Primate specific representatives were described also for several families of endogenous retroviruses (ERV).[13,14] In addition to provirus-like sequences, a great majority of ERV derived elements in the human genome are represented by solitary long terminal repeats (LTRs), presumably derived through recombination between the 5' and 3' LTRs of intact proviral elements. Different ERVs emerged in primate genomes at different evolutionary periods. Some of their families were formed 35–40 Mya, that is shortly after the prosimian and simian branchpoint,[15] whereas the others, like HERV-H and HERV-K families,[13] appeared in primate genomes considerably later (see Figs. 1 and 2).

Each of the above mentioned RE types includes fractions that were retropositionally active after the divergence of human and chimpanzee lineages. Almost all Alu copies absent from orthologous loci of non-human primate genomes are members of Y family. The most abundant human-specific Alu elements are dubbed Ya5 and Yb8 branch members.[16] Retropositionally active L1 elements form human specific fractions called Ta family.[17]

There are two kinds of species specificity of REs: (i) a whole group of the elements is present in one but absent from the other species; (ii) the group is present in both species but individual integration sites are different among species. The first type of specificity is rather rare. For example, SINE-R/SVA and MIR families were first suggested to be human-specific,[18,19] but later a large group of SINE-R.C2 elements was found in genomes of various apes.[20] The ancient activity of the families other than Alu SINE was also shown by advanced phylogenetic analysis.[21] Two human specific subfamilies of HERV-K(HML-2) elements will be described below.

Evolutionary young REs comprise roughly 8-15% of the human genome.[3,4] Along with chromosomal rearrangements and duplications of gene/pseudogene families, RE integrations are considered to be a major type of differences between human and great ape genomes.[22,23]

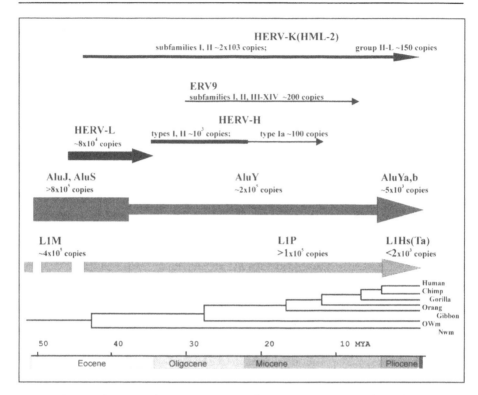

Figure 1. Timing and intensity of retroposition activity of REs. Lower part: a time scale in million years ago (Mya) and a corresponding primate evolution tree. Periods of particular REs expansion are marked by arrows. The ends of the arrows mark estimated times of the beginning and termination of various families' retropositions. The size of rectangles reflects the abundance of the corresponding REs subfamilies in the human genome.

Species specific subfamilies can be distinguished based on their structural features characteristic of a given, but not other, species. There are also a considerable number of species specific integrations of both the members of species specific subfamilies and those of older, not species specific, subfamilies the master genes of which were active before and after the speciation, being transposed after species divergence. The species specific integrations of both types can be considered as candidate causative factors of speciation.

## What Makes REs Possible Candidates for Evolutionary Pacemakers?

REs not only probably played an important role in the formation of the human genome structure, but they also could contribute much to the genome functioning. They can actively reshape the genome by causing genomic rearrangements, creating new genes and modifying the regulatory machinery of the existing genes thus being efficient pacemakers in evolution. Their ability to change genomes is reflected in etiology of various hereditary diseases.[24-26]

A major effect of L1 retropositions on the genome structure seems to be transduction of 3'-end adjacent non-L1 sequences and 'exon shuffling'.[11,27,28] According to the current estimation, the amount of human DNA transduced by L1 retropositions represents ~1% of the genome, a fraction comparable with that occupied by exons. Moreover, the L1 machinery is used for retroposition of other mobile elements such as Alu repeats. Recently, we demonstrated one

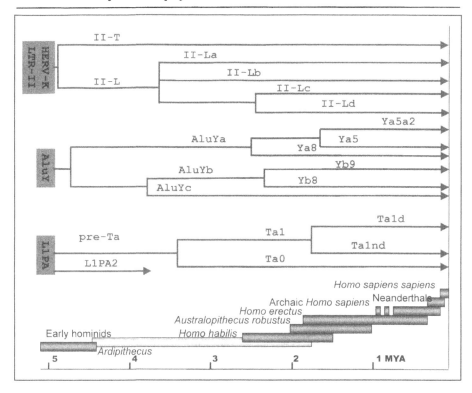

Figure 2. Phylogenetic trees of evolutionary young HERV-K(HML-2) LTR, AluY, and L1PA RE groups. Additive phylogenetic trees for consensus sequences of RE groups/subgroups were constructed using the Clustal software. Estimated times of the branching events are correlated with the time scale (Mya) in the lower part of the figure. Grey rectangles above the time scale denote approximate periods of existence of the corresponding hominid species and groups.

more mechanism of the L1 transducing activity—the recombination of L1 RNA with other RNAs followed by the integration of the recombinant cDNAs in the genome.[29]

Retroposed Alu repeats act not only as deleterious insertions but also as a source of additional codons, splice sites, and polyadenylation signals.[30] A large amount of Alu sequences detected within human transcripts suggest an involvement of these primate-specific SINEs in evolutionary modification of gene structures.[31] For example, an Alu element identified in orthologous loci of human, great apes, and Old World monkeys regulates the parathyroid hormone (PTH) gene via a negative calcium-response element; while another copy of Alu integrated in the human and ape gene coding for IgE receptor (FcɛRI-γ) defines tissue specificity of its expression. One more Alu copy specific for human and ape genomes forms a T-cell specific transcriptional enhancer in CD8 gene.[32]

"Newly" transposed endogenous retroviruses, and especially their solitary LTRs, are commonly considered as the effective modulators of neighboring genes expression. A large regulatory potential of LTRs is due to the presence in their structure of efficient Pol II promoters, enhancers, and several other sites recognizable by cellular transcription factors.[33] Current data on the influence of HERVs on cellular genes expression are reviewed in more detail in the Chapter 7 by Christine Leib-Mösch et al.

In general, the integration of an efficient regulatory module, like an LTR, in the preexisting regulation network of a gene involved in development could modify developmental trajectories thus causing morphological changes that might be then fixed by natural selection. Of course, such an integration can be imagined to interfere with a gene regulation system responsible for body size or certain aspects of brain development. However, such a straightforward impact seems too simple from the point of view of the known complexity of gene regulation, and at this point it is very like the explanation of how leopard became spotted given by R. Kipling in his famous stories for children. To give the hypothesis a real experimental background and to open a way to functional analysis of definite systems, one needs to perform a genome-wide comparison of the distribution in the human genome of those REs that differ between humans and other primates. The identification of these human specific RE integrations and of the genes the regulation of which could be modified by the inserted elements could provide a structural basis for revealing candidate evolutionary important insertions. It can be asked, why it should be human specific integrations and not those of chimpanzee. This is because hominid evolution seems to be much faster than that of pongids. Whereas pongids evolved rather slowly and remained phenotypically rather similar to the common ancestor of humans and pongids, the lineage of hominids is characterized by numerous and revolutionary phenotypic changes. Taking into account that the rates of molecular evolution in all hominoids were approximately the same, it means that hominids but not other hominoids have experienced some kind of a genetic regulatory revolution.

As described by Broude and Sverdlov in Chapter 2, to perform a genome-wide comparison one will need techniques allowing genome-wide analyses in a comprehensive and cost efficient way. It is also necessary to formulate the criteria of singling out the most promising candidate insertions:

1. These insertions must be young and integrated either immediately after the human and chimpanzee lineages divergence or close to the periods when revolutionary phenotypic changes occurred in hominids, that is around 5 Mya (*Ardipithecus* appearance), about 4.5 Mya (at the border of australophitecus appearance), around 2.5 Mya (early genus *Homo*), and 1.8 Mya (*Homo erectus* emergence).

2. They must be non-polymorphic indicating that they are essential for some universal features that we all share by virtue of being humans.

3. They must be located in the vicinity of the genes involved in the networks regulating developmental processes, in particular those determining body size and proportions and brain development.

The following sections will be devoted to the experimental identification of human specific RE integrations.

## Strategies and Approaches to the Genome-Wide Identification of Human-Specific RE Integrations

Theoretically, it might be possible to make a straightforward nucleotide-by-nucleotide comparison of completely sequenced genomes that would allow the identification of all types of differences including those in RE integration sites. However, it is currently a rather distant perspective. Modern strategies of a genome-wide search for human-specific RE integrations can be tentatively subdivided into two categories. The first one exploits the phylogenetic analysis of sequenced RE copies aimed at the detection of 'evolutionary young' branches. Low intrabranch divergence of RE sequences, absence of the branch from the genomes of related species, and insertional dimorphism of several representatives in human population can imply human specificity of the group as a whole. Here, further analyses of integrations in distinct loci are needed to verify the human specificity.

The second type of strategy is based on direct experimental comparison of human genomic DNA with DNAs of related primates by cloning and analysis of integration sites of REs under study. Both of the strategies received a great momentum from the draft sequence of the human genome, because it allows immediate assignment of identified human specific REs and their integration sites to precise genomic locations.

## *Phylogenetic and Functional Analysis of RE Families for Identification of Human Specific Subsets*

Being transpositionally active and the most abundant REs in the human genome, Alu repeats were chosen for the first attempts to construct their phylogenetic trees and to identify human-specific groups. Three groups of authors[34-36] have found some specific structural features common for a subset of Alu and differing it from other subsets which in turn have their own structural markers. Accordingly, Alu sequences were subdivided into three major classes: J, S and Y. The members of Y subset are characterized by the least intragroup divergence and therefore must have been evolved due to most recent propagation of one or a few muster genes. Further analysis of Alu Y sequences resulted in a concept of the evolution of recently integrated Alu subfamilies. Almost all of the Alu elements recently integrated in the human genome belong to one of several closely related subfamilies: Y, Yc1, Yc2, Ya5, Ya5a2, Ya8, Yb8, and Yb9, mostly being Ya5 and Yb8 branch members. As shown in Figure 2, Yb8 branch diverged from the ancestral Y family followed by the divergence of Ya8 and Ya5, the latter having given rise to Ya5a2 0.62 Mya.[26,16] The draft human genomic DNA sequence allows 'in silico' identification of the Alu elements "recently" integrated in the human genome. This approach was used by Batzer and co-workers[37,38] as a final stage of their long-term studies of Alu repeats. The authors searched through GenBank nr and htgs databases using the consensus sequences of Yb8 and Ya5 and identified, respectively, 1852 and 2640 members of these subfamilies in the human genome. A subset of 475 human loci containing the Alu insertions and suitable for designing pairs of unique primers was selected for subsequent PCR analysis of orthologous loci in genomes of human populations and non-human primates. All 231 Ya5 Alu family members analyzed in this way were shown to be human specific except for only one element present also in the orthologous locus of the chimpanzee genome. Similarly, only one of 244 AluYb8 subfamily members studied was found also in the orthologous locus of chimpanzee. Moreover, Alu-insertion dimorphism was detected within human population. This finding suggests that Alu elements were retropositionally active long after the human-chimpanzee divergence. Approximately 25 (58/231) and 20% (48/244) of the studied Ya5 and Yb8 Alu family members, respectively, were found to be polymorphic in European, African American, Asian, Egyptian, and Greenland Native human populations.[38]

A human specific subfamily of L1 (L1Hs) called Ta (transcribed, subset a) was first identified as a group of expressed elements with a high degree of sequence identity.[17] Later, Moran et al. developed an original cultured-cell assay for the detection of L1 retrotranspositions.[39] An improved technique[40] allowed the authors to perform selective screening for identification of active L1s just from a human genomic library. As a result, seven new retropositionally competent L1s from estimated 30-60 active L1 elements have been identified.[41] It is worth mentioning that 11 of 12 de novo L1 insertions found in these studies belonged to the Ta subfamily.[41] The current results of exhaustive search through GenEMBL database, phylogenetic analysis and massive locus-specific PCR-assay of human and non-human primate DNAs confirmed the human specificity of L1-Ta family (Fig. 2) and shed light on its evolutionary dynamics.[42,43] The family was subdivided into "Ta-0 and "Ta-1 subfamilies according to diagnostic positions within 3' UTR region of the Ta consensus sequence. The results of a PCR-assay using locus-specific primers revealed several Ta insertion dimorphisms in human population. It has been estimated that roughly 29% of Ta-0 insertions, 69% of Ta-1 insertions, and up to 90% of the

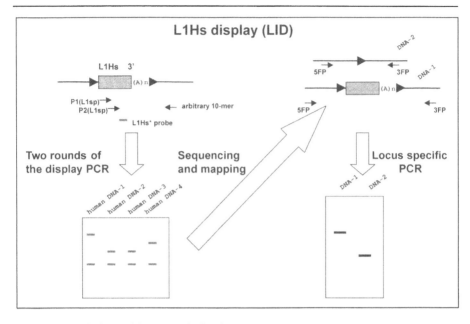

Figure 3. Principal scheme of the LID method. Left upper part. An L1-containing locus and positions of the primers used: straight line fragments, genome DNA; gray rectangle, an L1 element; black triangles, direct repeats of the integration site; short arrows, a 10mer random oligonucleotide and two primers (P1 and P2) against the 3'-end of the L1 consensus sequence; red bar, a hybridization L1 probe. Left lower part. Hybridization of PCR products with the L1 probe. Right part. Schemes of orthologous loci, positions of specific genomic primers 5FP and 3FP used for amplification of the sites of RE insertion, and results of locus-specific PCR. DNA1 and 2 correspond to loci with and without insertion. Other designations are the same as above.

youngest Ta-1d insertions are polymorphic in human.[42,43] Specific PCR amplification of an active subset of L1 elements allowed Swergold and co-workers[44,45] to develop a technique of whole genome identification of human-specific and/or polymorphic L1 insertions. Basically, the method (Fig. 3) includes nested PCR of 3'-end fragments of L1Hs-Ta together with adjacent fragments of cellular DNA, and differential display of the amplicons obtained on template DNAs of various individuals followed by hybridization with a labeled 3'UTR probe of L1. A random 10-mer universal primer was used to amplify the DNA fragments adjacent to L1 elements. The resulted fragments were cloned, sequenced and mapped within the human genome. The human specificity of the detected candidate L1 insertions was finally verified by locus-specific PCR-assay. The authors have successfully applied the technique to a large-scale screening of human population, and more than 30 new L1-Ta insertion polymorphisms were described. Although the genome-wide approach used demonstrated its effectiveness in simultaneous analysis of a variety of individual DNA samples, it has several limitations partially noticed by the authors. In particular, cross-hybridization of 3'UTR L1 probes systematically produces artifacts due to irreproducible DNA fragments primed by the arbitrary decanucleotide, and sometimes insufficient sensitivity and resolution of the display.

One of the first groupings of the ERV integration according to their age has been undertaken for HERV-K LTR mapped[46,47] on human chromosome 19. A set of 72 LTRs identified within clones of a chromosome-specific cosmid library was sequenced and used for the neighbor-joining tree construction. As a result, two major subfamilies including 16 distinct groups of HERV-K(HML-2) LTRs have been defined. The sequences of these groups have presumably originated from distinct 'master genes' that appeared at different periods of evolution.[47]

Two LTR groups named II-T and II-L are characterized by a very low degree of intragroup divergence, and the time of their master genes expansion was estimated at 8 and 4 Mya, respectively, which is close to the time of the hominid and pongid lineages divergence. Later, according to the results of locus specific PCR assay, several LTRs belonging to II-L group were detected in human orthologous loci but not in apes (see below). Medstrand and Mager[48] phylogenetically analyzed a set of HERV-K LTR sequences found in GenBank and detected the existence of a human specific HERV-K LTR group. The authors distinguished two young clusters of the LTRs in a bootstrapped phylogenetic tree (clusters 8 and 9). Further PCR analysis of the orthologous loci in primate genomes revealed the integrations of the LTRs belonging to cluster 8 in gorilla, chimpanzee, and human DNA, whereas the representatives of cluster 9 were found to be human specific or even polymorphic (in one case) in human population. Multiple alignment of the LTR sequences and selection of diagnostic positions allowed us (unpublished) to assign clusters 8 and 9 to II-T and II-L LTR groups, respectively. Thus, the results of both studies[46-48] suggest the existence of at least two recently active HERV-K master genes. One of these elements started its expansion in the ancestral genome and formed II-L/9 group of HERV-K LTRs after the human-chimpanzee divergence.

A recent analysis of available databases resulted in the identification of 22 HERV families independently acquired by the human genome.[14] The attempts to identify recently integrated HERV-related sequences using phylogenetic analysis were made predominantly for most abundant HERV families and their solitary LTRs. In particular, a phylogenetic analysis of the HERV-H family (also referred to as RTVL-H) revealed three subfamilies including the youngest one (Ia), the master gene of which was most probably formed due to a recombination event.[49] As estimated, the last expansion of the HERV-H family in the ancestral genome was 15-20 Mya. Some representatives of the Ia group were found in orthologous loci of the human and ape genomes. However, LTR sequences belonging to Ia subfamily were found also in the DNA of more ancient primates, African green monkey and marmoset. The time of the youngest HERV-H Ia copy insertion in the human genome was estimated at approximately 4.5 Mya.[13]

The ERV-9 family in the human genome includes about 40 proviral units and 4000 solitary LTRs. There are 14 branches of various ages in its phylogenetic tree, the youngest group of ERV-9 being 13 million years old.[50] With due regard to the expression of mRNAs from HERV-E family members in both normal and affected human tissues, an attempt for identification of human-specific group(s) among HERV-E sequences using a search of modern GDB with various HERV-E fragments was performed.[51] Newly identified family members showed 75 to 80% identity with the consensus sequence, and a number of HERV-E copies were found integrated in the close vicinity of cellular genes. Most probably, the expansion of the HERV-E family in the ancestral genome was ceased after the divergence of apes and Old World monkeys, that is about 20-25 Mya.

In general, phylogenetic analysis is a powerful instrument of identifying very recently formed subsets of REs or polymorphic insertions in a genome, although the identification of moderately transpositionally active, transposed for very long periods or rare REs using this approach is problematic. Obviously, to identify all human-specific RE insertions and to make definite conclusions on genome evolution, one needs an efficient technique for comprehensive genome-wide comparison.

## Genome-Wide Subtractive Hybridization of the Sequences Flanking RE Insertions

Unbiased techniques would allow one to directly experimentally analyze human specific integrations irrespective of whether they emerged as integrations of young RE families or of older, but still transpositionally active REs. We have recently proposed such a technique called Targeted Genomic Difference Analysis (TGDA).[52,53] The method includes: (i) a whole ge-

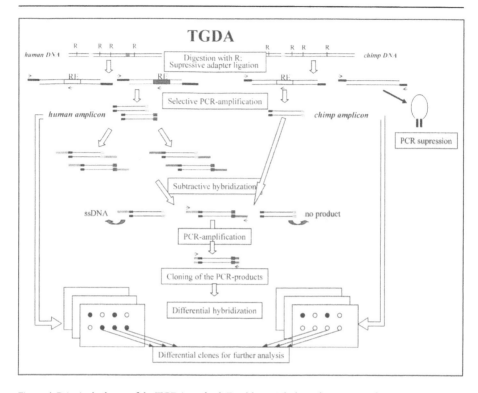

Figure 4. Principal scheme of the TGDA method. Double straight lines denote DNA fragments; letters RE indicate positions of identical (light ) or species specific (dark) RE insertions; R indicate positions of restriction sites; thin arrows mark positions of the PCR primers used. Black, green and red thick bars denote three different types of suppression adapters. A stem-loop PCR suppression structure of a single-stranded DNA fragment formed due to complementary adapter strains at both ends is schematized in the upper right corner. The lower part of the figure illustrates differential hybridization of two identical membranes containing clones of an arrayed DNA library with human (left) and chimpanzee (right) DNA probes.

nome amplification of the flanks adjacent to target interspersed repetitive elements in both genomic DNAs under comparison, and (ii) subtractive hybridization (SH) of the selected amplicons. The TGDA potential was demonstrated by the detection of differences in the integration sites of HERV-K retroviruses and related solitary LTRs between the human and chimpanzee genomes. Later, it has been successfully applied also to the detection of human specific integrations of the Ta subfamily of L1 elements.[54]

The technique of the SH is described in the Chapter 2 by Broude and Sverdlov. Briefly, the method is based on reannealing of two mixed genomic DNAs under comparison. Being digested to short fragments and mixed at large excess of one DNA (driver) over the other DNA (tracer), they are denatured and cooled to reanneal. During the reassociation most of the tracer DNA is hybridized to the excess driver DNA, except for the tracer specific target sequences ("differences") which form homoduplexes. The self-reassociated tracer fraction is enriched in these target fragments, as compared to the original tracer genome. These tracer specific sequences can be further cloned and analyzed.

Fortunately, to analyze human specific integrations of a certain RE, a rather complicated procedure of whole-genome subtraction is not needed. It is just necessary to subtract sequences flanking the sites of insertions. Therefore, the first stage of the TGDA technique (Fig. 4) is selective whole-genome amplification of the sequences flanking RE insertions in the two genomes under comparison. This procedure ensures efficient amplification of practically only the fragments that contain the target sequence, an HERV-K LTR in this example. It was used to prepare amplicons containing the DNA flanks of the LTR U3 parts both in human and chimpanzee DNAs. The human and chimpanzee DNA amplicons were used as tracer and driver, respectively. Some simple tricks were employed to ensure the selective amplification of the self-associated portion of the tracer after the reannealing step. The PCR products were cloned in *E. coli*, and the library obtained was shown to be highly enriched with the human sequences flanking LTR insertions present in human but not in chimpanzee DNA.

Since the selective amplification of the genome (simplification) in TGDA is targeted, the simplified fraction is not random but contains quite a definite portion of the genome. The complexity *C* of the simplified portion depends on the repetitive target content in the genome and the restriction endonuclease cutting frequency. With *N* target repeats in the genome and an average size of the amplified fragments of 256 bp (frequent cutter restriction endonuclease), $C=256 \times N$. In the case of HERV-K(HML-2) LTRs, the number of which in the human genome amounts to about 2000,[48] *C* is as low as $\sim 5 \times 10^5$, 6000-fold less than the complexity of the whole human genome. It results in a dramatic ($3.6 \times 10^7$) increase in the hybridization rate of the simplified vs. original genomic DNA.

Thus, TGDA enabled us to perform the first genome-wide experimental search for differences in LTR integrations between the human and chimpanzee genomes and to identify about 40 new human-specific LTRs.

The technique described is expected to be of universal use for analyses of the differences between related genomes in integration sites of various other REs, including Alu and L1 elements. For Alu or L1 targets represented roughly by $1 \times 10^6$ and $1 \times 10^5$ copies in the human genome, an increase in the rates of annealing should be 100 and 10000 fold, respectively, provided that the whole families are taken for the analysis. The youngest AluY and L1-Ta subfamilies are present in the human genome in about 3000 and over 500 copies, respectively. Therefore, one can expect approximately the same, or even better, kinetic characteristics when applying SH to selectively amplified flanks for each of these groups, as compared to HERV-K LTRs.

### Some Preliminary Results of the Analysis of the Sites of Human Specific Integrations of HERV-K(HML-2) LTRs

The TGDA technique has allowed us, for the first time, to perform a genome-wide analysis of the human specific integrations of the HERV-K(HML-2) family elements mostly represented by solitary LTRs. Further analysis of the integration sites in primate genomes confirmed that 37 loci of the human DNA contained LTR insertions absent from genomic orthologous loci of apes (Fig. 5).[52,53] The number of known human-specific LTRs was thus increased to about 60 elements of estimated[48] 2000 HERV-K(HML-2) LTRs per haploid human genome. The structures of the known human-specific LTRs were shown to be similar enough to derive a consensus human-specific (HS) LTR sequence (Fig. 6) that could be subdivided into four groups according to presence of a characteristic 12 nt deletion and diagnostic point mutations. As many as 142 LTRs of 98% and higher sequence similarity to the HS consensus have been found in nr and htgs GDB, and this set of LTRs was phylogenetically analyzed (Fig. 7). Part of the LTRs was assigned to II-T group with the master gene of predicted 16 million years age.[55]

Although the integrations of four LTRs of this group were human specific, many more representatives of the group were found in orthologous loci of the chimpanzee, gorilla, and even orangutan genomes. The other part of the LTR set belongs to II-L group which is estimated to be much younger, that suggests long term parallel retrotranspositions of two independent HERV-K retroelements. Moreover, the II-L group master gene gave rise to at least four active derivative variants, ancestors of four subgroups (II-La – II-Ld). The age of the master

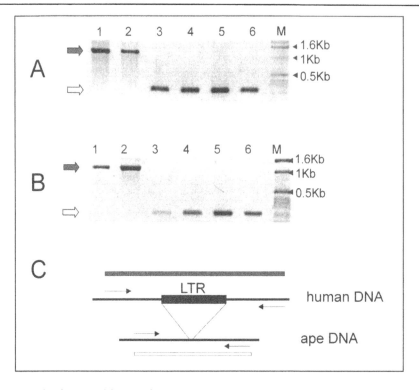

Figure 5. Results of PCR amplification of two individual LTR-containing loci AC018539 (A) and AC091895 (B) in various primate species. Red and white thick arrows indicate the predicted locations of LTR-containing and LTR-free products, respectively. Lanes 1-6, amplification of genomic DNAs: 1, 2 – two human individuals; 3, 4 – chimpanzee (*Pan troglodytes*); 5, 6 – gorilla (*Gorilla gorilla*), M, a 1kb ladder DNA length marker (Gibco-BRL). C) a scheme of locus specific PCR. Straight line fragments denote genomic orthologous loci; black rectangle, an LTR position; arrows, positions of unique genomic primers; red and white rectangles, PCR products. A color version of this figure is available at http://www.Eurekah.com.

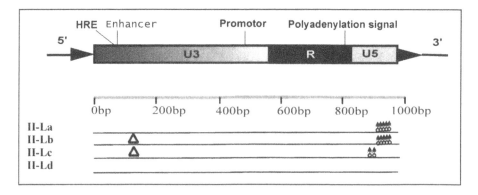

Figure 6. Human specific LTR consensus structures. Upper part: general scheme of LTR structure. Dark triangles mark short direct repeats of the integration site. Lower part schematizes consensus sequences of human specific LTR II-La, II-Lb, II-Lc, and II-Ld groups. Positions of a diagnostic 12-nucleotide deletion and diagnostic nucleotide substitutions are marked with an empty triangle and vertical arrows, respectively.

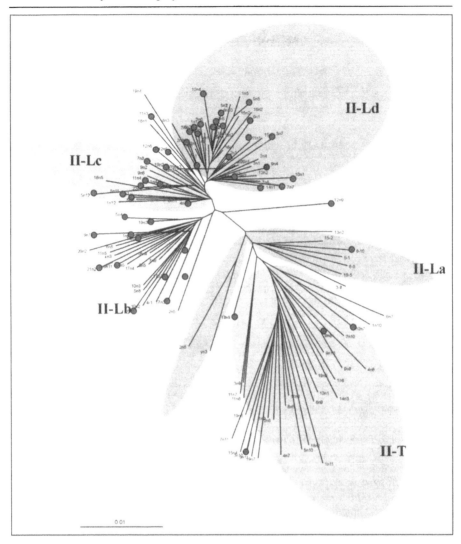

Figure 7. Additive unrooted tree for young HERV-K LTRs. The tree for 148 LTR sequences retrieved from nr GDB was constructed using the Clustal software. Separate branches of the corresponding LTR groups are shown as colored ovals. Red circles denote the LTRs whose human specific integrations are proved by locus-specific PCR. Lilac circles indicate the LTRs found in the orthologous loci both in the human and chimpanzee genomes. A color version of this figure is available at http://www.Eurekah.com.

genes of these subgroups can be estimated as ~10 to 3.5 Mya. Most of the LTRs shown to be absent from orthologous loci of the chimpanzee genome belong to the youngest II-Ld subgroup. The retropositions of the II-L group members continued far long after the human and chimpanzee lineages divergence, that was supported by a finding of three LTRs polymorphic in human population.[48,56]

To summarize, about 150 LTRs were inserted in the human genome due to recent activity of at least two HERV-K elements. Their retropositions across 5 million years after the divergence from pongids could affect many events of the hominid history starting from *Australopithecus*

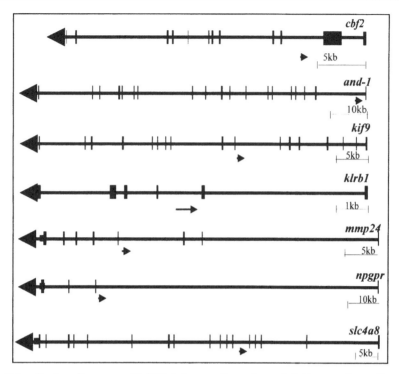

Figure 8. Localization of human specific LTRs in introns of genes (examples). Genes are denoted by arrows pointing the direction of transcription. Vertical bars indicate positions of exons. Short arrows mark positions of LTRs and point their U3-U5 directions.

through revolutionary changes in *Homo erectus* and to *Homo sapiens sapiens*. Much has to be done to answer the question of whether they were actually involved in these events, but first, it is necessary to determine which genes could be influenced by their integrations. We have analyzed the draft genome sequence and found that many truly human specific integrations are located in close vicinity to human genes, and frequently in gene introns, as shown in Figure 8 and Table 1. These integrations are good candidates for the involvement in gene regulation and, possibly, in evolutionary important events.

## Concluding Remarks

The current knowledge of hominoids and hominid evolution is contradictory and actively debated. This is quite understandable, because nothing is of more interest to people than the history of human origin and evolution. However, the traces left by human ancestors are too scattered and scanty to thoroughly track down the transformations which have finally led to our civilization. Furthermore, these fossils, however interesting and important they are in themselves, do not indicate why these transformations did occur. It is not enough to simply say that they were caused by mutations and natural selection. It would be interesting to know what mutations happened to be selected, how they changed the phenotype, what advantageous features were acquired due to these mutations etc. At this stage, we can only give very general answers to these questions as done by King and Wilson[5]: fast evolution of hominids was most probably due to changes in regulatory mechanisms. Clearly, there are a great many of regulatory systems that could be changed at *cis* or *trans* levels and thus

**Table 1. Human-specific solitary LTRs located in close vicinity with genes**

| Accession | Location[a] | Neighboring Genes | | Reference[b] |
|---|---|---|---|---|
| | | LTR position | Gene product (function) | |
| AC106051 | 4q13.3 | 1st intron of *GPR74* ; 2nd intron of *NPGPR* | G protein-coupled receptor 74; europeptide G protein-coupled receptor | 59 |
| AC007390 | 2p22.2 | 2nd intron of *CBF2* | CCAAT-box-binding transcription factor | 59 |
| AL121753 | 20q11.22 | 4th intron of *MMP24* | matrix metalloproteinase 24 | 59 |
| AC006432 | 12p13.31 | 2nd intron of *KLRB1* | killer cell lectin-like receptor 1 | 59 |
| AC008726 *M14123* | 5q33.3 | 5th intron of *SGCD* | sarcoglycan, delta (35kD dystrophin-associated protein | 48 |
| AC008648 | 5q35.1 | 30th intron of *kiaa0209* | predicted gene (D86924 mRNA) | 59 |
| AL352982 | 14q22.2 | 1st intron of *AND-1* | AND-1 protein | 59 |
| *AC107031* | 12q13.13 | 5th intron of *SLC4a8* | solute carrier family 4 | 59 |
| AC104447 | 3p21.31 | 9th intran of *KIF9* | kinesin protein 9 | 59 |
| AC002350 | 12q24 | 1st intron of *TA-PP2C* | T-cell activation protein phosphatase 2C | 48 |
| AL590377 | 9p22.2 | 13th intron of *FLJ20276* | hypothetical protein NM_017738 | 59 |
| AC074117 | 2p33.3 | intron of cDNA DKFZp434A163 | hypothetical protein AL110218 | 53 |
| AL590543 | 6q25.1 | introm of *FLJ12910* | ( hypothetical protein, mRNA - AK022972) | 59 |
| AC009132 | 16q23.1 | 25kb upstream of *FAAH* | fatty acid hydroxylase | 54 |
| AC006029 | 7q31.32 | 64kb upstream of *HYAL4* | hyaluronoglucosaminidase 4 | 53 |
| AL139404 | 10p12.1 | 60kb upstream of *SSH3BP1* | spectrin SH3 domain binding protein 1 | 53 |
| AL354855 | 9q34.13 | 47kb downstream of MGC12921 | hypothetical protein MGC12921 | 53 |
| AL139421 | 1p22.1 | 68kb upstream of DR1 | down-regulator of transcription 1 | 53 |
| AL135927 | 1q22 | 1,5kb downstream of *FLJ12287* | hypothetical protein NM_022367 similar to semaphorins | 53 |
| AL139022 | 14q23.3 | 7kb upstream of *FNTB* 36kb upstream of *GPX2* | farnesyltransferase gastrointestinal glutathione peroxidase 2 | 53 |
| AC025574 AC024884 | 12q13.3 | 8kb downstream of *MIP* | major intrinsic protein of lens fiber | 53 |
| AC007326 *AC008103,* *AF164609* | 22q11.21 | 2kb upstream of *PRODH* 27kb downstream of *DGCR6* | proline dehydrogenase DiGeorge syndrome critical region protein 6 | 57 |
| AC073898 *L47334* | 19q13.31 | 49kb upstream of *PVR* | poliovirus receptor | 47, 48 |

**Continued next page**

**Table 1. Continued**

| Accession | Location[a] | Neighboring Genes | | Reference[b] |
|---|---|---|---|---|
| AC048344 | 12p11.21 | 7kb upstream of *BICD1* | Bicaudal D homolog 1 | 59 |
| AL359644 | 9q33.2 | 59kb upstream of *EPB72* | erythrocyte membrane protein band 7.2 | 54 |
| AC012146 | 17p13.2 | 20kb downstream of *ZNF232* 49kb downstream of *KIF1C* | zinc finger protein 232 kinesin family member 1C | 53 |
| AL162412 | 9q21.12 | 2kb upstream of NT_023967.30 90kb upstream of *APBA1* | assembled gene amyloid beta A4 precursor protein-binding | 53 |
| AC099661 | 3q27.2 | 10kb upstream of *mPA-PLA1* 66kb upstream of *SENP2* | membrane-bound phosphatidic acid-selective protein sentrin-specific protease | 54 |
| AL359701 | 1p31.2 | 50kb downstream of *PDE4B* | phosphodiesterase 4B, cAMP -specific | 59 |
| AC002400 | 16p12,2 | 3kb upstream of *NDUFAB1* | NADH dehydrogenase (ubiquinone) 1, alpha/beta | 48 |
| AP001591 | 11q13 | 8kb upstream of *RARRES3* 14kb downstream of *LGALS12* | retinoic acid receptor responder (tazarotene) lectin, galactoside-binding, soluble, 12 | 48 |
| AP003385 | 11q13.3 | 3kb upstream of *NDUFV1* 17kb downstream of *GSTP1* | NADH dehydrogenase (ubiquinone) flavoprotein 1 glutathione transferase | 53 |
| U47924 | 12p13 | 13kb upstream of *Grcc9* | SPRY domain-containing SOCS box protein SSB-2 | 48 |
| AL450422 Z84493 | 3p21 | 16kb upstream of *CACNA2D2* | alpha 2 delta calcium channel subunit isoform I | 48 |

[a] Corresponding locus of human chromosome is indicated.
[b] Human-specific integration of LTRs in corresponding loci were reported.

possibly took part in evolutionary processes. Most probably, we will have to discard simple concepts like "this or that gene or regulatory element caused this or that phenotypic change" and to accept a more complicated model of evolution. In this model a crucial role in evolution is assigned to complex networks of flexible functional modules responsible for regulating a multitude of genes. However, we have no other way to analyze these complex systems than hypothesis driven approaches. Each of the hypotheses will have to be checked experimentally, so the progress in the science of human evolution will be very much dependent on the efficiency and capability of the techniques used for comparison of structural and functional properties of hominoid genomes. To verify the hypotheses, these techniques should at least provide unbiased genome wide analysis to identify the differences among the genomes. The whole human genome sequence available now and a coming sequence of the chimpanzee genome[57] will highly facilitate the comparative analyses and revealing the evolutionary important elements by means of bioinformatics. It would be probably still too expensive and time consuming to test each difference for being truly species-specific. Therefore genome-wide comparative analyses will remain of value. One of the hypotheses, which seems to be worth of checking, suggests a possible role of REs in changing *cis* nodes of some regulatory networks and thus causing evolutionary significant modifications of phenotypes. The techniques reviewed above allow one to perform genome-wide analyses also for this particular type of interference with evolution. New techniques will probably combine SH with other methods of large-scale genome analysis, such as DNA microarrays. However restricted may be the information obtained by all these techniques, it is indispensable for understanding the molecular mechanisms of human evolution and role of REs therein.

## Acknowledgments

The work partially supported by INTAS-01-0759, the Russian Foundation for Basic Research 02-04-48614 and 2006.200054 grants.

## References

1. Haile-Selassie Y. Late Miocene hominids from the Middle Awash, Ethiopia. Nature 2001; 412:178-181.
2. Foley R. The context of human genetic evolution. Genome Res 1998; 8:339-347.
3. International Human Genome Sequencing Consortium. Initial sequensing and analysis of the human genome. Nature 2001; 409:860-921.
4. Venter JC, Adams MD, Myers EW et al. The sequence of the human genome. Science 2001; 291:1304-1351.
5. King MC, Wilson AC. Evolution at two levels in humans and chimpanzees. Science 1975; 188:107-116.
6. Smit AF. The origin of interspersed repeats in the human genome. Curr Opin Genet Dev 1996, 6:743-748.
7. Jurka J. Repeats in genomic DNA: mining and meaning. Curr Opin Struct Biol 1998; 8:333-337.
8. Prak ET, Kazazian HH Jr. Mobile elements and the human genome. Nature Rev Genet 2000; 1:134-144.
9. Britten RJ. Evolution of Alu retroposons. In: Jacson MS, Strachan T, Dover G, eds. Gemone Evolution. Oxford: Human BIOS Scientific Publishers Ltd., 1996:211-228.
10. Batzer MA, Arcot SS, Phinney JW et al. Genetic variation of recent Alu insertions in human populations. J Mol Evol 1996; 42:22-29.
11. Kazazian HH Jr, Moran JV. The impact of L1 retrotransposons on the human genome. Nature Genet 1998; 19:19-24.
12. Ostertag EM, Kazazian HH Jr. Biology of mammalian L1 retrotransposons Annu Rev Genet. 2001; 35:501-538.
13. Johnson WE, Coffin JM. Constructing primate phylogenies from ancient retrovirus sequences. Proc Natl Acad Sci USA 1999; 96:10254-10260.
14. Tristem M. Identification and characterization of novel human endogenous retrovirus families by phylogenetic screening of the human genome mapping project database. J Virol 2000; 74:3715-3730.
15. Benit L, Lallemand JB, Casella JF et al. ERV-L elements: a family of endogenous retrovirus-like elements active throughout the evolution of mammals. Virol 1999; 73:3301-3308.
16. Batzer MA, Deininger PL. Alu repeats and human genomic diversity. Nature Rev Genet 2002; 3:370-379.
17. Skowronski J, Fanning TG, Singer MF. Unit-length line-1 transcripts in human teratocarcinoma cells. Mol Cell Biol 1988; 8:1385-1397.

18. Ono M, Kawakami M, Takezawa T. A novel human nonviral retroposon derived from an endogenous retrovirus. Nucleic Acids Res 1987; 15:8725-8737.

19. Smit AF, Riggs AD. MIRs are classic, tRNA-derived SINEs that amplified before the mammalian radiation. Nucleic Acids Res 1995; 23:98-102.

20. Kim HS, Takenaka O. Phylogeny of SINE-R retroposons in Asian apes. Mol Cell 2001; 12:262-266.

21. Gilbert N, Labuda D. CORE-SINEs: eukaryotic short interspersed retroposing elements with common sequence motifs. Proc Natl Acad Sci USA 1999; 96:2869-2874.

22. Hacia JG. Genome of the apes. Trends Genet 2001; 17:637-645.

23. Gagneux P, Varki A. Genetic differences between humans and great apes. Mol Phylogenet Evol 2001; 18:2-13.

24. Meischl C, Boer M, Ahlin A et al. A new exon created by intronic insertion of a rearranged LINE-1 element as the cause of chronic granulomatous disease. Eur J Hum Genet 2000; 8:697-703.

25. Schwahn U, Lenzner S, Dong J et al. Positional cloning of the gene for X-linked retinitis pigmentosa 2. Nature Genet 1998; 19:327-332.

26. Deininger PL, Batzer MA. Alu repeats and human disease. Mol Gen Metab 1999; 67:183-193.

27. Moran JV, DeBerardinis RJ, Kazazian HH Jr. Exon shuffling by L1 retrotransposition. Science 1999; 283:1530-1534.

28. Kazazian HH Jr, Goodier JL. LINE drive. retrotransposition and genome instability. Cell 2002; 110:277-280.

29. Buzdin A, Ustyugova S, Gogvadze E et al. A new family of chimeric retrotranscripts formed by a full copy of U6 small nuclear RNA fused to the 3' terminus of L1. Genomics 2002; 80:402-406.

30. Rowold DJ, Herrera RJ. Alu elements and the human genome. Genetica 2000; 108:57-72.

31. Brosius J. RNAs from all categories generate retrosequences that may be exapted as novel genes or regulatory elements. Gene 1999; 238:115-134.

32. Hamdi HK, Nishio H, Tavis J et al. Alu-mediated phylogenetic novelties in gene regulation and development. J Mol Biol 2000; 299:931-939.

33. Sverdlov ED. Retroviruses and primate evolution. BioEssays 2000; 22:161-171.

34. Britten RJ, Baron WF, Stout DB et al. Sources and evolution of human Alu repeated sequences. Proc Natl Acad Sci USA 1988; 85:4770-4774.

35. Jurka J, Smith T. A fundamental division in the Alu family of repeated sequences. Proc Natl Acad Sci USA 1988; 85:4775-4778.

36. Quentin Y. The Alu family developed through successive waves of fixation closely connected with primate lineage history. J Mol Evol 1988; 27:194-202.

37. Roy-Engel AM, Carroll ML, Vogel E et al. Alu insertion polymorphisms for the study of human genomic diversity. Genetics 2001; 159:279-290.

38. Carroll ML, Roy-Engel AM, Nguyen SV et al. Large-scale analysis of the Alu Ya5 and Yb8 subfamilies and their contribution to human genomic diversity. J Mol Biol 2001; 311:17-40.

39. Moran JV, Holmes SE, Naas TP et al. High frequency retrotransposition in cultured mammalian cells. Cell 1996; 87:917-927.

40. Sassaman DM, Dombroski BA, Moran JV et al. Many human L1 elements are capable of retrotransposition. Nature Genet 1997; 16:37-43.

41. Kimberland ML, Divoky V, Prchal J et al. Full-length human L1 insertions retain the capacity for high frequency retrotransposition in cultured cells. Hum Mol Genet 1999; 8:1557-1560.

42. Boissinot S, Chevret P, Furano AV. L1 (LINE-1) retrotransposon evolution and amplification in recent human history. Mol Biol Evol 2000; 17:915-928.

43. Myers JS, Vincent BJ, Udall H et al. A comprehensive analysis of recently integrated human Ta L1 elements. Am J Hum Genet 2002; 71:312-326.

44. Sheen FM, Sherry ST, Risch GM et al. Reading between the LINEs: human genomic variation induced by LINE-1 retrotransposition. Genome Res 2000; 10:1496-1508.

45. Ovchinnikov I, Troxel AB, Swergold GD. Genomic characterization of recent human LINE-1 insertions: evidence supporting random insertion. Genome Res 2001; 11:2050-2058.

46. Vinogradova T, Volik S, Lebedev Y et al. Positioning of 72 potentially full size LTRs of human endogenous retroviruses HERV-K on the human chromosome 19 map. Occurrences of the LTRs in human gene sites. Gene 1997; 199:255-264.

47. Lavrentieva I, Khil P, Vinogradova T et al. Subfamilies and nearest-neighbour dendrogram for the LTRs of human endogenous retroviruses HERV-K mapped on human chromosome 19: physical neighbourhood does not correlate with identity level. Hum Genet 1998; 102:107-116.

48. Medstrand P, Mager DL. Human-specific integrations of the HERV-K endogenous retrovirus family. J Virol 1998; 72:9782-9787.

49. Anderssen S, Sjottem E, Svineng G et al. Comparative analyses of LTRs of the ERV-H family of primate-specific retrovirus-like elements isolated from marmoset, African green monkey, and man. Virology 1997; 234:14-30.

50. Costas J, Naveira H. Evolutionary history of the human endogenous retrovirus family ERV-9. Mol Biol Evol 2000; 17:320-330.

51. Taruscio D, Floridia G, Zoraqi GK et al. Organization and integration sites in the human genome of endogenous retroviral sequences belonging to HERV-E family. Mamm Genome 2002; 13:216-222.

52. Buzdin A, Khodosevich K, Mamedov I et al. A technique for genome-wide identification of differences in the interspersed repeats integrations between closely related genomes and its application to detection of human-specific integrations of HERV-K LTRs. Genomics 2002; 79:413-422.

53. Mamedov I, Batrak A, Buzdin A et al. Genome-wide comparison of differences in the integration sites of interspersed repeats between closely related genomes. Nucleic Acids Res 2002; 30:e71.

54. Buzdin A, Usyugova S, Gogvadze E et al. Genome-wide targeted search for human specific and polymorphic L1 integrations. Hum Genet 2003; 112:527-533.

55. Lebedev YB, Belonovitch OS, Zybrova NV et al. Differences in HERV-K LTR insertions in orthologous loci of humans and great apes. Gene 2000; 247:265-277.

56. Turner G, Barbulescu M, Su M et al. Insertional polymorphisms of full-length endogenous retroviruses in humans. Curr Biol 2001; 11:1531-1535.

57. Normile D. Genomics. Chimp sequencing crawls forward. Science 2001; 29:2297.

58. Buzdin A, Usyugova S, Khodosevich K et al. Human-specific subfamilies of HERV-K(HML-2) long terminal repeats: three master genes were active simultaneously during branching of hominoid lineages. Genomics 2003; 8:149-156.

# CHAPTER 9

# Genome-Wide Analysis of Human Gene Expression:

## Application to the Expression of Human Endogenous Retroviruses

Tatyana V. Vinogradova

## Abstract

This chapter reviews the state-of-the-art of the approaches to the investigation of human and other complex transcriptomes. The techniques of differential cDNA libraries screening, subtractive hybridization, serial analysis of gene expression, DNA microarrays and differential display are considered, as well as the application of these techniques to analysis of the expression patterns of human endogenous retroviruses in various tissues, including cancerous tissues. A conclusion is made that none of the present techniques is fully adequate for the complexity of the human transcriptome, and that principally new techniques are needed to decipher the transcriptome in its dynamics.

## Introduction: What Is the Function of a Genomic Constituent?

The sequencing of a variety of genomes allowed, for the first time, to estimate the real scale of the gene content in the genomes. It made it also clear how far we are from understanding genome functioning when it is an integral part of a very complex cellular life maintaining machinery. As soon as a researcher in the postgenomic era decides to turn from structural analysis to functional studies, his life immediately becomes Sisyphus' work.

A pregenomic functional paradigm came down to answers to the three following questions:

1. What is the biochemical function of the element (phosphatase? kinase? transcription factor? etc.)
2. Which partners participate in the element functioning?
3. What is the physiological meaning of the molecular events in which the element is involved?

Attempts to answer these questions have relied upon the analysis of individual genes and their role in specific cells or tissues. Genome analysis and parallel attempts to functionally annotate newly identified genes have lifted the curtain and revealed a tremendous complexity of functional organization of even simple genomic constituents. They appeared to be involved not in a single, but in multiple functions, sometimes even contrary, depending on the cellular state. The functions can be different in different tissues or at different stages of development and can be changed upon receiving specific signals by the cell. The problems of complexity are touched upon in Chapter 1 of this issue. All these problems are also characteristic of retroelements (REs). Let us consider an example of a solitary long terminal repeat (LTR) of endogenous retrovirus (ERV). One and the same LTR can play a role of:

- an enhancer to boost gene transcription
- a hormone responsive element to mediate the hormone action
- a promoter to initiate gene transcription
- a terminator of transcription
- a recombination hot spot
- a target for signals of chromatin remodeling
- a source of antisense RNA which silences the gene transcribed in the opposite direction.

This list can probably be extended. In various particular cell types an LTR can realize from none to a few of the above potentials, different in different cells.[1,2] On the other hand, various LTRs scattered across the genome and very similar in their primary structures, can function differently even in the same cells. Each LTR can be involved in its own network of cellular pathway(s) and thus participate in different physiological functions.

Realizing all the complexity of functional analysis, we could still try to somehow unravel this tangle. Unless principally new approaches to such complex systems are developed, there is no other way as systematic functional study of every gene, regulatory element and other functional units of the genome. It is a rather simple principle known and popular in Russia as "The eyes fear, but the hands do". There is another useful principle which a karate teacher taught his pupils, "When you do not know what to do – take a step forward". In this review we will try to consider the ways of how to take a step forward in the problem of functional role of retroelements, and in particular human endogenous retroviral LTRs. This step forward is the analysis of the participation of REs, mainly LTRs, in cellular transcription processes. The review is devoted mainly to the technical aspects of the approaches to this problem, whereas the research aspect is discussed by Christine Leib-Mösch et al in Chapter 7.

## Detection of Gene Expression and Comparative Analysis of the Expression Using Cross-Hybridization of the Samples Under Comparison

It must be admitted that although there are in silico methods for detection of various regulatory sites in sequenced genomes, they are still far from providing reliable prediction of the regulatory role of various genomic constituents. Moreover, no experimental technique for reliable genome-wide identification of such regulatory sites as enhancers, promoters or insulator elements has been proposed so far. Although there are few approaches to isolation of transcription factor binding sites and some hypo- or hypermethylated CpG islands possibly related to promoters,[3] they are actually not real regulatory sites which require binding of factor combinations and are not necessarily associated with CpG islands. By now, the only genome-wide approach to the solution of the problem of transcription regulation is based on detection of transcripts and comparison of their content in various tissues. Particular genomic constituents of interest can be further fished out in this or that way.

The complete set of RNA transcripts produced by the genome at any given time and both qualitatively and quantitatively characteristic of a particular cell type is often referred to as cell transcriptome or expression profile. The expression profile is a major determinant of a cellular phenotype[4] because RNA production is the first step in the chain of events on the way from gene to phenotype. Differences in gene expression are responsible for phenotypic differences and relevant to cellular responses to environmental changes. The transcriptome is highly dynamic and changes both in the course of normal cellular processes, such as DNA replication and cell division, and due to numerous variations in cellular environment. Clearly, the information on expression profiles of various cells in variable environments is indispensable for unraveling the mechanisms of the genome guided cellular functions. Furthermore, such a knowledge can help to determine the causes of diseases, to find what gene products might have therapeutic uses themselves or as targets for therapeutic intervention, and to understand how drugs and drug candidates work in cells and organisms.

There are five basic methods currently used to search for transcribed genes and identify differentially transcribed genes: construction and comparative analysis of cDNA libraries, DNA arrays including microarrays (microchips), subtractive hybridization (SH), differential display (DD) and serial analysis of gene expression (SAGE). These techniques are being actively developed and became standard tools for researchers. However, none of the existing techniques of searching for differentially transcribed sequences can be considered ideal or definitely superior to other (related) techniques. The choice of the appropriate technique is to a great extent determined by the properties of the biological system under study and by the goals and objectives of the research.

Below follows a very brief description of the experimental techniques. The use of existing EST databases and other bioinformatic tools for analysis of transcripts is beyond the scope of the present review.

## Construction and Analysis of cDNA Libraries

Historically the earliest, this method enables one to prepare libraries of fragmented or full-size cDNAs corresponding to the total pool of cellular transcripts.[5,6] In the beginning, construction of cDNA libraries required large amounts of starting material. RNA reverse transcription followed by polymerase chain reaction (RT-PCR)[7] allowed to adapt the method for much lesser amounts of the material, a very essential improvement for working with microdissected tumors, biopsies and embryonic material.[7-9]

The frequency of occurrence of a particular cDNA in a library is roughly proportional to the abundance of the transcript in the cell, though considerable deviations from the linear correlation are quite usual. This feature greatly impedes the search for rare transcripts and requires to analyze too many clones making the procedure clumsy and laborious. Therefore, the methods for constructing normalized cDNA libraries were developed.[10,11] A normalized cDNA library is different from a "usual" one in that each transcript in it is more or less equally abundant. The normalization eliminates many copies of highly abundant cDNAs thus facilitating the identification of rare transcripts.[12] But on the other hand, it also masks the differences in various RNAs' content, an important element of the expression profile.

Non-normalized libraries can be used to identify highly and moderately expressed genes and to compare the expression of such genes in different cells or tissue samples.[13] As an example, to identify genes potentially suitable for diagnostics of ovarian cancer, 21500 clones from 5 cDNA libraries obtained from ovarian tissues and cell cultures were analyzed. These clones were hybridized with labeled first strand cDNA from 10 ovarian tumors and six normal tissues. Genes of 134 clones were overexpressed in at least five of the 10 tumors, and among them a new gene which might be used as a marker for this type of cancer.[14]

The serious disadvantages of this approach include laboriousness and poor reliability of comparative analysis not allowing to identify differences in the abundance of rare transcripts. The method provides an idea of relative transcription levels only from the frequencies of occurrence of particular transcripts among the library clones, and this is often misleading.

Nevertheless, the method allows one to rather efficiently detect and analyze the total pool of transcripts characteristic of a given cell or tissue type. It was successively applied to analysis of full-size transcripts from the human brain.[15]

Construction and analysis of cDNA libraries were also successfully used in search of new families of endogenous retroviruses. For instance, the transcripts of the HERV-K family were detected using a DNA fragment of a Syrian hamster type A retrovirus (intracisternal A particle, IAP) gene as a probe for screening of a fetal human liver gene library.[16] Similarly, hybridization of multiple sclerosis-associated retrovirus (MSRV) with a cDNA library from placenta has enabled one to reveal the transcripts of a new human endogenous retrovirus family, named HERV-W.[17] Hybridization analysis allowed also to isolate an ERV-9 endogenous retroviral sequence from a human embryonic carcinoma cDNA library.[18]

Transcripts containing ERV sequences were searched for by screening of various cDNA libraries using fragments of the retroviral sequence as hybridization probes, as illustrated by some examples below. HERV-K LTR related cDNA clones were detected by screening a human placenta cDNA library with a HERV-K LTR probe.[19] The ability of human endogenous retroviral HERV-K LTRs to provide transcriptional processing signals for nonviral sequences was demonstrated by detection of chimeric cDNAs containing polyadenylation signals originating from HERV-K-T47D-related LTR fused to cellular transcripts.[20,21] The expression of novel HERV-containing mRNAs in human T cells was shown by screening of a cDNA library derived from PHA treated T cells using fragments of a HERV-H LTR as hybridization probes.[22]

## DNA Arrays and Microarrays

Fragments of DNA spotted on a porous membrane in a certain order (DNA arrays) were long in use for hybridizational detection of homologous sequences in different biological samples[4] (and refs. therein). Hybridization probes used with arrays were usually radioactively labeled, and the spots bound to complementary sequences could be detected by autoradiography.

Recently, a powerful, based on DNA arrays technique of genome-wide analysis with various applications was developed and called DNA microarrays (DNA microchips) technology. The use of glass as substrate and fluorescence for detection, together with the achievements in synthesizing and depositing nucleic acids on glass slides at very high densities, have allowed the miniaturization of nucleic acid arrays. Presently, DNA arrays with as many as 250,000 different oligonucleotide probes or 10,000 different cDNAs per square centimeter can be produced. This technology progressed by leaps and bounds, and though it is still not a routine tool in most laboratories and someone even joked that there are more reviews on DNA microarrays than experimental papers, this time will undoubtedly come soon. How to work with DNA microarrays is described in detail in Chapter 2, and for additional information a number of other reviews can be recommended.[4,23-26] The following paragraph is just a brief overview of the DNA microarrays potential in comparative analysis of gene expression in various tissues.

Figure 1 illustrates basics of the technique. Gene representative sequences (probes) are spotted on microscopic slides. Either total RNA or mRNA for analysis is converted into cDNAs which can be further labeled with fluorescent reporter dyes. In the figure these reporters are marked as red and green. Differently colored dyes, mostly Cy3 and Cy5, are used for two different cDNA sets to be compared. The two sets can be prepared e.g., for normal and tumor tissue, for two different stages of the cell cycle etc. The use of two different dyes enables one to detect the relative transcript abundance for each cDNA set. In two-color hybridization strategy two cDNA samples are co-hybridized to a microarray. After washing, the array is scanned at two different wavelengths to estimate the content of Cy3 and/or Cy5 in each spot. Red color in the figure marks the location of the probes that hybridize efficiently only to one cDNA set, while green indicates the probes that hybridize only to another one cDNA set. Yellow indicates the probes that hybridize equally to the cDNA of both sets. Laser excited fluorescence of the dyes is measured using a detector, e.g., a confocal laser microscope. Since the emission spectra for two dyes differ, their fluorescence can be measured separately.

Monochrome images from the scanner are imported into a software processing package in which hybridization data are presented as logarithm of a Cy3/Cy5 content ratio in various spots. Significant deviations of the Log from 0 (no change) are indicative of increased (>0) or

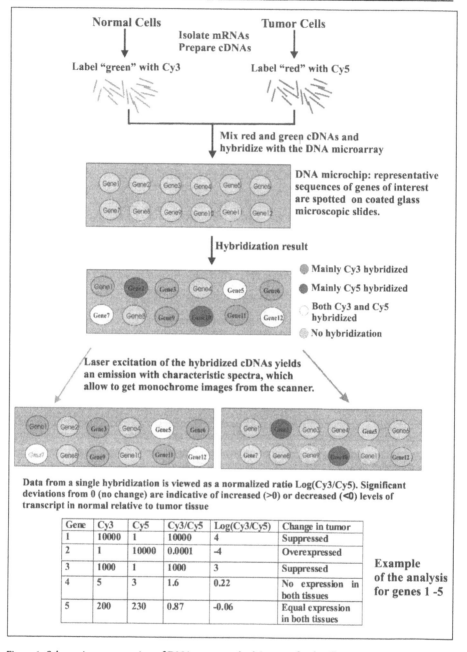

Figure 1. Schematic representation of DNA arrays method (see text for detail).

decreased (<0) specific mRNA content in a Cy3 labeled sample as compared to that labeled with Cy5.

This seemingly rather simple principle requires extremely precise, high-tech equipment and sophisticated software for reliable and accurate interpretation of great amounts of data obtained in tens of thousands of parallel gene expression analyses. Discussion of these tools and problems of reliability and reproducibility of the modern microarray technology are beyond the scope of this chapter and can be found in the reviews mentioned above.

The array-based global gene expression monitoring of tens of thousands of different genes has opened new horizons in functional analyses:

## Genes Behaving Similarly under Various Conditions Can Have Related Functions

It is now possible to compare the expression profiles of the same type of cells under different conditions to see the differences and similarities in the expression of individual genes. A special statistical strategy of such an analysis known as hierarchical clustering[23] allows one to group genes according to the extent of similarity in their expression behavior under different conditions.

It appeared that genes can be grouped into clusters comprising the genes similarly up- and down-regulated when the cellular environment changes. Other clustering methods were also used to treat gene expression data (e.g., see Self Organizing Maps[27]) but the hierarchical clustering is probably most popular due to its logical simplicity.

It seems quite logical to suggest that genes with similar expression behavior are likely to be related functionally. Overrepresentation of some particular gene types within a cluster can provide a way for broad functional classification of large groups of genes. Moreover, detection of genes with unknown functions within a cluster could shed light on their functions. The co-regulated genes are likely involved in related biological or biochemical processes—the 'guilt-by-association' concept which can be very fruitful.

A good evidence in favor of the validity of this concept has been provided in the now classical research by Eisen et al[23] which described the hierarchical clustering algorithm. The authors analyzed 2,467 genes of budding yeast *S. cerevisiae* functionally annotated at that time in the *Saccharomyces* Genome Database. The expression of the genes under various conditions was analyzed by microchip hybridization, and it was convincingly demonstrated that genes of unrelated sequence but similar function tightly cluster together. For example a large cluster of 126 genes strongly down-regulated in response to stress included genes encoding ribosomal proteins (112 genes) and other proteins involved in translation. This result was not unexpected because yeast responds to favorable growth conditions by increasing the production of ribosomes through transcriptional regulation of genes encoding ribosomal proteins. Another example of grouping functionally related genes were mitochondrial protein synthesis genes that were also expressed concordantly along with a number of genes involved in respiration. These and other data are an experimental background to the extension of this concept to genes with unknown functions.

## Different Cell Types Can Be Grouped According to Similarity in Their Expression Profiles under Different Conditions

The best example of such grouping is provided by the analysis of normal and cancerous cells.[28,29] These pioneer works demonstrated that it was possible to distinguish acute myeloid leukemia from acute lymphoblastic leukemia by gene expression profiling. The authors concluded that gene expression profiling can provide a new tool for analysis of tumor pathology, including cells of origin, stage, grade, clinical course and response to treatment. Another example of tumor classification by microarray based gene expression monitoring is given by a work of Ross et al.[30] The authors studied 60 cancer cell lines with cDNA microarrays and demonstrated that global characteristics of the carcinomas' expression patterns were similar to those of their normal origin cells, even though there were multitudes of expression variations in cancer cell lines. Gene expression profiles can be applied also for clinical purposes. The first unambiguous case of successful using gene expression profiling to discern otherwise indistinguishable tumors was described for the diffuse large B-cell lymphoma (DLBCL).[31] The clinical outcome of the lymphoma is different for different patients. An analysis of gene expression patterns with hierarchical clustering of lymphomas revealed the existence of two molecularly distinct forms of DLBCL. Patients with one of them had a significantly better overall survival rate as compared to that with the other form. This demonstrated a principal possibility of identification of previously undetectable subtypes of cancer, thus opening the way to better diagnostics and possibly therapy of cancer.

The potentials of such a classification of cells are certainly not restricted to cancer. It can be applied to any problem where it is needed to know global changes in gene expression underlying developmental processes like organogenesis or embryogenesis, as recently demonstrated with *Caenorhabditis elegans*.[32] Large scale application of gene expression profiling could give an unprecedented view into genome organization, gene functioning, and complex functional gene networks responsible for regulation in individual cells and multicellular communities.

## Subtractive Hybridization

A subtractive hybridization technique (SH) allows one to directly isolate sequences (targets) present solely in one of two related genomes/transcriptomes, without any preliminary knowledge of the genome sequences. In analysis of complex genomes this method has serious limitations due to high complexity (diversity) of the material to be analyzed (see Chapter 2). However, it was successively applied to analysis of small genomes (e.g., bacterial genomes) or changes in the cellular RNAs abundance due to regulatory processes in the cell genome.[33, 34] The technique was also successfully used to compare transcriptomes of tumor and normal cells.[35,36]

Subtractive hybridization is based on reassociation of two DNA sets to be compared, in particular cDNA sets. Being digested to short fragments and mixed at a large excess of one cDNA (defined as *driver*) over another one (defined as *tracer*), they are denatured and cooled to reanneal. During the reassociation most of the tracer DNA is hybridized to the excess driver DNA. The self-reassociated tracer is enriched in reassociated target fragments, as compared to the original tracer cDNA.

The next step is a separation of targets from other molecules using chromatography on hydroxyapatite,[37] avidin-biotin binding,[38-40] and other techniques. Despite of successful identification by this method of various important transcripts, SH was generally inefficient regarding rare transcripts and required large starting quantities of RNA. In a classical version of SH the abundance of each chosen target is proportional to the abundance of the corresponding transcript in the RNA pools of cells or tissues under comparison. Therefore, dominating transcripts of high and medium expression level hinder the identification of low abundant (rare) transcripts.

With the advent of polymerase chain reaction (PCR) technology,[41] it became possible to selectively amplify the subtracted fraction of cDNA retaining a representative set of all or almost all genes. PCR allowed one to abandon physical separation of single-stranded (targets) and double-stranded cDNA[42] and to select the targets by specific PCR amplification, tracer and driver being double-stranded. RT-PCR made it possible to strongly reduce the amounts of starting material.

A further important improvement of SH was an introduction of a normalization stage to equalize the abundance of various cDNAs within the target population, that is to equalize concentrations of low and highly abundant transcripts.[10,11] To normalize libraries, a pool of RNA to be analyzed is denatured, and then reannealed during a limited time period within which only highly abundant sequences can be renatured. This enables one to eliminate or significantly reduce the representation of highly abundant molecules thus increasing the specific concentration of low and medium abundant molecules further used for subtraction. The normalization stage made SH efficient in searching for rare differential transcripts. However, in this approach the frequency of occurrence of the isolated targets does not correlate with their differential expression in the cell, and transcripts 10 or even 100-fold different in the levels of expression will be represented almost equally. In contrast to SH, the methods of differential display (DD) and microarray allow to at once estimate relative levels of expression for transcripts under study.

It is believed that using SH one can reliably identify differential transcripts if their expression levels differ more than 5-fold. Currently available methods of SH do not provide sufficient enrichment of low abundant transcripts,[43-45] especially in the presence of much more abundant transcripts. In the latter case particularly low abundant transcripts will most probably be lost.

Apparently, the efficiency of SH is markedly decreased due to non-specific formation of poly(dA)/(dT) hybrids. It was shown that long poly(A) tracts present in most expressed mRNAs is a serious problem in subtractive reactions. Long poly(dT) regions of tracer cDNA, generated from the RNA of interest, randomly hybridize with long poly(dA) regions of driver cDNA resulting in the loss of transcripts. This loss particularly affects low abundant mRNAs.[46]

The method of SH was greatly improved by using the effect of PCR suppression (Fig. 2). This effect results in selective amplification of only those DNA molecules which contain a target sequence complementary to one of the primers used. All the other molecules are either not amplified at all or amplified with very low specificity. For PCR suppression, DNA fragments are ligated to suppression adapters (40 nucleotide (nt) long oligonucleotides of high GC content), specially constructed (see Fig. 2) to form self-complementary termini capable of forming intramolecular "stem-loop" structures. The PCR amplification of such DNA fragments will be suppressed if a single primer complementary to the 5'-end of the adapter is used because the formation of "stem-loop" structures proceeds more readily than the binding of the primer. However, if this primer is used in pair with a primer against the target, complementary to a single-stranded loop, then the DNA fragments of interest will be amplified because this primer binds the target within the loop and initiates the DNA synthesis. The resulting DNA fragments lack self-complementary termini and are capable of exponential amplification with the two primers. DNA fragments lacking the target sequence will not be amplified.

This suppression subtractive hybridization (SSH) technique combines normalization and subtraction in a single procedure (Fig. 2). Predominant PCR amplification of target sequences by using PCR suppression ensures higher enrichment in targets. The enrichment by SSH may be as high as 1,000-fold in one round of subtractive hybridization.[47] The SSH technique is applicable to many molecular genetic and positional cloning studies for identification of genes involved in disease etiologies, developmental, tissue-specific, or other differentially expressed genes.[48] Currently it is the most available and simple method of subtractive hybridization.

Enrichment of a cDNA mixture obtained after subtraction using double-stranded cDNA hybridization is determined by the tracer/driver ratio but is actually much lower than this, and therefore SH is carried out in a number of rounds. Nonetheless, additional screening like differential hybridization is needed to reveal differentially expressed genes in the enriched libraries. Identification of all (up and down) differentially transcribed genes requires two separate subtraction experiments, and it is impossible to compare more than two samples at once. An important weakness of SH is cross-hybridization of genes or any other sequences with extended regions of high identity, e.g., repetitive genomic elements within both mature and immature transcripts.[49] Due to the formation of imperfect heteroduplexes, the cross-hybridization can lead to the elimination of target molecules that are partially homologous with driver molecules. For example, among the eliminated molecules can be members of a gene family differentially expressed in the cells under comparison when particular cells express non-identical spectra of this family genes. As a result, differential expression of the family members will be either undetected at all or seriously misestimated. This drawback is characteristic of all other methods based on hybridization, including microarray. It can be partially overcome by using short DNA fragments for hybridization.[47]

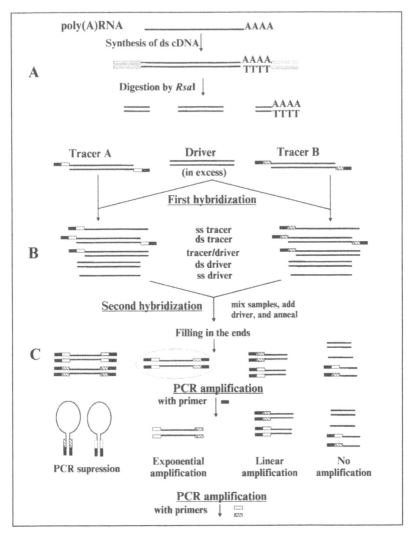

Figure 2. Schematic representation of Suppression Subtractive Hybridization technique (SSH). A) Tracer and driver ds cDNAs were prepared from two mRNA samples under comparison and separately digested with a restriction enzyme (here *Rsa*I). B) Two separate tracer populations were coupled with different suppression adapters (A and B). The suppression adapters (40 nt long oligonucleotides of high GC content) consisted of two parts: external (black-rectangle), identical for the both adapters, and internal (white and hatched rectangles), different between the adapters. Driver cDNA had no adapters. The first hybridization was carried out separately for tracers A and B with an excess of driver. The cDNAs were melted, and then annealed during a limited time period within which only highly abundant sequences could be reannealed. During this step the contents of various cDNAs within the ss tracer population (stage of normalization) were equalized. In the second round the two mixtures were combined, a fresh portion of the driver was added and the hybridization was continued. C) After filling in the protruding ends of ds reannealed DNA by DNA polymerase, differentially expressed tracer sequences are supposed to have different adapters on their 5'- and 3'-ends (encircled). The first PCR with a primer identical to the external parts of the adapters 1 and 2 results in exponential amplification of only these tracer sequences. Due to the PCR suppression effect, amplification of most tracer molecules' homoduplexes with a "stem-loop" structure is inhibited. Finally, PCR amplification with internal primers was performed to further reduce the content of any background PCR products.

This feature of SH can seriously distort the interpretation of the data on ERV and other REs expression.

## Methods of Detection and Comparison of Transcripts Avoiding Denaturation-Renaturation Steps

### SAGE (Serial Analysis of Gene Expression) As a Tool for Unbiased Transcriptome Analysis

Serial analysis of gene expression (SAGE) has provided a rapid and comprehensive approach for elucidation of quantitative gene expression patterns that does not depend on the prior availability of transcript information and can be used to identify and quantify new, as well as known, genes.[50,51]

SAGE is based on two basic ideas. First, a short nucleotide sequence tag (9 to 10 bp) can uniquely identify a transcript, provided it is isolated from a certain position within the transcript. Second, concatenation of short sequence tags allows efficient analysis of transcripts in a serial manner by sequencing of multiple tags within a single clone.

Figure 3 shows the main steps of the technique exemplified by one of its latest versions applicable to small cell populations.[51] Poly(A) RNA is directly isolated from cell lysates using oligo(dT) coated beads and reverse transcribed to cDNA. A frequently cutting anchoring enzyme, for example *Nla*III, cleaves cDNA molecules, leaving the 3' end of the cDNA attached to the beads. This process provides a unique site on each transcript that corresponds to the restriction site located closest to the poly(A) tail. The cDNA is then divided in halves, and each of them is ligated via the anchoring restriction site to different linkers (1 and 2), each containing identical type IIS restriction nuclease recognition site. Type IIS restriction endonucleases (tagging enzymes) cleave the cDNA at a defined distance up to 20 bp from their asymmetric recognition sites. The cleavage releases short pieces of the cDNA attached to the linkers (SAGE tags).

Two pools of released tags are then mixed, ligated to each other and PCR amplified with primers P1 and P2 specific to linkers 1 and 2, respectively. In the amplicons the tags are ligated tail to tail to form ditags. Apart from amplification, this step also provides orientation and punctuation of the tag sequences. Due to low probability of any two tags to be coupled in more than one ditag even for abundant transcripts, repeated ditags potentially produced by biased PCR can be disregarded practically without affecting final results. Cleavage of the PCR products with the anchoring enzyme allows to isolate ditags which are further concatenated by ligation, cloned, and sequenced. In the concatenated product the ditags are separated by a 4 bp site characteristic of the anchoring enzyme. Sequencing of contcatemer clones reveals the identity and abundance of each tag, and thus provides quantitation and identification of cellular transcripts. Absolute abundances are calculated by dividing the observed abundance of any tag by the total number of tags analyzed.

For several years, SAGE has been used for a comprehensive analysis of a variety of different tissue samples, each usually consisting of millions of cells. This approach has recently been extended to permit analysis of gene expression in substantially fewer cells (reviewed in ref. 51), thereby allowing analysis of heterogeneous tissues or microanatomical structures.

The tags obtained with the original SAGE version are sufficiently long (14-15 bp) to reliably identify cDNAs but too short to precisely map genes in the genome. To increase the length of the tags, a SAGE version termed longSAGE was proposed[52] (reviewed in ref. 53). The use of *Mme*I restriction endonuclease instead of *Bsm*FI allows to obtain 21-bp long SAGE tags. This modification made possible to unambiguously identify any tag within the genome sequence and therefore to assign particular transcripts to genes or open reading frames in the genome.

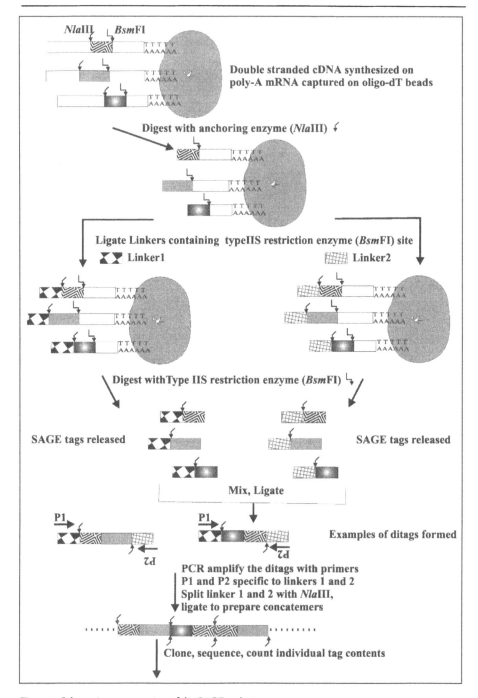

Figure 3. Schematic representation of the SAGE technique.

Rapid sequencing of many tags with a modern automated sequencer enables monitoring of the activity of thousands of genes in a relatively short period of time. The technique permits not only to identify the genes active in a given cell sample but also to simultaneously measure the abundance of their transcripts.

The studies of human cancer cells with SAGE[54,55] allowed to get an image of the human transcriptome. An analysis of 3.5 million transcripts from 19 normal and diseased human tissues has identified the expression of about 84,000 transcripts and revealed that more than 43,000 genes can be expressed in a single cell type. The latter number exceeds the modern estimates, and the reasons for this discrepancy remain to be clarified. About 1000 genes were estimated to be expressed in all the investigated cells, and expression of 40 genes was enhanced in cancerous cells as compared to their normal counterparts. It should be noted that the abundance of various transcripts in the cells was very divergent—from 0.3 to about 9500 copies per cell. This study has produced a heretofore unavailable global picture of gene expression patterns.

SAGE has been successfully used to characterize transcriptomes of various organs such as kidney, liver and thyroid, as well as for gene expression profiling of a number of human diseases. Nearly in all cases, several known and uncharacterized genes were found to be up- or down-regulated (reviewed in refs. 53, 56).

A challenging finding concerning mapping of highly expressed genes in the human genome with SAGE has been reported recently[57] (reviewed in ref. 56). An analysis of 2.5 million SAGE tags derived from 12 different normal and cancerous human tissue types revealed that highly expressed genes seem to cluster in particular chromosomal regions. These regions were therefore named 'regions of increased gene expression' or RIDGEs. These highly transcribed regions correlated with gene-rich regions suggesting that some common factors may determine both gene density in a chromosome region and their level of transcription.

SAGE determines absolute expression levels that strongly facilitates comparisons between numerous SAGE libraries. Due to this feature, SAGE has been selected as the major platform technology for the Cancer Genome Anatomy Project (CGAP). A large number of SAGE libraries generated from diverse cancer and normal tissues in many laboratories are accessible at the National Center for Biotechnology Information web site (http://www.ncbi.nlm.nih.gov/ SAGE). The web site SAGE Genie (http://cgap.nci.nih.gov/SAGE) was constructed recently allowing to monitor tissue-specific expression of a gene or tag of interest in color-coded symbolic organs. The current data set includes more than 5 million SAGE tags derived from 114 different cell types.[56] There are also other databases available, for example http://www.sagenet.org. Science magazine maintains a SAGE site http://sageke.sciencemag.org.

Despite many evident advantages, SAGE is a time consuming and rather expensive method that restricts its (as well as DNA microarray) wide use in the laboratories involved in functional genomics.

## Differential Display

The differential display method (DD) has no problem of homologous tracts in the primary structure of transcripts. It has a number of apparent advantages as compared to the methods described previously: the assay is rapid, simple, and much more sensitive than with other techniques. The method is based on simple and widely used technologies and can be therefore easily reproduced by many researchers. DD allows to detect even a slight (2-3 fold) difference in the abundance of transcripts, to use small amounts of starting RNA and amplify rare transcripts. It enables identification of both inducible and repressed genes, as well as comparison of more than two stages or different conditions simultaneously.

### DDRT-PCR (Differential Display Reverse Transcription PCR) Technique

A technique for identification and comparison of mRNAs expressed in various cells or under altered conditions was first described in 1992.[58,59] The procedure included three basic steps: 1) reverse transcription of RNA isolated from various cell populations with a set of degenerated anchored oligo(dT) primers to get a cDNA pool; 2) PCR amplification of part of the cDNA pool with an initial anchored oligo(dT) primer and an upstream arbitrary primer. Short sequences can be further visualized due to radioactive labeling of cDNA during amplification; 3) Separation of the amplified fragments by gel electrophoresis. The samples under comparison can be separated in parallel gel lanes that allows one to directly discern differences after autoradiography.

DDRT-PCR was tested with various cell populations.[58] The method yields a set of cDNA fragments for each of the samples to be compared. The sets are analyzed in denaturing poly-acrylamide gel to compare the distribution of bands in side-by-side lanes. cDNA from differential bands (that is differing in intensity between two samples) is eluted, amplified and cloned. Since the resolution of gel electrophoresis is rather low and permits to compare only a limited number of fragments with a length from 100 to 600 bp, the initial DDRT-PCR version used 12 anchored primers to RNA poly(A) tail and a set of 10-nt arbitrary primers to increase the number of transcripts analyzed. It was calculated that for complete assessment of RNA from mammal cells one needs to use 20 arbitrary primers and 12 anchored primers. Although this number is sufficient to amplify all mRNAs of the cell, the method is still too laborious and time consuming, that stimulated further elaboration of the technique.

Many improvements to the method were suggested. In particular, it was shown that 4 anchored primers instead of 12 is sufficient, and 200 µg of total RNA instead of poly(A)RNA is enough to get major information on the transcriptome of a given cell type.[60] Further improvements were made in various directions: modification of primers,[61-63] their extension to increase reproducibility,[64] decreasing starting RNA amounts, reamplification, convenient analysis of cDNA fragments,[61,65] elaboration of screening,[66] and many others.

A detailed modern protocol of DDRT-PCR is described in a review by McClelland et al.[67] Many new genes were identified by this methods, among them e.g., *PTI-1*, an oncogene associated with prostate cancer,[68] and two novel MAD family members.[69] Whereas the main advantage of the method is its simplicity in implementation, a principal disadvantage is that due to low specificity of arbitrary priming[62] there is a large number of false positives amounting up to 50-75%.[44,60,70] Even duplicate experiments with the selection of only reproduced bands does not solve the problem. Amplification and cloning of a single band very often gives more than one fragment because it is difficult to avoid an admixture of closely running fragments when excising the band, or due to insufficient resolution of the fragments in the gel. Moreover, the use of unspecific arbitrary primers for reamplification often leads to the appearance of unspecific bands from one band.[71,72] Sometimes it is impossible to single out the required fragment from the amplified group, that makes necessary laborious checking of all the revealed differential bands by common methods like Northern blot, RT-PCR or RNase protection.

Another serious disadvantage of using arbitrary primers is that one and the same gene can be represented on a gel by more than one fragment amplified with a particular arbitrary primer. Moreover, one and the same gene can be represented in pools obtained by PCR with different arbitrary primers, in other words samples can be overlapped. As mentioned above, complete assessment of total RNA pool requires a large number of arbitrary primers.[73] However, the more primers are used, the higher probability that already revealed cDNAs will be reidentified as new.[74,75]

One of the major questions being debated in literature is that about the capability of DD to identify rare transcripts. Model experiments with addition of exogenous RNA as a target demonstrated that it was not detected by DDRT-PCR if its relative content was less than 1%.[76] Other experiments provide more optimistic estimates of the method sensitivity.[44] All these complications and problems had a consequence that after a relatively short peak of popu-larity DDRT-PCR was practically not used anymore for comprehensive analysis of cell transcriptomes.

## Systematic Differential Display

As mentioned above, the main source of errors with DD is unspecificity of arbitrary primers. To avoid this problem, a new technique called systematic differential display (SDD) was suggested.[77-81] This technique principally differs from the classical DDRT-PCR version in two points. First, instead of original DDRT-PCR "random sampling", an alternative strategy of gel separated band patterns (fingerprints) production which may be called "systematic sampling" is used to analyze mRNA pools. This modification ensures the correspondence of each particular mRNA species to a unique fragment on fingerprints. Second, amount ratios of the cDNA fragments amplified by SDD reflect those of transcripts in the starting RNA pools, whereas in the classical DDRT-PCR these ratios can be seriously distorted because the intensity of a band in fingerprints strongly depends on the arbitrary primer match to the corresponding cDNA.

All the proposed SDD versions provide differential displays of only 3'-ends of cDNA supposed to represent the total mRNA pool. To enable the selection of such fragments, at the initial stage of cDNA synthesis 3'-ends are linked to a specific adapter[79,81] or biotin.[78,80] After this, double stranded cDNA is cut with a restriction endonuclease, the restriction sites are ligated to an adapter, and 3'-end cDNA fragments are selected by PCR amplification or using biotin binding. Each transcript of the initial RNA is thus represented by a cDNA fragment of definite length corresponding to the distance between the poly(A) tail and the restriction site.

The first stages of SDD in the version suggested by Ivanova et al[78] and named Gene Expression Fingerprinting (GEF) are as above (Fig. 4). Then, 5'-termini of the fragments are labeled and the mixture is digested with a new restriction enzyme. The labeled 5'-terminal fragments are separated by PAGE. The remaining 3'-terminal fraction is separated by biotin binding to streptavidin beads, the 5'-termini of this fraction are again labeled, split off with one more restriction enzyme and separated in PAGE. The operation is repeated each time with a new restriction endonuclease. In such a way, firstly simplified sets of fragments can be obtained suitable for separation on PAGE, and secondly each pool of fragments represents an mRNA fraction most probably not overlapping with the others.

A principally very similar technique, the Restriction Landmark cDNA Scanning (RLCS), was proposed by Suzuki et al.[80] However, at the last stages of RLCS initial restriction fragments obtained from the total cDNA are separated in an agarose gel. After this the agarose lane containing the separated fragments is treated with another restriction enzyme, and its content separated in the second dimension in PAGE (Fig. 4). Since only 3'-ends of the fragments carry the radioactive label, each cDNA is represented on a two-dimensional gel by a single spot. Up to 500-1,000 cDNA fragments could be resolved by two-dimensional electrophoresis. Naturally, an analysis of 1000 transcripts can not be considered comprehensive, since there are 10 times more transcripts in the cell. Minor modifications of the method involving carefully planned use of restriction enzymes (as in ref. 74) will allow one to obtain non-overlapping patterns, and 10-15 two-dimensional gels will be sufficient to cover all RNA species. Surely, this is a rather complicated technique, and the main complication is two-dimensional gel electrophoresis, when reproducible patterns are a matter of art. Moreover, it is much more difficult to compare the distribution of spots on 2D-gels than in side-by-side lanes.

3'-end cDNA fragments selected by PCR amplification in the method proposed by Prashar and Weissman and named READS (Restriction Endonucleolytic Analysis of Differentially expressed Sequences)[79] are also not separated into subpopulations. It makes impossible to analyze total RNA pool comprising 10,000 RNA species. Therefore, READS can not be considered a true systematic differential display. However, the method can be easily improved by introducing an additional stage of amplification with anchored oligo(dT) primers and thus by subdividing of the DNA fragments under study into several subpopulations. READS gives many false positives because the only step for selection of the target molecules is the amplification with a 5' PCR primer and a "heel" primer, the latter being short and therefore able to unspecifically anneal with non-target sequences.

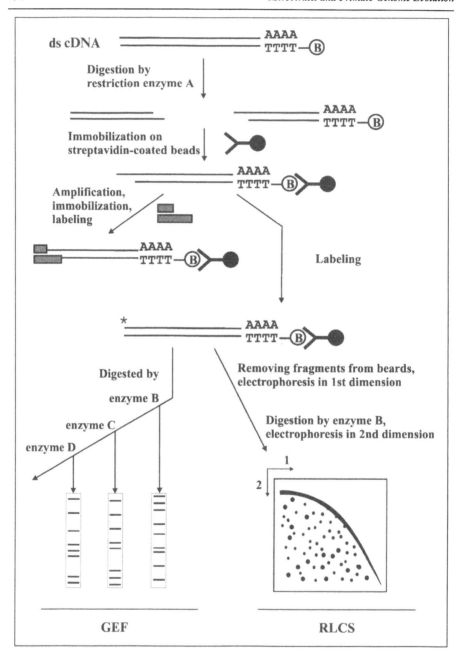

Figure 4. Schematic representation of Gene Expression Fingerprinting (GEF) and Restriction Landmark Scanning (RLCS) techniques (modified from refs. 78, 80).

    In the method proposed by Kato, subpopulations are obtained by amplification with an anchored oligo(dT) primer and a primer against the corresponding ligated adapter (Fig. 5).[77] To this end cDNA is hydrolyzed with a class IIS restriction endonuclease, *Fok*I. This enzyme

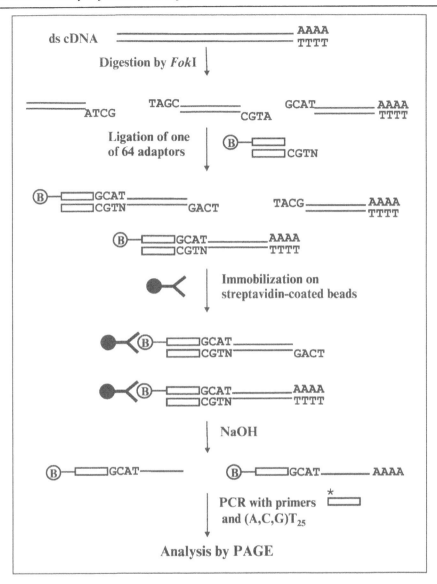

Figure 5. Schematic representation of SDD based on cDNA 3'-fragments generated by class II restriction enzymes (modified from ref. 77). N in protruding tetranucleotides of the adapters symbolizes a redundant mononucleotide.

produces 3'-terminal fragments with a poly(A) stretch and a 4-nucleotide 5'-overhang with unknown sequence. These fragments are ligated to a biotinylated adapter with protruding ends of 4 nucleotides in length. Streptavidin granules are then used to select only fragments with biotinylated adapters. The selected fragments are released from the granules and amplified with a primer against the ligated adapter and a T-primer anchored with one nucleotide. The experiment is repeated with each of 64 adapters corresponding to all possible combinations of protruding trinucleotides. Two more IIS enzymes and three anchored oligo(dT) primers are used

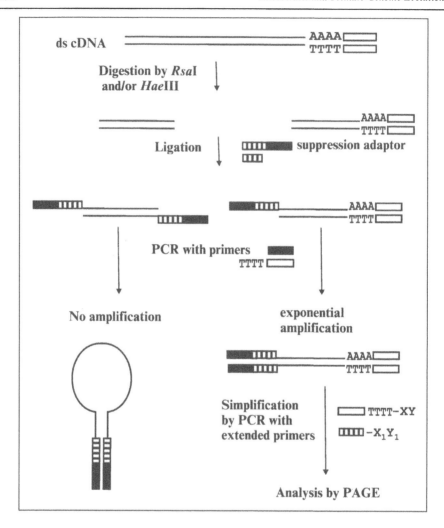

Figure 6. Schematic representation of Ordered Differential Display (ODD) (modified from ref. 81).

to produce in total 576 non-overlapping 3'-cDNA fragment subpopulations. Such a strong simplification allows to analyze even very complex RNA pools.

However, the method is quite laborious. The same author proposed a less laborious modification, however not giving non-overlapping subpopulations.[74]

The most simple version of SDD is ODD (Ordered Differential Display) which uses the effect of PCR suppression (Fig. 6).[81] Here ds cDNAs are digested with *Rsa*I/*Hae*III restriction enzymes, and the resulting DNA fragments are ligated with a suppression adapter. Representative pools of 3'-end restriction fragments of cDNAs are then selectively amplified using the PCR suppression effect with an oligo(dT) primer and a primer specific to the outer part of the suppression adapter. Simplified subsets of the amplicon obtained suitable for comparison by PAGE are further amplified with an oligo(dT) primer and a primer complementary to the inner part of the adapter, both primers being extended by two randomly picked bases at their 3'-ends. As a result, the population of 3'-end cDNA fragments is subdivided into 192 subsets which are displayed on a sequencing gel. The whole mRNA pool may thus be systematically investigated.

The use of SDD is justified when a comprehensive study of RNA samples is required. Similar studies but only for known transcripts can be also done using microarray technology, and its use for solving these problems is now more and more increasing. However, one should not forget serious limitations of microarray technology as a method based on hybridization.

## Targeted Differential Display

It is often desirable to analyze a particular portion of the transcripts having definite structural features in common, e.g., a common motif. Here one can use selective PCR amplification of this portion with primers recognizing this common motif. Such a special approach is considerably less expensive than microarrays and probably is a technique of choice for this particular case.

The first version of such a targeted differential display (TDD) was used to identify genes containing specific domains, like MADS-box genes,[82] genes with zinc finger homology[83,84] or with a common nucleotide motif,[85,86] as well as genome repetitive sequences,[1,87] and other families.[88]

As a rule, for selective PCR amplification TDD uses a target primer complementary to constant regions of the sequences and an oligo(dT) primer with anchors or without them. The primers should meet specific requirements. In particular, target primers should have very high identity level with the target domain and should be moderately degenerated. This condition enables amplification of target cDNA fragments at high temperatures thus decreasing the number of false positives. Since the probability of multiple binding of the target primer with different sites of one and the same cDNA is minimized, identical RNA molecules are all represented by fragments of equal length. In this regard TDD can be considered as a particular case of SDD.

TDD applications demonstrated that the method permits not only to detect variations in transcription levels of known gene families' members but also to reveal new members of these families.

## Targeted Differential Display Analysis of Human Endogenous Retroviral LTRs Incorporated in Cellular Transcripts

One more modification of SDD called Selective Differential Display of RNAs containing Interspersed Repeats (SDDIR) was developed by Vinogradova et al.[1] The method was used to compare the HERV-K(HML-2) LTR content in the total RNA of testicular germ cell tumors and their normal tissue counterparts. SDDIR has enabled us to get an overview of individual LTRs within the cell transcriptomes and to reveal differences in expression profiles of the LTRs in normal and tumor tissues. An unexpectedly large number of the LTRs was found to be transcribed, the content of many of the transcripts being different in normal and tumor tissues.

SDDIR involves two principal stages: (i) selective RT-PCR amplification of a total cellular RNA subset containing a certain type of repetitive elements, and (ii) side-by-side display of the amplicons derived from the tissues under comparison by means of gel electrophoresis in parallel lanes. The use of SDDIR to elucidate transcriptional status of specific repetitive elements implies the analysis of total RNA because many important regulatory elements can apparently be absent from mRNA as a result of processing.

The protocol of SDDIR includes the following steps:

1. Preparation of PCR amplicons corresponding to the LTR containing transcripts (Fig. 7A, step 1). For this purpose the cDNA first strand was synthesized from the total RNA using PL-1 primer complementary to the coding strand of the LTRs to selectively obtain amplicons containing only the LTRs transcribed in the U3->R->U5 direction and only the LTR U3 parts. The synthesis was performed with SuperScript II reverse transcriptase, in the presence of the second primer, SmrtRD. SuperScript II allows to synthesize long cDNAs and, having reached the 5'-end of RNA, adds a few deoxycytidine (dC) nucleotides to the newly synthesized cDNA. The 3'-end of SmrtRD oligonucleotide anneals with the (dC) tail, reverse transcriptase switches templates and replicates the oligonucleotide. The resulting cDNA

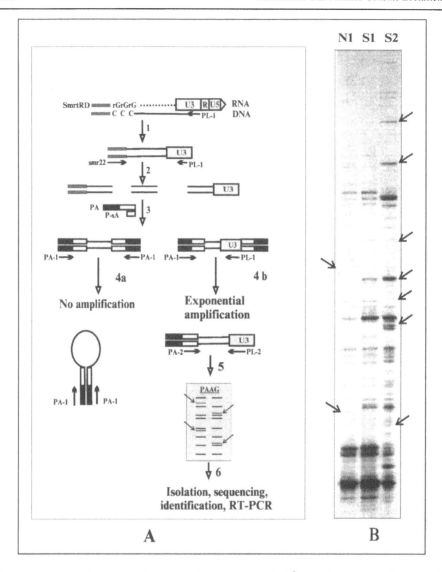

Figure 7. A) Schematic representation of Selective Differential Display of RNAs containing Interspersed Repeats (SDDIR) used for the analysis of RNAs containing HERV-K LTRs. 1-6, various stages of the procedure: (1) selective PCR amplification of the LTR containing RNAs, (2) restriction of the amplicons with a restriction endonuclease and (3) ligation of stem-loop forming adapters, (4) PCR amplification with a primer identical to an external part of the adapter and an LTR specific primer, (5) nested PCR and PAGE, and (6) isolation and analysis of the fragments. Horizontal arrows marked as SmrtRD, smr22, PL-1, and PA-1 indicate positions and directions of the primers utilized for reverse transcription and/or PCR amplification. The black-and-white rectangles correspond to the adapters which form terminal inverted repeats (TIRs) after ligation to the restriction fragments and filling in the protruding single-stranded 5'-termini. A stem-loop structure formed by the TIRs at the denaturation stage of the PCR is shown in the left lower part of the Figure. Grey rectangles mark U3, R, and U5 parts of LTRs. The arrows on the polyacrylamide gel indicate differential bands. B) Results of SDDIR for two seminomas and normal testicular parenchyma. Arrows indicate differential bands. S1, S2 – two different seminoma samples; N1 – normal testicular parenchyma obtained from a patient with S1 seminoma.

    contains both a sequence complementary to SmrtRD oligonucleotide and the sequence of the PL-1 primer and can be PCR amplified with an smr22 and PL-1 pair of primers.

2. Digestion of the ds cDNA amplicons with a frequent cutter restriction endonuclease (Fig. 7A, step 2) to represent each transcript by only one short cDNA restriction fragment. The choice of AluI in this case was determined by the peculiarities of the restriction sites distribution within the LTR consensus sequence.

3. Ligation of the fragments obtained to stem-loop forming PA adapters (Fig. 7A, step 3) which leave 5'-single-stranded protruding termini after the ligation. The ss-termini were filled in to double-stranded with Taq DNA-polymerase resulting in the terminal inverted repeats formation.

4. Selective PCR amplification of the LTR-containing fragments using PL-1 and PA-1 primers (Fig. 7A, step 4a). The amplification of all other fragments is suppressed by double stranded stems formed by the terminal inverted repeats (Fig. 7A, step 4b) at annealing during PCR. The stability of the stems is enhanced due to high GC content of the 40 bp long PA adapter preventing PA-1 primer binding. The PCR suppression was shown to ensure the efficient amplification of only those individual genomic fragments that contain the target sequences. The amplicons obtained were re-amplified with nested PL-2 and PA-2 primers.

5. Comparison of the final amplification products spectra obtained for different tissues using side-by-side PAGE separation (Fig. 7A, step 5).

6. Isolation of differential bands, their re-amplification with the same primers, sequencing and analysis by comparison with existing databases (Fig. 7A, step 6).

SDDIR was used to compare LTR-containing transcripts from seminomas and their counterpart normal tissue. Two seminomas from different individuals and one normal parenchyma were analyzed. An example of the differential display is presented in Figure 7B. It can be seen that the expression patterns of seminoma and normal testicular parenchyma are distinctly different, some of the transcripts being up-regulated whereas the others down-regulated in the tumor as compared to the normal tissue. Some of the differential transcripts were common for both tumors but many of the differences were specific for one or another tumor. Bands on displays for particular tissues had different intensities, probably due to different abundance of individual LTR-containing transcripts.

Isolation of the material from differential bands, their PCR amplification and sequencing of about 60 differential fragments obtained demonstrated that all of them contained LTR fragments at one of their termini, thus confirming that a subset of the LTR-containing fragments was specifically amplified. Such a high specificity was achieved due to PCR suppression that was shown to ensure the efficient amplification of only those individual DNA fragments that contain the target sequence.[89]

The differential transcription revealed by SDDIR was additionally tested by RT-PCR for 12 randomly sampled transcripts. Based on the known cDNA sequences, unique primers were designed for the PCR amplification. The test confirmed the SDDIR data for 11 transcripts. In one case RT-PCR revealed no difference in transcription levels in normal and tumor tissues. The data demonstrated that the specificity of the technique (i.e., the percentage of correctly identified cDNAs) was about 90%.

RT-PCR analysis of 12 randomly picked up transcripts allowed to evaluate their amounts relative to β-actin mRNA. It was concluded that the transcripts fall into a group of low abundant RNAs with copy numbers of 0.5–15 per cell.

SDDIR avoids the use of arbitrary primers which cause non-specific background and are responsible for low reproducibility of classical differential display approaches. The method also seems simpler than the protocol called RFLP-coupled domain directed differential display applied to the isolation of the MADS-box genes.[82] Due to the PCR suppression effect one can amplify repeat-containing RNA fragments practically at one go, because amplification of the fragments free of repeats is suppressed.

Therefore, SDDIR seems to be a highly specific tool for analysis of repeat-containing RNAs. The technique allows one to achieve almost 100% specific amplification of repeat-containing clones and about 90% specific identification of differentially expressed repeat-containing clones even for low abundant (0.1-0.5% of β-actin RNA) RNAs.

## Concluding Remarks

The human genome sequencing was a tremendous scientific achievement that opened new horizons to further development of life sciences, this time to an understanding how an orchestra of a huge number of genes is conducted and functions both at the level of single cells and multi-cellular organisms. However, a long way towards a deep understanding of coordinated regulation of gene expression is still ahead. First steps on this way are inevitably connected with accumulation of comprehensive catalogues of genes that are switched off and on, or up and down regulated in the processes of cell development, differentiation, cell responses to environmental changes etc. The EST data base (http://www.ncbi.nlm.nih.gov/dbEST), Transfac data base (http://transfac.gbf.de/homepage/databases/transfac/transfac.html), Gene ontology (http://www.geneontology.org) and others are the first attempt to compile such catalogues. Clearly, this task is by far more complex than just genome sequencing, however laborious is the latter. An enormous complexity of the task becomes more evident if to remember that the total number of cells in the adult human body is about $10^{14}$, and $10^{16}$ cell divisions are required to reach this number, and that there are 200 major types of somatic cells, and that it is very difficult to obtain stem cells in quantities sufficient for analysis, and that some cells appear during development for only a very limited period of time, and some regulatory transcripts exist in less than 1 copy per cell, etc., etc. It became also evident that even all the existing techniques will hardly allow to resolve this task. Therefore new methods and principally new approaches to the investigation of human and other transcriptomes are needed. It is known that there is a cyclic interplay between science and engineering: science discovers new phenomena, which are then used to build new instruments which in turn stimulate new science, and so on. The advent of genomic era has put so many questions that they can not be answered with the currently available technologies. On the other hand, even these technologies give so much data that one does not know how to handle them.

While it is currently unclear how to answer these questions, let us hope that a great technical and informational revolution will open new prospects for the life science research.

### Acknowledgments

The author thank Boris O. Glotov and Eugene D. Sverdlov for critical reading of the manuscript and valuable comments. We also apologize to all colleagues whose works were not cited due to a broad scope of this review and limited space. This work was partially supported by INTAS 991-1143, Physico-Chemical Biological Program of the Russian Academy of Sciences, and the Russian Foundation for Basic Research 2006.20034 grants.

## References

1. Vinogradova T, Leppik L, Kalinina E et al. Selective Differential Display of RNAs containing interspersed repeats: analysis of changes in the transcription of HERV-K LTRs in germ cell tumors. Mol Genet Genomics 2002; 266:796-805.
2. Vinogradova TV, Leppik LP, Nikolaev LG et al. Solitary human endogenous retroviruses-K LTRs retain transcriptional activity in vivo, the mode of which is different in different cell types. Virology 2001; 290:83-90.
3. Pollack JR, Iyer VR. Characterizing the physical genome. Nature Genet 2002; 32(Suppl):515-521.
4. Lockhart DJ, Winzeler EA. Genomics, gene expression and DNA arrays. Nature 2000; 405:827-836.
5. Burrell MM. Construction of cDNA libraries. Methods Mol Biol 1996; 58:199-209.
6. Suzuki Y, Sugano S. Construction of full-length-enriched cDNA libraries. The oligo-capping method. Methods Mol Biol 2001; 175:143-153.

7. Belyavsky A, Vinogradova T, Rajewsky K. PCR-based cDNA library construction: general cDNA libraries at the level of a few cells. Nucleic Acids Res 1989; 17:2919-2932.

8. Endege WO, Steinmann KE, Boardman LA et al. Representative cDNA libraries and their utility in gene expression profiling. Biotechniques 1999; 26:542-548, 550.

9. McCarrey JR, Williams SA. Construction of cDNA libraries from limiting amounts of material. Curr Opin Biotechnol 1994; 5:34-39.

10. Ko MS. An 'equalized cDNA library' by the reassociation of short double- stranded cDNAs. Nucleic Acids Res 1990; 18:5705-5711.

11. Patanjali SR, Parimoo S, Weissman SM. Construction of a uniform-abundance (normalized) cDNA library. Proc Natl Acad Sci USA 1991; 88:1943-1947.

12. Bonaldo MF, Lennon G, Soares MB. Normalization and subtraction: two approaches to facilitate gene discovery. Genome Res 1996; 6:791-806.

13. Ji H, Liu YE, Jia T et al. Identification of a breast cancer-specific gene, BCSG1, by direct differential cDNA sequencing. Cancer Res 1997; 57:759-764.

14. Schummer M, Ng WV, Bumgarner RE et al. Comparative hybridization of an array of 21,500 ovarian cDNAs for the discovery of genes overexpressed in ovarian carcinomas. Gene 1999; 238:375-385.

15. Ohara O, Nagase T, Ishikawa K et al. Construction and characterization of human brain cDNA libraries suitable for analysis of cDNA clones encoding relatively large proteins. DNA Res 1997; 4:53-59.

16. Ono M. Molecular cloning and long terminal repeat sequences of human endogenous retrovirus genes related to types A and B retrovirus genes. J Virol 1986; 58:937-944.

17. Blond Jl, Beseme F, Duret L et al. Molecular characterization and placental expression of HERV-W, a new human endogenous retrovirus family. J Virol 1999; 73:1175-1185.

18. La Mantia G, Maglione D, Pengue G et al. Identification and characterization of novel human endogenous retroviral sequences prefentially expressed in undifferentiated embryonal carcinoma cells. Nucleic Acids Res 1991; 19:1513-1520.

19. Simon M, Haltmeier M, Papakonstantinou G et al. Transcription of HERV-K-related LTRs in human placenta and leukemic cells. Leukemia 1994; 8(Suppl 1):S12-17.

20. Baust C, Seifarth W, Germaier H et al. HERV-K-T47D-Related long terminal repeats mediate polyadenylation of cellular transcripts. Genomics 2000; 66:98-103.

21. Baust C, Seifarth W, Schon U et al. Functional activity of HERV-K-T47D-related long terminal repeats. Virology 2001; 283:262-272.

22. Kelleher CA, Wilkinson DA, Freeman JD et al. Expression of novel-transposon-containing mRNAs in human T cells. J Gen Virol 1996; 77:1101-1110.

23. Eisen MB, Spellman PT, Brown PO et al. Cluster analysis and display of genome-wide expression patterns. Proc Natl Acad Sci USA 1998; 95:14863-14868.

24. Holloway AJ, van Laar RK, Tothill RW et al. Options available—from start to finish—for obtaining data from DNA microarrays II. Nature Genet 2002; 32 Suppl:481-489.

25. Slonim DK. From patterns to pathways: gene expression data analysis comes of age. Nature Genet 2002; 32 Suppl:502-508.

26. Chuaqui RF, Bonner RF, Best CJ et al. Post-analysis follow-up and validation of microarray experiments. Nature Genet 2002; 32(Suppl):509-514.

27. Tamayo P, Slonim D, Mesirov J et al. Interpreting patterns of gene expression with self-organizing maps: methods and application to hematopoietic differentiation. Proc Natl Acad Sci USA 1999; 96:2907-2912.

28. Liotta L, Petricoin E. Molecular profiling of human cancer. Nature Rev Genet 2000; 1:48-56.

29. Golub TR, Slonim DK, Tamayo P et al. Molecular classification of cancer: class discovery and class prediction by gene expression monitoring. Science 1999; 286:531-537.

30. Ross DT, Scherf U, Eisen MB et al. Systematic variation in gene expression patterns in human cancer cell lines. Nature Genet 2000; 24:227-235.

31. Alizadeh AA, Eisen MB, Davis RE et al. Distinct types of diffuse large B-cell lymphoma identified by gene expression profiling. Nature 2000; 403:503-511.

32. Reinke V. Functional exploration of the C. elegans genome using DNA microarrays. Nature Genet 2002; 32 Suppl:541-546.

33. Byers RJ, Hoyland JA, Dixon J et al. Subtractive hybridization—genetic takeaways and the search for meaning. Int J Exp Pathol 2000; 81:391-404.

34. Morrish DW, Linetsky E, Bhardwaj D et al. Identification by subtractive hybridization of a spectrum of novel and unexpected genes associated with in vitro differentiation of human cytotrophoblast cells. Placenta 1996; 17:431-441.
35. Sager R. Tumor suppressor genes in the cell cycle. Curr Opin Cell Biol 1992; 4:155-160.
36. Louro ID, Bailey EC, Ruppert JM. Suppression subtractive hybridization for identification and functional analysis of tumor suppressor genes. Methods Mol Biol 2003; 222:453-462.
37. Hedrick SM, Cohen DI, Nielsen EA et al. Isolation of cDNA clones encoding T cell-specific membrane-associated proteins. Nature 1984; 308:149-153.
38. Duguid JR, Dinauer MC. Library subtraction of in vitro cDNA libraries to identify differentially expressed genes in scrapie infection. Nucleic Acids Res 1990; 18:2789-2792.
39. Sargent TD, Dawid IB. Differential gene expression in the gastrula of Xenopus laevis. Science 1983; 222:135-139.
40. Davis MM, Cohen DI, Nielsen EA et al. Cell-type-specific cDNA probes and the murine I region: the localization and orientation of Ad alpha. Proc Natl Acad Sci USA 1984; 81:2194-2198.
41. Saiki RK, Gelfand DH, Stoffel S et al. Primer-directed enzymatic amplification of DNA with a thermostable DNA polymerase. Science 1988; 239:487-491.
42. Hubank M, Schatz DG. Identifying differences in mRNA expression by representational difference analysis of cDNA. Nucleic Acids Res 1994; 22:5640-5648.
43. Milner JJ, Cecchini E, Dominy PJ. A kinetic model for subtractive hybridization. Nucleic Acids Res 1995; 23:176-187.
44. Wan JS, Sharp SJ, Poirier GM et al. Cloning differentially expressed mRNAs. Nature Biotechnol 1996; 14:1685-1691.
45. Ermolaeva OD, Lukyanov SA, Sverdlov ED. The mathematical model of subtractive hybridization and its practical application. Proc Int Conf Intell Syst Mol Biol 1996; 4:52-58.
46. Wang SM, Fears SC, Zhang L et al. Screening poly(dA/dT)- cDNAs for gene identification. Proc Natl Acad Sci USA 2000; 97:4162-4167.
47. Diatchenko L, Lau YF, Campbell AP et al. Suppression subtractive hybridization: a method for generating differentially regulated or tissue-specific cDNA probes and libraries. Proc Natl Acad Sci USA 1996; 93:6025-6030.
48. Diatchenko L, Lukyanov S, Lau YF et al. Suppression subtractive hybridization: a versatile method for identifying differentially expressed genes. Methods Enzymol 1999; 303:349-380.
49. Lin CT, Sargan DR. A method for generating subtractive cDNA libraries retaining clones containing repetitive elements. Nucleic Acids Res 1997; 25:4427-4428.
50. Zhang L, Zhou W, Velculescu VE et al. Gene expression profiles in normal and cancer cells. Science 1997; 276:1268-1272.
51. Velculescu VE, Vogelstein B, Kinzler KW. Analysing uncharted transcriptomes with SAGE. Trends Genet 2000; 16:423-425.
52. Saha S, Sparks AB, Rago C et al. Using the transcriptome to annotate the genome. Nature Biotechnol 2002; 20:508-512.
53. Hermeking H. Serial analysis of gene expression and cancer. Curr Opin Oncol 2003; 15:44-49.
54. Velculescu VE. Essay: Amersham Pharmacia Biotech & Science prize. Tantalizing transcriptomes— SAGE and its use in global gene expression analysis. Science 1999; 286:1491-1492.
55. Velculescu VE, Madden SL, Zhang L et al. Analysis of human transcriptomes. Nature Genet 1999; 23:387-388.
56. Ye SQ, Usher DC, Zhang LQ. Gene expression profiling of human diseases by serial analysis of gene expression. J Biomed Sci 2002; 9:384-394.
57. Caron H, van Schaik B, van der Mee M et al. The human transcriptome map: clustering of highly expressed genes in chromosomal domains. Science 2001; 291:1289-1292.
58. Liang P, Pardee AB. Differential display of eukaryotic messenger RNA by means of the polymerase chain reaction. Science 1992; 257:967-971.
59. Welsh J, Chada K, Dalal SS et al. Arbitrarily primed PCR fingerprinting of RNA. Nucleic Acids Res 1992; 20:4965-4970.
60. Liang P, Averboukh L, Pardee AB. Distribution and cloning of eukaryotic mRNAs by means of differential display: refinements and optimization. Nucleic Acids Res 1993; 21:3269-3275.
61. Martin KJ, Pardee AB. Principles of differential display. Methods Enzymol 1999; 303:234-258.
62. Zhao S, Ooi SL, Pardee AB. New primer strategy improves precision of differential display. Biotechniques 1995; 18:842-846, 848, 850.

63. Afonina I, Zivarts M, Kutyavin I et al. Efficient priming of PCR with short oligonucleotides conjugated to a minor groove binder. Nucleic Acids Res 1997; 25:2657-2660.

64. Jurecic R, Nachtman RG, Colicos SM et al. Identification and cloning of differentially expressed genes by long- distance differential display. Anal Biochem 1998; 259:235-244.

65. Wang X, Feuerstein GZ. Direct sequencing of DNA isolated from mRNA differential display. Biotechniques 1995; 18:448-453.

66. Corton JC, Gustafsson JA. Increased efficiency in screening large numbers of cDNA fragments generated by differential display. Biotechniques 1997; 22:802-804, 806, 808.

67. McClelland M, Honeycutt R, Mathieu-Daude F et al. Fingerprinting by arbitrarily primed PCR. Methods Mol Biol 1997; 85:13-24.

68. Shen R, Su ZZ, Olsson CA et al. Identification of the human prostatic carcinoma oncogene PTI-1 by rapid expression cloning and differential RNA display. Proc Natl Acad Sci USA 1995; 92:6778-6782.

69. Topper JN, Cai J, Qiu Y et al. Vascular MADs: two novel MAD-related genes selectively inducible by flow in human vascular endothelium. Proc Natl Acad Sci USA 1997; 94:9314-9319.

70. Carulli JP, Artinger M, Swain PM et al. High throughput analysis of differential gene expression. J Cell Biochem Suppl 1998; 31:286-296.

71. Callard D, Lescure B, Mazzolini L. A method for the elimination of false positives generated by the mRNA differential display technique. Biotechniques 1994; 16:1096-1097, 1100-1093.

72. Li F, Barnathan ES, Kariko K. Rapid method for screening and cloning cDNAs generated in differential mRNA display: application of northern blot for affinity capturing of cDNAs. Nucleic Acids Res 1994; 22:1764-1765.

73. Matz MV, Lukyanov SA. Different strategies of differential display: areas of application. Nucleic Acids Res 1998; 26:5537-5543.

74. Kato K. RNA fingerprinting by molecular indexing. Nucleic Acids Res 1996; 24:394-395.

75. McClelland M, Mathieu-Daude F, Welsh J. RNA fingerprinting and differential display using arbitrarily primed PCR. Trends Genet 1995; 11:242-246.

76. Bertioli DJ, Schlichter UH, Adams MJ et al. An analysis of differential display shows a strong bias towards high copy number mRNAs. Nucleic Acids Res 1995; 23:4520-4523.

77. Kato K. Description of the entire mRNA population by a 3' end cDNA fragment generated by class IIS restriction enzymes. Nucleic Acids Res 1995; 23:3685-3690.

78. Ivanova NB, Belyavsky AV. Identification of differentially expressed genes by restriction endonuclease-based gene expression fingerprinting. Nucleic Acids Res 1995; 23:2954-2958.

79. Prashar Y, Weissman SM. Analysis of differential gene expression by display of 3' end restriction fragments of cDNAs. Proc Natl Acad Sci USA 1996; 93:659-663.

80. Suzuki H, Yaoi T, Kawai J et al. Restriction landmark cDNA scanning (RLCS): a novel cDNA display system using two-dimensional gel electrophoresis. Nucleic Acids Res 1996; 24:289-294.

81. Matz M, Usman N, Shagin D et al. Ordered differential display: a simple method for systematic comparison of gene expression profiles. Nucleic Acids Res 1997; 25:2541-2542.

82. Fischer A, Saedler H, Theissen G. Restriction fragment length polymorphism-coupled domain-directed differential display: a highly efficient technique for expression analysis of multigene families. Proc Natl Acad Sci USA 1995; 92:5331-5335.

83. Stone B, Wharton W. Targeted RNA fingerprinting: the cloning of differentially-expressed cDNA fragments enriched for members of the zinc finger gene family. Nucleic Acids Res 1994; 22:2612-2618.

84. Johnson SW, Lissy NA, Miller PD et al. Identification of zinc finger mRNAs using domain-specific differential display. Anal Biochem 1996; 236:348-352.

85. Broude NE, Chandra A, Smith CL. Differential display of genome subsets containing specific interspersed repeats. Proc Natl Acad Sci USA 1997; 94:4548-4553.

86. Dominguez O, Ashhab Y, Sabater L et al. Cloning of ARE-containing genes by AU-motif-directed display. Genomics 1998; 54:278-286.

87. Lavrentieva I, Broude NE, Lebedev Y et al. High polymorphism level of genomic sequences flanking insertion sites of human endogenous retroviral long terminal repeats. FEBS Lett 1999; 443:341-347.

88. Birch PR. Targeted differential display of abundantly expressed sequences from the basidiomycete Phanerochaete chrysosporium which contain regions coding for fungal cellulose-binding domains. Curr Genet 1998; 33:70-76.

89. Chenchik A, Diachenko L, Moqadam F et al. Full-length cDNA cloning and determination of mRNA 5' and 3' ends by amplification of adaptor-ligated cDNA. Biotechniques 1996; 21:526-534.

## CHAPTER 10

# Phylogeny of Human Endogenous and Exogenous Retroviruses

**Aris Katzourakis and Michael Tristem**

## Abstract

T he human genome contains a number of human endogenous retrovirus (HERV) families, each resulting from a single germ-line infection. Here we combine information on all previously described HERV families, and further search the human genome sequencing project databases to add taxa to taxon poor areas of the retroviral phylogeny. We use the highly conserved retroviral reverse transcriptase (RT) motif to construct phylogenies and provide the most comprehensive phylogeny reported to date, consisting of 352 HERVs from 31 HERV families. This phylogeny was used to determine the relationships of HERVs to exogenous retroviruses, and to endogenous retroviruses from other taxa. We confirm that most HERV families are not closely related to exogenous retroviruses. Of the 31 families, 23 are assigned to Class I, 4 to Class II and 4 to Class III. Furthermore, we characterize seven novel HERV families and further characterize two for which only partial information was previously available. Three of the HERV proviruses described here are particularly interesting as they may represent some of the oldest integrations into the primate lineage, with one element having integrated between 62 and 100 millions of years ago.

## Introduction

The human genome contains a wide variety of endogenous retroviruses (HERVs) and, until recently, they were typically identified by homology to exogenous mammalian retroviruses using Southern hybridization, via PCR, or even by chance during analysis of human loci. However, with the advent of the human genome sequencing project, it is not surprising that they are now being largely identified through the use of computer assisted searches of nucleotide databases such as GenBank.[1,2] The HERVs have been broadly subdivided into three classes, on the basis of sequence similarity to exogenous retroviruses (see Chapter 5). Class I HERVs display sequence similarity to the gammaretroviruses (mammalian type C retroviruses), and class II to betaretroviruses and alpharetroviruses (mammalian type B and D, and avian type C retroviruses). The more recently identified class III HERVs are distantly related to the spumaviruses.[2] These three major groupings have, in turn, been further subdivided into families on the basis of sequence similarity, and on the similarity of their primer binding sites (PBSs) to host tRNAs. For example, the HERV-L family contains a PBS similar to tRNA$^{Leu}$, whereas tRNA$^{Trp}$ primes the HERV-W family. It should be noted that the use of the term 'family' to describe these HERV lineages is not used in its strict taxonomic sense, given that the *Retroviridae* as a whole have been assigned family status. Rather, it is used to refer to a group of

*Retroviruses and Primate Genome Evolution*, edited by Eugene D. Sverdlov.
©2005 Eurekah.com.

HERVs that, probably, arise from a single cross-species infection of the human genome. This nomenclature is problematic for a number of reasons. Clearly distinguishable groups of retroviruses sometimes utilize the same PBS, such as the HERV-K elements, that (on the basis of sequence divergence) may form up to ten distinct groups.[3] Furthermore, some highly divergent retroviruses are primed by the same tRNA, such as HERV-R, ERV-9 and HERV-R type (b), which are all primed by tRNA$^{Arg}$, but form three distinct clades in the HERV phylogeny.[1] Another difficulty is that available sequence information has often been derived from different regions of the retroviral genome (sometimes not including the region with obvious PBS homology). Furthermore, it is also difficult to distinguish genuinely monophyletic HERV families, and those that only appear monophyletic due to the relative lack of sequence information from nonhuman endogenous retroviruses. HERV families may appear to be monophyletic only because related viruses in other hosts have yet to be identified, while in fact they are derived from more than one germ-line infection.

A more systematic framework for the classification of retroviruses is now feasible, both because of the isolation of endogenous retroviruses from many different vertebrate taxa,[2,4,5] and due to the large amount of sequence data that is becoming available as a result of the human genome sequencing project. As of October 2001 (at the time we initiated the searches described below), almost two billion nucleotides, accounting for 60% of the human genome were in the finished format within the 'nr' (nonredundant) database. An additional two billion nucleotides were also available in unfinished format from the high throughput databases (htgs).[6,7] The first study in which novel retroviral families were identified, by computer-based screening of 7% of the human genome, reported 22 HERV families, six of which were novel and four of which were previously only partially characterized.[1] Benit et al[2] identified a further four new families based on more recent screening of the human genome using Env rather than reverse transcriptase (RT).

Here we present an updated phylogenetic study of the HERVs, (based on the build 28 version of the human genome) using all the HERV families identified in the study of Tristem[1] together with the novel families identified by Bénit[2] and a potential new family suggested by Andersson.[3] Rather than adding new HERV sequences to all of the existing families, we concentrated on identifying novel HERVs in taxon-poor areas of the tree. We also set out to identify and characterize additional members of HERV families which were (i) weakly supported by bootstrap analysis in previous studies (indicating the potential presence of more than one family in that region of the tree) or (ii) known from only a single (or small number) of elements.

We used a conserved section of the *pol* gene encoding RT to search the human genome database for HERVs with which to build phylogenies. It has been shown that trees derived from RT display much higher bootstrap scores than trees derived from the *env* gene,[2] possibly due to the greater constraint imposed by conservation of RT enzymatic function. This, therefore, allows more robust phylogenies to be constructed.

The construction of ever more inclusive phylogenies from both endogenous and exogenous retroviruses provides a very important source of information for the study of retroviral evolution as a whole. Phylogenies carry direct imprints of the processes that have generated them, namely lineage birth and extinction. Such phylogenies have been used previously to infer information about horizontal transmission and coevolution.[5,8] Retroviral phylogenies can also provide insights into the evolution of their hosts, by serving as clade markers.[9,10] Furthermore the potential of retroviruses to infer evolutionary events in the genomes in which they are contained is only now beginning to be realized, with studies such as that by Hughes and Coffin,[11] who used retroviral phylogenies to infer genomic rearrangements. Retroviruses may even reveal information about key events in the evolution of modern lineages.[12]

# Identification and Phylogenetic Reconstruction of Endogenous Retroviruses in the Human Genome

## *Identification of HERVs within Sequence Data Banks*

The novel sequences used in this study were obtained from the EMBL/GenBank/DDBJ databases at the end of October 2001. The databases were screened using an amino acid query sequence consisting of domains 1 to 7 of the RT protein as described by Xiong and Eickbush.[13] The prototypic element from each taxon-poor HERV family reported by Tristem[1] (those clades containing less than five members), was used in a TBLASTN database search.[14] The new HERV families identified by Bénit[2] and Andersson[3] were also included in the search.

Novel elements, found to encode most of the appropriate region of the RT protein, were added to the alignment previously produced by Tristem.[1] Elements were added sequentially, in decreasing order of similarity to the element used in the TBLASTN search. This was done until a known member of another family was found. Space constraints and the length of time required to build the resultant phylogenies meant that, generally, a maximum of five taxa were added to each family. Hence the number of elements presented in the phylogenies below is not exhaustive.

When more than one cosmid contained identical HERV sequences, we kept only one of the sequences (occasionally the same region of the genome has been sequenced several times and has been entered into the databases under more than one accession number).

## *Alignment and Phylogenetic Reconstruction*

The novel elements identified using the method described above were aligned to known members of the same family by eye. This is simplified by the conserved nature of the RT protein and its domain structure. The full dataset contained 352 taxa, 117 of which were identified during this study, with 167 HERVs and 68 nonhuman endogenous retroviruses taken from other studies.[1-3] The total number of sequences in this alignment is less than the total number of HERVs that have been sequenced to date, as we selectively added sequences to groups with few members, and did not attempt to include all the HERVs. Neighbor-Joining trees of the whole dataset were constructed using PAUP[15] and the protpars matrix.[16] Bootstrap support scores were obtained from 1000 replicates.

Maximum parsimony trees were generated from a reduced version of the dataset due to the long computation times required. Neighbor-Joining trees indicated several well-supported branches containing numerous members derived from a single family. When these branches showed at least 95% bootstrap support, they were pruned, removing all but three of the taxa. This reduced the number of taxa in the dataset from 352 to 177. We employed the searching strategy described by Quicke[17] on this reduced dataset, using an unordered matrix. We initially performed 20,000 random additions with tree bisection and reconnection (TBR) while holding one tree in memory during each replicate. The shortest tree was subsequently used in a heuristic search with TBR until 10,000 equally parsimonious trees were identified and this pool of trees was used to reweight the data matrix using the rescaled consistency index. The last tree identified in the previous step was used as the starting tree for a heuristic search with the reweighted matrix, and the last equally parsimonious tree was again used in a heuristic search with all the characters equally weighted. This led to no further reduction in tree length. A strict consensus of the 10,000 equally parsimonious shortest trees saved in the heuristic search was computed. This is a sample of all the equally parsimonious trees possible for this dataset. Bootstrap support scores were computed from 100 replicates, with 100 random additions with TBR while holding one tree in memory for each replicate.

### Paired Long Terminal Repeats (LTR) Identification, Primer Binding Sites and Genomic Organization

Paired LTRs were located using the BLAST to BLAST program[18] to compare two 10 kb regions upstream and downstream of the *pol* sequence used in our phylogenies. Paired structures separated by between 2 and 10 kb, that were between 300 and 1000 bp in length, were then investigated for the presence of characteristic LTR features, such as short inverted repeats, promoter and polyadenylation signals. We also searched for short identical repeats bordering the LTRs, as well as a polypurine tract (PPT) immediately preceding the 3' LTR.

The percentage divergence between the two paired LTRs was determined using the method employed by Tristem.[1] The entire alignable length was used, excluding regions containing deletions. We corrected the divergence figures for multiple mutations at the same site, back mutations and convergent substitutions using the Kimura 2-parameter model,[19] as implemented in PAUP.[15] Two estimates of actual time of integration were provided, using two alternative estimates of the mutation rate of the human genome. The faster rate was $2.1 \times 10^{-9}$ nucleotides per synonymous site per year[20] and the slower rate was $1.3 \times 10^{-9}$.[21] Both estimates rely on molecular clocks calibrated from the divergence date of human and Old World monkeys of 27.5 million years ago (Mya).[22] RepeatMasker (Smit and Green, unpublished results) was used to check that sequences identified as LTRs did not correspond to nonretroviral repetitive elements such as short interspersed nuclear elements (SINEs).

When an LTR pair was identified, we attempted to identify the PBS region, as HERV families have often been classified according to the similarity of their PBS to specific types of host tRNA. This was done by comparing the sequence immediately 3' of the 5' LTR against a tRNA sequence database.[23] The genomic organization of the prototypic member of each family identified in this report was determined by locating certain motifs that are typically found in retroviral proteins. These include the major homology region (MHR) and Cys-His box in Gag, the protease (PR) and Integrase (IN) motifs in Pol, and the transmembrane domain (TM) in Env. This search was performed using the TBLASTN option in the BLAST to BLAST program[18] against the Gag, Pol and Env proteins of exogenous retroviruses, and then searching for conserved motifs by eye. The exogenous retroviruses used were gibbon ape leukemia virus (GaLV), Jaagsiekte sheep virus, mouse mammary tumor virus (MMTV) and walleye dermal sarcoma virus (WDSV), depending on the phylogenetic placement of the provirus investigated. In addition, the provirus was translated in all three frames and used to search the conserved domain database[24] with reverse position specific BLAST, a variant of the PSI-BLAST program.[14]

## HERV Lineages within Retroviral Phylogeny

The phylogenies shown in Figures 1 and 2 (based on Maximum parsimony and Neighbor-Joining analyses respectively), showed a number of distinct well-supported HERV lineages within the retroviral phylogeny. This is consistent with previous results[1,2] and occurs because all members of a particular HERV family are likely to be derived from only a few rounds of viral replication. Thus they remain relatively closely related, even though they may have been endogenized for long periods.[25]

Despite this, it is still difficult to determine the actual number of HERV families present within a phylogeny, due to sampling biases. These biases lead to the apparent monophyly of what are actually multiple HERV families, because counterparts of each family remain unidentified in nonhuman hosts. This means that simply counting the number of HERV clades within a phylogeny will lead to an underestimate of the number of HERV families present.

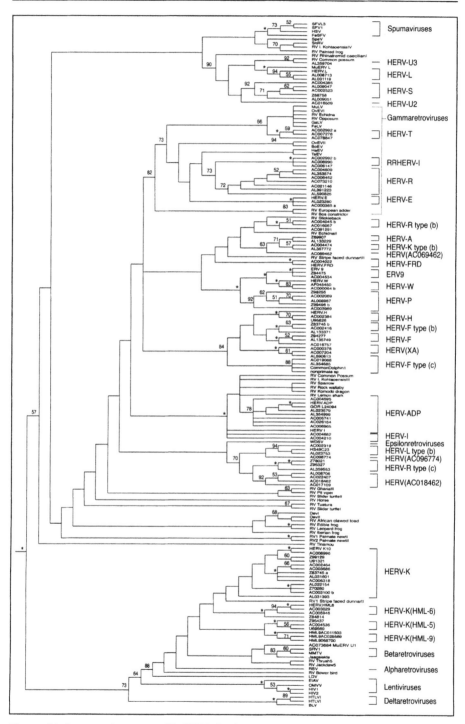

Figure 1. Maximum parsimony tree of the HERV families. Strict consensus of 10,000 equally parsimonious trees, based on domains 1 to 7 of RT. Numbers before nodes indicate bootstrap values where greater than 50%, asterisks indicate bootstrap values above 95%.

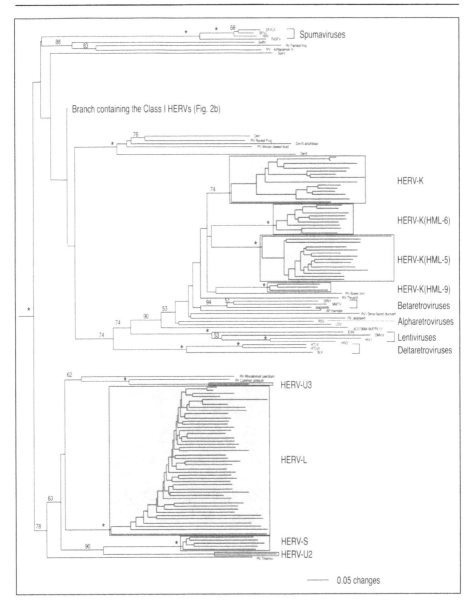

Figure 2A. Neighbour joining tree, with numbers before nodes indicating bootstrap values where greater than 50%, and asterisks indicating bootstrap values above 95%. The grey boxes indicate HERV families. Bootstrap scores within HERV families are not shown.

In order to estimate the actual number of HERV families, we made the assumptions used by Tristem.[1] First, HERV clades containing elements with more than one PBS homology were assumed to have been derived from separate horizontal transmission events, and were therefore split into separate families. Second, HERVs which encoded the same PBS but which were polyphyletic with respect to viruses derived either from nonhuman hosts, or with viruses with alternate PBS homologies, were regarded as separate families. Finally, HERV families that were

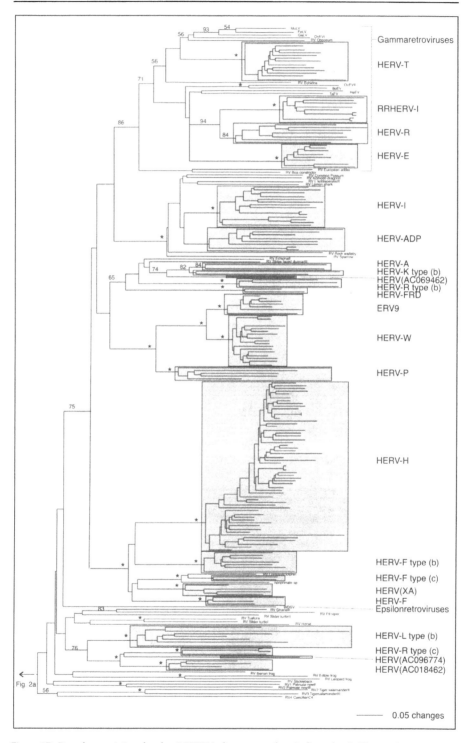

Figure 2B. Branch containing the class I HERVs, bootstrap values indicated as in Figure 2A.

either paraphyletic to nonhuman viruses, or to HERVs with alternate PBS homologies, were also assumed to be independently derived.

We infer the presence of 31 HERV families in the human genome on the basis of the above criteria, this being the most extensive survey of HERV diversity to date. Tristem[1] originally identified 22 families, and Bénit[2] proposed a further 3 families. We provide evidence for six more families, one of which has been suggested by Andersson.[3] Two of these were identified by splitting the HERV(Z69907) family into three families. HERV(Z69907) was previously known from two members, located on cosmids Z69907 and AC004474, which clustered with robust bootstrap support in Neighbor-Joining trees but were, nevertheless, still separated by long branch lengths.[1] Figures 1 and 2 show that there are now five HERV elements clustering in the region of the phylogenies where HERV(Z69907) is located. We were able to find paired LTRs and hence characterize the PBS affinity of two proviruses in this clade, henceforth known as HERV-K type (b) (AL133229) and HERV-A (AL357772). Although bootstrap support is weak, we tentatively assign the partially characterized AC004474 provirus to the HERV-K type (b) family, and the uncharacterized Z69907 provirus (both from Tristem[1]) to the HERV-A family. The provirus on AC069462 is paraphyletic to these two families. We could not determine the PBS affinity for this HERV, so we named this family HERV(AC069462). Thus the HERV(Z69907) family has been split into HERV-K type (b), HERV-A and HERV(AC069462).

A further three families were identified by splitting the HERV(HS49C23) family from Tristem[1] into four families by both the addition of taxa, and subsequently, determining the PBS homology for two of the proviruses. The HERV(HS49C23) family was previously known from five elements that clustered together with relatively weak bootstrap support, and displayed relatively long branch lengths. The Neighbor-Joining phylogeny shown in Figure 2 now contains 15 elements in the region of the tree where HERV(HS49C23) is located, and they cluster together with 76% bootstrap support. One sub-branch of this clade contained seven of the 15 elements, including HERV(HS49C23), clustered together with 100% bootstrap support. The three taxa used in the parsimony analysis also showed a strong bootstrap score of 94%. A PBS with sequence similarity to tRNA$^{Leu}$ was determined for the provirus on cosmid AC087240, so we termed this clade HERV-L type (b). The position of the provirus on cosmid AC096774 varied depending on the method of analysis, being paraphyletic to HERV-L type (b) using parsimony, but paraphyletic to HERV-R type (c) using Neighbor-Joining. Due to the uncertainty surrounding its phylogenetic placement, we tentatively assigned this element into a new family, termed HERV(AC096774) (we were unable to find any LTRs for this element). The other seven of the 15 elements located in this region clustered into two well supported clades. Two elements within one of the clades contained paired LTRs and a PBS with 17 out of 17 matches to tRNA$^{Arg}$. Hence this family was designated HERV-R type (c). The other clade was named HERV(AC018462) as we were again unable to find any paired LTRs. We consider this a separate family because its topology also varied in the different analyses (it was paraphyletic to HERV(AC096744) in Neighbor-Joining, but a sister taxon to HERV-R type (c) in maximum parsimony). All the elements in this region of the tree were highly defective, and were mostly separated by very long branch lengths. It is therefore likely that the addition of more taxa to this region of the tree will lead to the identification of further HERV families.

The existence of the sixth family, HERV-K(HML-9), was originally suggested by Andersson,[3] but was not previously shown to represent a distinct family based on phylogenetic criteria. HERV-K(HML-9) is separated from the other class II HERVs by RV bower bird in the tree shown in Figure 2, and further investigation revealed that it has a different PBS affinity to many of the HERV-K(HML) families, having tRNA$^{Cys}$ PBS rather than tRNA$^{Lys}$.

In addition to these six new families, our phylogenies confirmed the presence of well supported retroviral clades with multiple copies in some families that were previously known only from one member: HERV-R (7 copies), HERV-R type (b) (9 copies), HERV-F (3 copies),

HERV-T previously known as HERV-S71,[2] (13 copies) and HERV(XA) (4 copies). We also identified more members for families previously known from only a small number of representatives: HERV-K(HML-5) (14 copies), HERV-K(HML-6) (9 copies), HERV-ADP (8 copies) and HERV-F type (b) (7 copies). We did not find any additional members of the HERV- FRD family, which is known from only two members.

## Characterization of New Families

### HERV-K type (b)

LTRs were identified for a member of the family within cosmid AL357772. The element is approximately 9.9 kb in length, and the distance between the 3' end of the 5' LTR and the start of *pol* was unusually long at approximately 3.8 kb. This suggests that the AL357772 element contains a 1.7 kb insertion that is not present within the Z69907 provirus. However, we were unable to find any obvious database matches to this 1.7 kb region. The provirus within cosmid AL357772 could be aligned to the provirus on Z69907 across 3.3 kb with 77% nucleotide similarity. The alignment was not continuous throughout the length of the provirus, being fragmented in *gag* and only matching a 300-nucleotide region at the end of *env*. A putative MHR and a Cys-His motif were identified. These motifs are present within most retroviral *gag* genes, suggesting that HERV-K type (b) encodes Gag, although it has multiple stop codons and frameshift mutations. The provirus also appears to contain a *pol* gene, encoding regions of RT, IN and RNaseH, though no region with similarity to PR could be identified. A TM motif, a characteristic of the retroviral *env* gene was also identified. The TM motif was, however, interrupted by an Alu element, identified using RepeatMasker (Smit & Green, unpublished results). The LTRs were 485 nucleotides (nt) in length, and contained characteristic promoter and polyadenylation motifs. The observed divergence between the paired LTRs was 18.6%, and the corrected figure was 21.6%. The estimated dates of integration for the two different rates of mutation that we used were 51 and 83 million years ago. The PBS showed 14 of 18 matches to human lysine tRNA, making it the only class I HERV to be primed by this tRNA, and hence care must be taken to avoid confusion with the class II HERV-K families.

### HERV-A

This family contains the previously described HERV(Z69907) element. The prototypic HERV-A provirus is located on cosmid AL133229. It has a smaller genome (approximately 6.2 kb) than its sister family, HERV-K type (b). A Cys-His box was identified indicating the presence of the *gag* gene, although there was no obvious indication of an MHR. The *pol* gene appeared to have a typical retroviral structure, containing motifs associated with PR, RT and IN. HERV-A does not appear to have an *env* gene, with the 3' LTR being located immediately downstream of *pol*. In contrast, the provirus within cosmid Z69907 contains a defective *env* gene, suggesting that *env* has been deleted in the prototypic HERV-A element. The LTRs were 485 nt in length being the same length as the LTRs on HERV-K type (b), and polyadenylation and promoter motifs were identified. The putative LTR identified by Tristem[1] for HERV(Z69907) is unlikely to be correct as it showed no homology to the LTRs on AL133229. The observed divergence between the LTRs was 20%, and the corrected divergence was 23.8%, giving an estimated date of integration of between 57 and 92 million years ago. The PBS showed 17 of 18 matches to a bovine tRNA$^{Ala}$.

### HERV(AC069462)

No LTRs could be identified for HERV(AC069462) so the PBS affinity and integration date of this provirus could not be determined. Putative MHR and Cys-His box motifs were found, indicating the presence of *gag*. The *pol* gene, while displaying multiple stop-codons and frameshift mutations, encoded motifs associated with all the major products encoded by *pol*, namely PR, RT, RNaseH and IN. There was no obvious indication of an *env* gene.

## HERV-L type (b)

We found paired LTRs for only one member of this clade (which contains the original prototypic element HS49C23), within the provirus on cosmid AC087240. This provirus is approximately 9 kb long. The 5' LTR is 439 bp in length, and is bordered by the short inverted repeats TG and CA. The 3' LTR is partial, aligning with the 5' LTR over 340 bp, and is preceded by a PPT of 15 bp in length. The PBS immediately follows the 5'LTR and displays 16 out of 18 matches to human leucine tRNA. This makes it the only known class I provirus with a PBS homology to tRNA$^{Leu}$ and we therefore name this family HERV-L type (b). A putative Cys-His box was identified but there was no indication of an MHR. An amino acid motif showing sequence similarity to IN is present within *pol* but there was no obvious indication of PR. We were also unable to find recognizable motifs within *env* on the prototypic HERV-L type (b) but another member of the family (on cosmid AC048351) contained both the TM domain characteristic of *env* and a PR motif.

The new prototypic HERV-L type (b) element could be aligned to the provirus on cosmid HS49C23 over 5.6 kb with 79% nucleotide identity. The LTRs showed a divergence of 21.5% across the 379 bp that could be aligned, and the corrected divergence was 25.9%. This corresponds to an estimated integration date of between 62 and 100 million years ago, making this one of the earliest known HERV integration events; almost certainly present in the primate lineage since its origin.

## HERV-R type (c)

Paired LTR-like structures, 586 nt in length, were identified in two members of this family. The first pair, present on cosmid Z95327, showed 8.2% divergence and 8.75% corrected divergence, corresponding to an estimated date of integration of between 21 and 34 million years ago. A Cys-His box was identified in *gag*. IN and RNaseH motifs were also present within *pol*, although there was no obvious indication of a PR protein. The provirus also encodes Env, as a TM domain was identified. A PBS displaying 17 of 17 matches to human arginine tRNA was located adjacent to the 5' LTR, as was a 20 bp PPT immediately upstream of the 3' LTR. The two LTRs were unusually far apart, at over 16 kb, and the distance between the 3'end of the 5'LTR and the start of RT was 8.5 kb, suggesting that there has been a large insertion around the *gag* region. Furthermore, the 3'LTR contained a 1 kb insertion. The 5'LTR of Z95327 was used (via a BLAST search) to identify paired LTRs within a second member of the family present within cosmid AL359553 (a comparison between the LTRs derived from Z95327 and AL359553 indicated 84% identity). The element was 8.6 kb in length, hence lacking the large insertion 5' of RT present in Z95327 and had an identical PBS to the element within cosmid Z95327.

## HERV(AC018462)

No LTRs were found for the members of this family, and the elements comprising it could not be aligned over any great length, with the notable exception of the elements derived from cosmids AC018462 and AC022407. These elements showed 90% identity across 11 kb. The retroviral *pol* gene was present in both elements, containing the PR, IN and RNaseH motifs, but there was no evidence of *gag* or *env* genes. Furthermore, the lack of LTRs raises the possibility that the two elements represent a segmental duplication of a defective element. We failed to find any motifs associated with genes other than *pol* in any of the elements comprising this clade.

### HERV(AC096774)

The only member of this family is highly defective, only containing a *pol* gene (IN, RNaseH, and RT motifs were identified). Sequences related to long interspersed nuclear element (LINE) and SINE elements were found both upstream and downstream of the defective *pol* gene, suggesting that the provirus has been degraded by nonERV related transposable element insertions.

## Further Characterization of Previously Reported Families

### HERV-K(HML-9)

We identified an additional member of the class II HERVs, HERV-K(HML-9), based on our phylogenetic criteria. To date the characterization of this family has been based on 244 nt from the RT region. We identified complete LTRs for two members, with a third member containing only a partial 3' LTR. The 5' and 3' LTRs of AC011503 were 421 and 420 nt long respectively, and their similarity (6.4% divergence, 6.8% corrected) suggested a relatively recent integration date of between 16 and 26 Mya. This element was found to be 8.5 kb in length, and showed a genomic organization similar to HERV-K and HERV-K(HML-6). There appeared to be no large deletions in *gag* or *pol*, with *gag* encoding a Cys-His box and an MHR motif and *pol* displaying PR, RT, IN and RNaseH motifs. A reading frame showing similarity to dUTPase was located between the *gag* and *pol* genes. The *env* gene displayed similarity to MMTV, HERV-K and Jaagsiekte *env* in BLAST based searches. The other element with complete LTRs (within cosmid AC025569) is likely to have had a similar integration date. Both elements contained a PBS with 16 out of 18 matches to bovine tRNA$^{Cys}$. However, for continuity we have retained the original name HERV-K(HML-9) for this family, rather than renaming it HERV-C.

### HERV-F type (c)

We identified an additional member of the HERV-F type (c) family, first described by Bénit[2] on cosmid AL354685. This family was further characterized as only a section of the RT and TM regions have been investigated to date.[2] We found this family to cluster with nonprimate endogenous retroviruses with strong bootstrap support in both Neighbor-Joining and maximum parsimony trees. The provirus was approximately 8 kb long and bordered by short paired LTRs 376 and 370 bp in length, showing the characteristic inverted repeats TG and CA. The LTRs showed a divergence of 5.9% (6.2% corrected) corresponding to integration dates of between 15 and 24 Mya. The PBS showed 18 of 18 matches to a phenylalanine tRNA. The provirus showed typical retroviral organization, displaying the MHR and Cys-His box motifs within *gag*, the PR, IN and RNaseH motifs of *pol* and the TM domain of *env*. In addition, no appreciable deletions were found in any of these genes, and there were large uninterrupted coding regions of 2109 bp in *pol*, 1413 bp in *gag* and 1413 bp in *env*, suggesting that this provirus may still be capable of producing some retroviral protein products.

## Relationships between HERVs and Exogenous Retroviruses

The congruence between the phylogenies of HERV families, and the existing division of HERVs into the three classes, varied between the two phylogenetic reconstruction methods. Of the 31 families in our phylogenies, 23 belong to class I, 4 to class II and 4 to class III HERVs. The class I HERVs are most closely related to the gammaretroviruses, such as feline leukemia virus (FeLV) and gibbon ape leukemia virus (GaLV) (see Introduction and Chapter 5). Several HERV families appear to belong to this retroviral genus, namely RRHERV-I, HERV-R, HERV-E and HERV-T (previously known as HERV-S71).[1] In particular, HERV-T is closely related to FeLV and GaLV. A clade containing the four HERV families as well as nonhuman retroviruses, clustered together with strong bootstrap support (82% in maximum parsimony trees, 86% in neighbor joining trees). All the other HERV families in class I are more closely related to other HERV families, or to endogenous retroviruses derived from nonhuman taxa, than they are to the gammaretroviruses. The families derived from what was previously HERV(HS49C23) (now families HERV-L type (b), HERV-R type (c), HERV(AC096774) and HERV(AC018462)) are particularly unusual in that they are most closely related to a clade containing nonmammalian vertebrate retroviruses such as RV-Pit Viper and RV-Slider Turtle.

The class II HERVs are classified as being most similar to the betaretroviruses, such as simian retrovirus type 1 (SRV-1) or MMTV, or to alpharetroviruses such as Rous sarcoma virus (RSV). The clade containing the class II HERVs was placed as a sister taxon to the betaretroviruses (SRV-1 and MMTV) in Neighbor-Joining analyses. The avian leukosis viruses like RSV appear to be more distantly related to the HERV class II viruses. No endogenous retroviruses of the exogenous lentivirus or deltaretrovirus clades were identified.

Class III HERVs are distantly related to the spumaviruses, such as human spumavirus (HSV). The spumaviruses appear to be more closely related to endogenous retroviruses derived from nonmammalian vertebrates, clustering together with the RV-Painted Frog and caecilian retroviruses (RV-I KohtaoensisIV). This spumavirus/nonmammalian ERV clade clusters with the class III families in the parsimony tree although with less than 50% bootstrap support. In the Neighbor-Joining analyses the spumaviruses were placed in a clade containing endogenous retroviruses from other species, the whole clade forming an outgroup to the other retroviruses in the phylogeny.

This study is consistent with previous reports suggesting that HERV classification on the basis of relatedness to exogenous retroviruses needs to be revised[1,2] but probably not until all the HERV families in the human genome have been identified. The existing classifications should therefore not, (with the exception of the one or two cases mentioned above), be thought of as implying close relationships between HERVs and exogenous retroviruses. However, it remains possible that exogenous retroviruses that are more closely related to HERV families remain to be identified.[26]

It is likely that the number of HERV families identified in the human genome so far is an underestimate for a number of reasons. Firstly, some families have a sufficiently low copy number that they might not have been detected in the fraction of the human genome that has been studied so far. We have not identified any further members for HERV-FRD (despite our search being based on 60% of the human genome), raising the possibility that a number of HERV families may have very low copy numbers. Two of the new families that we present in this report are also known from only one member. As a result, it is possible that other families with a single or a very small copy number remain to be identified in the human genome. Second, heterochromatin, which accounts for approximately 8% of the human genome, and centromeric regions, are very poorly represented in the sequence databases.[27] One of the main reasons that these regions are hard to sequence is that they contain a large amount of repetitive sequence. As a result, they may contain a higher proportion of retroviral sequence than the euchromatic chromosomal regions, and may contain families that have not yet been discovered. Finally, the sampling of nonhuman viruses in several regions of the tree is relatively poor, compared to the HERVs. It is therefore possible that a number of the families that appear monophyletic in this phylogeny are actually polyphyletic, but that they have not been split due to the lack of nonhuman data. For example, Andersson[3] has suggested that HERV-K contains 10 distinct groups, based on sequence divergence. In this study, we have split HERV-K into four groups based on our phylogenetic criteria: HERV-K, HERV-K(HML-5), HERV-K(HML-6) and HERV-K(HML-9). It is possible that additional nonhuman retroviral sequences will split the class II HERVs into more families.

When the PBS affinity of a HERV cannot be determined, a variety of other designations have been used, such as a nearby gene (e.g.HERV-ADP), an amino acid motif present within the sequence, (e.g., HERV-FRD) or the cosmid number of the prototypic member. It is now obvious that many distantly related families share the same tRNA primer, examples being class I and II HERVs both being primed by tRNA$^{Lys}$ and Class I and III HERVs being primed by tRNA$^{Leu}$. This has resulted in the rather cumbersome use of an arbitrary differentiation between the families based on an alphabetical character, for example the three families that are primed by arginine are referred to as HERV-R type (a), (b) and (c). Once all the families in the human genome have been identified, it would be preferable to use a system of classification based on phylogeny. As with any classification system, this one may also have to change, particularly as the sampling of endogenous retroviruses from nonhuman taxa may never be fully complete, and the emergence of new sequences may further split HERV families.

## Ancient and Modern HERV Families

We have identified three HERV proviruses in this study that may represent some of the oldest integration events into the primate lineage. The HERV-K type (b) prototypic element probably integrated between 51 and 83 Mya, the HERV-A prototypic element is between 57 and 92 million years old, and the HERV-L type (b) prototype is between 62 and 100 million years old. Whilst all three of these elements may well be present in the same genomic locations in many different primate species, it is possible that the HERV-L type (b) element has been present since the origins of the primate lineage. This is based on fossil evidence (discussed in Chapter 3) for the date of divergence between strepsirrhines (lemurs and lorises) and haplorhines (tarsiers and anthropoids) which is thought to have occurred approximately 57 Mya.[28] However, it should be noted that more recent work, taking into account the incompleteness of the fossil record, suggests an older date for the origin of primates of 81.5 Mya.[29] This would mean that the HERV-L type (b) element could have integrated into the primate lineage after the divergence of strepsirrhines and haplorhines (depending on the actual rate of neutral substitution). Furthermore, the hominid lineage is thought to have undergone a slowdown in its rate of mutation (see refs. 20, 30 and discussion in Chapter 4). As a result, it is possible that the ages of particular retroviral insertions are being overestimated. Dates of integration such as the older figure of 100 Mya are even more prone to problems of calibration, as mutation rates in early primates are unlikely to have been the same as those in extant species.

We provide evidence for the existence of four families (HERV-L type (b), HERV-R type (c), HERV(AC096774) and HERV(AC018462)) in the region of the phylogenies that was previously thought to contain a single family, HERV(HS49C23)[1] All the elements in this region of the tree were very degraded, coding for few short ORFs and, where LTRs could be identified, they gave the oldest dates of integration relative to other HERV families. In addition, the branch including all four of these families is the most basal Class I HERV branch, and is actually more closely related to the exogenous epsilonretroviruses (e.g., WDSV) than to the gammaretroviruses in the maximum parsimony trees (though not in the neighbor-joining trees). The ancient integration dates of the families, coupled with their basal positions, raise the possibility that they represent the remains of an early period of HERV evolution. They show little similarity to exogenous retroviruses, other HERV families or even to each other. It is therefore possible that they represent multiple families that integrated over a time-scale of tens of millions of years. Thus, they only appear monophyletic due to the lack of other retroviruses in this region of the tree. It is possible that the addition of more HERV sequences (or retroviral sequences from other taxa) will further split this branch, but the degraded nature of the elements described so far suggests that we may be nearing the 'look back time' limit for detecting ancient HERV families.

**Table 1. Previously identified families used in this study**

| Family | Genomic Organization | Original Reference | |
|---|---|---|---|
| HERV-H | LTR-*gag-pol-env*-LTR | Mager & Henthorn,[48] Hirose et al[49] | K01891, D11079 |
| HERV-L | LTR-*gag-pol*-DUT-LTR | Cordonnier et al[50] | X89211 |
| HERV-K | LTR-*gag*-DUT-*pol-env*-LTR | Ono et al[51] | M14123 |
| HERV-W | LTR-*gag-pol-env*-LTR | Perron et al[52], Blond et al[36] | AF009668 |
| HERV-E | LTR-*gag-pol-env*-LTR | Repaske et al[53] | M10976 |
| HERV-I | LTR-*gag-pol-env*-LTR | Maeda & Kim[54] | M92067 |
| ERV-9 | LTR + part *gag-pol-Δenv*-LTR* | La Mantia et al[55] | X57147 |
| HERV-K(HML-5) | LTR-*Δgag*-DUT-*pol-env*-LTR | Medstrand & Blomberg[56] | U35161 |
| HERV-K(HML-6) | LTR-*gag*-DUT-*pol-env*-LTR | Mesdstrand et al[57] | HSU60268 HSU60270 |
| HERV-FRD | LTR-ORF?-*gag-pol-env*-LTR | Seifarth et al[58] | U27240 |
| RRHERV-I | LTR-*Δpol-env*-LTR* | Kannan et al[59] | M64936 |
| HERV-R | LTR + part *gag* + part *pol* + *env*-LTR* | O' Connell et al[60] | M12140 |
| HERV-R type (b) | LTR-*gag-pol-env*-LTR | Tristem [1] | AC004045 |
| HERV-ADP | LTR-*gag-pol-env*-LTR | Lyn et al[61] | L14752 |
| HERV-P | LTR + part *pol* + LTR* | Kröger and Horak[62] | AC002069 |
| HERV-T | LTR-*gag-Δpol-env*-LTR | Werner et al[63] | M32788 |
| HERV(XA) | LTR-*gag-Δpol-Δenv*-LTR | Widegren et al[64] | U29659 |
| HERV-S | LTR-*gag-pol-env*-LTR | Tristem,[1] | AC004385 |
| HERV-F | LTR-*gag-pol-env*-LTR | Tristem, [1] | Z94277 |
| HERV.U2 | LTR-*pol-env*-LTR* | Bénit et al[2] | AC016509 |
| HERV.U3 | LTR-*pol-env*-LTR* | Bénit et al[2] | AL359704 |
| HERV-F type (b) | LTR-*gag-Δpol-Δenv*-LTR | Tristem[1] | AC002416 |
| HERV-F type (c) | *pol-env*\* | Bénit et al[2] | AC020617 |
| HERV-K(HML-9) | part *pol*\* | Andersson et al[3] | AF016001 |
| HERV(Z69907) | *gag-pol-Δenv*-LTR* | Tristem[1] | Z69907 |
| HERV(HS49C23) | *Δgag-pol-Δenv*\* | Tristem[1] | HS49C23 |

*Entries followed by an asterisk indicate that the proviral sequence has not been fully characterized, and indicate the sequence that is known.

The majority of the elements described here are defective, containing multiple stop-codons and frameshifts, with a number of elements containing large insertions and deletions. This is in line with previous studies.[1,25,31] The HERV-K(HML-2) family, a member of the Class II group has received a lot of attention, as it appears to contain large almost intact reading frames in *gag*, *pol* and *env*. One member of this family carries only a single inactivating point mutation in the *pol* gene.[32] More recent work has shown that this point mutation is polymorphic in humans, with two out of seven individuals carrying the intact virus.[33] Beyond HERV-K(HML-2), no proviruses intact in all three retroviral genes have been found, although a complete *env* gene has been found in HERV-H[34,35] and HERV-W.[36] The HERV-F type (c) element that we describe on cosmid AL354685 is the most intact Class I provirus described to date, with a complete *env* gene and few mutations in *gag* and *pol*; *gag* contains a single stop codon while *pol* contains a frameshift and a stop codon.

The results of the research described in this chapter are summarized in Tables 1 – 3.

***Table 2. HERV families described in this study***

| Family | Accession Number | Position Orientation | Primer tRNA | Genomic Organization | Chromosome Location |
|---|---|---|---|---|---|
| HERV-K type (b) | AL357772 | 40560(-) | Lys | LTR-ins1-*gag*-Δ*pol*-*env*ᵃ-LTR | 9p22 |
| HERV-A | AL133229 | 39710(+) | Ala | LTR-*gag*-*pol*-LTR | 20q12 |
| HERV (AC069462) | AC069462 | 86659(+) | ND | *gag*-*pol* | 3p25 |
| HERV-L type (b) | AC087240 | 13315(+) | Leu | LTR-??-Δ*pol*-??-LTR | 12p13 |
| HERV (AC018462) | AC018462 | 123477(+) | ND | *pol* | 2p13 |
| HERV (AC096774) | AC096774 | 143980(+) | ND | Δ*pol* | 4q31+1 |
| HERV-R type (c) | AL359553 | 52520(-) | Arg | LTR-*gag*-Δ*pol*-env-LTR | 1q11 |
| HERV-K(HML-9) | AC011503 | 99975(-) | Cys | LTR-*gag*-DUT-*pol*-*env*ᵇ-LTR | 19p12 |
| HERV-F type (c) | AL354685 | 44152(+) | Phe | LTR-*gag*-*pol*-env-LTR | Xq21 |

a- *env* gene contains an Alu insertion
b- the provirus on AC025569 did not contain an *env* gene, and the provirus on AC068700 contained reading frames coding for an undetermined protein.
ND-None detected

## Concluding Remarks: Prospects for HERV Phylogenetics

The phylogenies that we present are based on the highly conserved RT gene. As such, it is important to note that they represent the evolutionary history of that gene and, by extension, that of the retroviruses which contain it. Exogenous retroviruses have been known to recombine,[37] and past recombination events between endogenous and exogenous retroviruses have been described previously.[38] The role of recombination between HERVs has not yet been fully assessed, but a recent phylogenetic study comparing trees derived using the relatively labile retroviral *env* gene, with those derived from the RT gene, revealed that although the two phylogenies are broadly congruent, there are differences. For example, the HERV-F type (b) family, which appears closely related to the HERV-H family in phylogenies based on RT, clusters with the HERV-E/HERV-R/RRHERV-I clade in the trees derived from the *env* gene.[2] No single gene carries the full evolutionary history of the organism that carries it and, to get a complete picture of the evolutionary history of the HERVs, one would have to consider the evolutionary history of all of their genes. This is problematic with the HERVs, as the *gag* gene is highly divergent and can only be aligned over the relatively short MHR and Cys-His box. Furthermore, a number of retroviral families entirely lack the *env* gene. Nevertheless, the RT region of *pol* genes and *env* genes could be used together to infer the evolutionary history of retroviruses that carry both of these genes using reconciled trees, following the approach set out by Page.[39,40]

The production of ever more complete retroviral phylogenies due to the completion of the human genome sequencing opens up the possibility of using more powerful tests of macroevolutionary hypotheses than has been possible to date.[41-44] These tests can be used to reveal information about the long-term evolutionary biology and retrotranspostional dynamics of endogenous retroviruses, by comparing these phylogenies against null models in order to describe the branching process. The existence of a time scale facilitated by the molecular nature of

***Table 3. Families made redundant in this study and their replacements***

| Redundant Family Name | Replacement Family Name |
| --- | --- |
| HERV(Z69907) | HERV-A |
| | HERV-K type (b) |
| | HERV(AC069462) |
| HERV(HS49C23) | HERV-L type (b) |
| | HERV(AC096774) |
| | HERV-R type (c) |
| | HERV(AC018462) |

HERV data, and the presence of LTRs allowing the estimation of integration dates, allows the use of tests that can determine the demographic history of HERVs.[45-47] We are currently using data-mining tools and HERV data, in combination with these recently developed tests, to better understand the evolution of these elements in the primate lineage.

## Acknowledgements

This work was funded by the Wellcome Trust, a NERC studentship to A.K, and by the Onassis foundation. We thank R. Gifford and L. Bénit for unpublished sequence data, and J. Martin for comments on the manuscript.

## References

1. Tristem M. Identification and characterisation of novel human endogenous retrovirus families by phylogenetic screening of the human genome mapping project database. J Virol 2000; 74:3715-3730.
2. Benit L, Dessen P, Heidmann T. Identification, phylogeny, and evolution of retroviral elements based on their envelope genes. J Virol 2001; 75:11709-11719.
3. Andersson ML, Lindeskog M, Medstrand P et al. Diversity of human endogenous retrovirus class II-like sequences. J Gen Virol 1999; 80:255-260.
4. Herniou E, Martin J, Miller K et al. Retroviral diversity and distribution in vertebrates. J Virol 1998; 7:5955-5966.
5. Martin J, Herniou E, Cook J et al. Interclass transmission and phyletic host tracking in murine leukaemia virus related retroviruses. J Virol 1999; 73:2442-2449.
6. Beck S, Sterk P. Genome scale DNA sequencing: Where are we? Curr Opin Biotechnol 1998; 9:116-120.
7. Beck S, Sterk P. Genome monitoring table, www2.ebi.ac.uk/genomes/mot/.2001.
8. van der Kuyl AC, Dekker JT, Goudsmit J. Distribution of baboon endogenous virus among species of african monkeys suggests multiple ancient cross-species transmissions in shared habitats. J Virol 1995; 69:7877-7887.
9. Johnson WE, Coffin JM. Constructing primate phylogenies from ancient retrovirus sequences. Proc Natl Acad Sci USA 1999; 96:10254-10260.
10. Cook JM, Tristem M. 'SINEs of the times' - transposable elements as clade markers for their hosts. Trends Ecol Evol 1997; 12:295-297.
11. Hughes JF, Coffin JM. Evidence for genomic rearrangements mediated by human endogenous retroviruses during primate evolution. Nature Genet 2001; 29:487-489.
12. Sverdlov ED. Retroviruses and primate evolution. BioEssays 2000; 22:161-171.
13. Xiong Y, Eickbush TH. Origin and evolution of retroelements based upon their reverse transcriptase sequences. EMBO J 1990; 9:3353-3362.
14. Altschul SF, Madden TL, Schaffer AA et al. Gapped BLAST and PSI-BLAST: A new generation of protein database search programs. Nucleic Acids Res 1997; 25:3389-3402.

15. Swofford DL, PAUP*. Phylogenetic analysis using parsimony and other methods. 4 ed. Sunderland, MA: Sinauer Associates, 1998.
16. Felsenstein J. PHYLIP (Phylogeny inference package). 3.5c ed. Seattle: Department of Genetics, University of Washington, 1993.
17. Quicke DLJ, Taylor J, Purvis A. Changing the landscape: A new strategy for the estimation of large phylogenies. Syst Biol 2001; 50:60-66.
18. Tatusova TA, Madden TL. BLAST 2 SEQUENCES, a new tool for comparing protein and nucleotide sequences. FEMS Microbiol Lett 1999; 174:247-250.
19. Kimura M. A simple method for estimating evolutionary rates of base substitutions through comparative studies of nucleotide sequences. J Mol Evol 1980; 16:116-120.
20. Li W-H, Tanimura M. The molecular clock runs more slowly in man than in apes and monkeys. Nature 1987; 336:93-96.
21. Miyamoto MM, Goodman M. DNA systematics and evolution of primates. Annu Rev Ecol Syst 1990; 21:197-220.
22. Purvis A. A composite estimate of primate phylogeny. Philos Trans R Soc Lond Ser B-Biol Sci 1995; 348:405-421.
23. Sprinzl M, Horn C, Brown M et al. Compilation of tRNA sequences and sequences of tRNA genes. Nucleic Acids Res 1998; 26:148-153.
24. Marchler-Bauer A, Panchenko AR, Shoemaker BA et al. CDD: A database of conserved domain alignments with links to domain three-dimensional structure. Nucleic Acids Res 2002; 30:281-283.
25. Boeke JD, Stoye JP. Retrotransposons, endogenous retroviruses, and the evolution of retroelements. In: Retroviruses. Coffin JM, Hughes SH, Varmus HE, eds. New York: CSHL Press, 1997:343-435.
26. Lower R, Lower J, Kurth R. The viruses in all of us: Characteristics and biological significance of human endogenous retrovirus sequences. Proc Natl Acad Sci USA 1996; 93:5177-5184.
27. Lander ES, Linton LM, Birren B et al. Initial sequencing and analysis of the human genome. Nature 2001; 409:860-921.
28. Gingerich PD, Uchen MD. Time of origin of primates. J Hum Evol 1994; 27:443-445.
29. Tavaré S, Marshall CR, Wil O et al. Using the fossil record to estimate the age of the last common ancestor of extant primates. Nature 2002; 416:726-729.
30. Bailey WJ, Fitch DHA, Tagle DA et al. Molecular evolution of the psi-eta-globin gene locus - gibbon phylogeny and the hominoid slowdown. Mol Biol Evol 1991; 8:155-184.
31. Wilkinson DA, Mager DL, Leong JC. Endogenous human retroviruses. In:The retroviridae,vol III. Levy JA, ed. NY: Plenum Press, 1994:465-535.
32. Mayer J, Sauter M, Racz A et al. An almost-intact human endogenous retrovirus K on human chromosome 7. Nature Genet 1999; 21:257-258.
33. Reus K, Mayer J, Sauter M et al. Genomic organization of the human endogenous retrovirus HERV- K(HML-2.HOM) (ERVK6) on chromosome 7. Genomics 2001; 72:314-320.
34. Lindeskog M, Mager DL, Blomberg J. Isolation of a human endogenous retroviral HERV-H element with an open env reading frame. Virology 1999; 258:441-450.
35. Jern P, Lindeskog M, Karlsson D et al. Full-length HERV-H elements with env SU open reading frames in the human genome. AIDS Res Hum Retrovir 2002; 18:671-676.
36. Blond JL, Beseme F, Duret L et al. Molecular characterization and placental expression of HERV-W, a new human endogenous retrovirus family. J Virol 1999; 73:1175-1185.
37. Zhang JY, Sapp CM. Recombination between two identical sequences within the same retroviral RNA molecule. J Virol 1999; 73:5912-5917.
38. Benson SJ, Ruis BL, Fadly AM et al. The unique envelope gene of the subgroup J avian leukosis virus derives from ev/J proviruses, a novel family of avian endogenous viruses. J Virol 1998; 72:10157-10164.
39. Page RD. Maps between trees and cladistic-analysis of historical associations among genes, organisms, and areas. Syst Biol 1994; 43:58-77.
40. Page RD. GeneTree: Comparing gene and species phylogenies using reconciled trees. Bioinformatics 1998; 14:819-820.
41. Harvey PH, Leigh Brown AJ, Maynard Smith J et al. New uses for new phylogenies. Oxford: Oxford University Press, 1996.
42. Mooers A, Heard SB. Evolutionary process from phylogenetic tree shape. Q Rev Biol 1997; 72:31-54.

43. Nee S. Inferring speciation rates from phylogenies. Evolution 2001; 55:661-668.
44. Purvis A. Using interspecific phylogenies to test macroevolutionary hypotheses. In: Harvey PH, Leigh Brown AJ, Maynard Smith J, Nee S, eds. New uses for new phylogenies. Oxford: OUP, 1996:153-168.
45. Nee S, Barraclough TG, Harvey PH. Temporal changes in biodiversity: Detecting patterns and identifying causes. In: Gaston KJ, ed. Biodiversity: A biology of numbers and difference. Oxford: Blackwell Science, 1996:230-252.
46. Pybus OG, Harvey PH. Testing macro-evolutionary models using incomplete molecular phylogenies. Proc R Soc Lond Ser B-Biol Sci 2000; 267:2267-2272.
47. Paradis E. Detecting shifts in diversification rates without fossils. Am Nat 1998; 152:176-187.
48. Mager DL, Henthorn PS. Identification of a retrovirus-like repetitive element in human DNA. Proc Natl Acad Sci USA 1984; 81:7510-7514.
49. Hirose Y, Takamatsu M, Harada F. Presence of env genes in members of the RTLV-H family of human endogenous retrovirus-like elements. Virology 1993; 192:52-61.
50. Cordonnier A, Casella JF, Heidmann T. Isolation of novel human endogenous retrovirus-Like elements with foamy virus-related pol sequence. J Virol 1995; 69:5890-5897.
51. Ono M, Yasunaga T, Miyata T et al. Nucleotide sequence of human endogenous retrovirus genome related to the mouse mammary tumor virus genome. J Virol 1986; 60:589-598.
52. Perron H, Garson JA, Bedin F et al. Molecular identification of a novel retrovirus repeatedly isolated from patients with multiple sclerosis. Proc Natl Acad Sci USA 1997; 94:7583-7588.
53. Repaske R, Steele PE, Oneill RR et al. Nucleotide sequence of a full-length human endogenous retroviral segment. J Virol 1985; 54:764-772.
54. Maeda N, Kim HS. Three independent insertions of retrovirus-like sequences in the haptoglobin gene-cluster of primates. Genomics 1990; 8:671-683.
55. La Mantia G, Maglione D, Pengue G et al. Identification and characterization of novel human endogenous retroviral sequences preferentially expressed in undifferentiated embryonal carcinoma cells. Nucleic Acids Res 1991; 19:1513-1520.
56. Medstrand P, Blomberg J. Characterization of novel reverse-transcriptase encoding human endogenous retroviral sequences similar to type A and type B retroviruses - differential transcription in normal human tissues. J Virol 1993; 67:6778-6787.
57. Medstrand P, Mager DL, Yin H et al. Structure and genomic organization of a novel human endogenous retrovirus family: HERV-K (HML-6). J Gen Virol 1997; 78:1731-1744.
58. Seifarth W, Skladny H, Kriegschneider F et al. Retrovirus-like particles released from the human breast-cancer cell-line T47-D display type-B-related and type-C-related endogenous retroviral sequences. J Virol 1995; 69:6408-6416.
59. Kannan P, Buettner R, Pratt DR et al. Identification of a retinoic acid-inducible endogenous retroviral transcript in the human teratocarcinoma-derived cell-line PA-1. J Virol 1991; 65:343-6348.
60. O'connell C, O'brien S, Nash WG et al. Erv3, a full-length human endogenous provirus - chromosomal localization and evolutionary relationships. Virology 1984; 138:225-235.
61. Lyn D, Deaven LL, Istock NL et al. The polymorphic ADP-ribosyltransferase (NAD+) pseudogene-1 in humans interrupts an endogenous pol-like element on 13q34. Genomics 1993; 18:206-211.
62. Kroger B, Horak I. Isolation of novel human retrovirus-related sequences by hybridization to synthetic oligonucleotides complementary to the tRNAPro primer-binding site. J Virol 1987; 61:2071-2075.
63. Werner T, Brackwerner R, Leib-Mösch C et al. S71 is a phylogenetically distinct human endogenous retroviral element with structural and sequence homology to simian sarcoma virus (SSV). Virology 1990; 174:225-238.
64. Widegren B, Kjellman C, Aminoff S et al. The structure and phylogeny of a new family of human endogenous retroviruses. J Gen Virol 1996; 77:1631-1641.

# CHAPTER 11

# Evolutionary Aspects of Human Endogenous Retroviral Sequences (HERVs) and Disease

**Jonas Blomberg, Dmitrijs Ushameckis and Patric Jern**

## Abstract

Endogenous retroviruses (ERVs) are remnants of retroviral infections. ERVs preserve functions of exogenous retroviruses to various extents. ERVs are both parasites and symbionts. Although the most pathogenic elements are eliminated by selection, some pathogenicity may remain. Some recently endogenized elements of mice and cats are known to cause disease. The situation in humans is less certain. Diseases where a role for ERVs has been discussed are multiple sclerosis, schizophrenia, diabetes, systemic lupus erythematosus, seminoma, malignant melanoma, preeclampsia and azoospermia. Several pathogenic mechanisms have been implicated: Antigenic mimicry, immune dysregulation, receptor interference, growth stimulation by cis- or transactivation, loss of physiological functions mediated by retroviral genes, and gene loss by illegitimate recombination are among them.

In most cases, much work remains before a pathogenic mechanism is established. The biology of HERVs must be better understood in order to understand their role in human disease.

## Introduction

Virus-induced disease is a special case of the interaction between host and virus. An endogenous retrovirus (ERV) is part "virus", part "selfish gene"(see below), part "host gene", and part "sequence". The least pretentious aspect is the last one. In order to avoid overinterpretation we will therefore in the following use the term "endogenous retroviral sequences" (ERVs). This term solely reflects the structural similarity between sequences of exogenous (infectious) retroviruses (XRVs) and ERVs. ERVs occur not only in all vertebrates[1,2] but also the whole animal kingdom and in plants where they often are dominating genetic components. The interactions of ERVs with the host are manifold, and go beyond host-virus interactions as they are usually discussed. Endogenization of an XRV is an example of lateral gene transfer, and ERVs contribute in several ways to genetic diversity.[3,4] In an evolutionary context, disease equates negative selection. As the persistence of ERVs in a lineage can stretch over several hundred million years there is ample time to establish a stable host-viral sequence interaction, with a minimum of negative selection, i.e., disease. A fundamental question therefore is: Do ERVs induce disease?

The question is justified because of its medical importance. In general, mobile genetic elements (like ERVs) mainly cause disease by transposition.[5] It is a special case of genetic disease, i.e., disease arising from mutation in the host genome.

*Retroviruses and Primate Genome Evolution*, edited by Eugene D. Sverdlov.
©2005 Eurekah.com.

In a broad biological context, the answer is "yes, ERVs can cause disease". In an intensely studied species, mice, recently integrated ERVs like the AKR murine leukemia virus, and intensely transposing ERVs like intracisternal type A particle sequences (IAPs), are major causes of disease (for a review see ref. 6). In the former case, production of infectious virus leads to transduction of oncogenes and immune deficiency. In both cases transposition can knock out important genes and/or activate oncogenes. Consequently, the expected selection neutrality of ERVs may take a rather long time to develop.

The "pacification" of a retroviral gene is not straightforward. Retroviruses carry with them a number of *cis-* and *trans-*acting mechanisms optimized to suit a free-living exogenous virus, making them potentially dangerous to the host even after eons, when the integrated viral gene ("provirus", or "virogene") has been severely damaged. Eukaryotic cells probably have been exposed to retroelements since their very beginning. More than half of the human genome has probably undergone reverse transcription before incorporation.[3,7] Thus, many defence mechanisms against damage from newly acquired retroelements must exist. Some are known. Methylation[8] and inhibitory RNA[9] are two of them. Previous ERV integrations may protect against new ones in several ways.[10] However, in principle there are two major causes of disease; either i. there are adverse environmental influences, i.e., a conflict of interest between an invading microbe and the host, occasionally being negative for the host, or ii. there are imperfections in the host machinery itself, such disease being of a degenerative or chance nature. The subject of this essay has elements of both. We will discuss disease from the aspect of the interplay between host and retrovirus as a price for the evolutionary flexibility provided by the ERV.

We will also give a brief review of the status of the HERV-disease field. The provisional hybrid HERV nomenclature proposed by Andersson et al[11] will be used with slight modifications. Briefly, it is "HERV-X(clone or sequence name, group)", where X is the tRNA specificity of the primer binding site, and group a cluster name based on sequence similarity, if defined. This nomenclature is an attempt to join the previous PBS-based nomenclature[12] to a sequence similarity based one.[13]

## Animal ERVs and Disease

The first findings were made in chicken, mice and cats. During the 1930s a factor was found in some inbred mice that could transmit breast cancer (Mouse Mammary Tumor virus, MMTV).[14] It is now known to be a Betaretrovirus. During the 1970s it was shown that both MMTV and mouse leukemia virus (MLV; belonging to the Gammaretrovirus genus) exist both in infectious (exogenous) and hereditary (endogenous) forms. The former is contagious between individuals in the same generation (horizontal spread), the latter between subsequent generations (vertical spread). The pathogenic role of ERVs can be elaborate. In MLV leukemogenesis there is a stepwise sequence of recombinations with ERVs leading to the pathogenic virus.[6] In both the MMTV and MLV examples ERV expression is positively correlated with disease.

The disease in mice from Lake Casitas in California is another example. These wild mice consist of two populations.[15] One population gets a degenerative neurological disease, a myelitis. It is caused by a pathogenic variant of MLV. After infection early after birth, the infection becomes chronic, with viremia for several months. The other mouse population is resistant against the disease. The resistant mice have a genetic property (FV4), which is a defective endogenous MLV sequence. This apathogenic ERV produces an envelope glycoprotein which protects against the exogenous virus by binding to and blocking the same cell surface receptor that is used by the pathogenic exogenous MLV. The example shows i. that infection just after birth followed by a chronic viremia eventually can lead to disease, and ii. that an endogenous retroviral sequence can protect against an exogenous retrovirus. In this case ERV expression is negatively correlated with disease. The epidemiology of the disease is determined both by heredity and infection after contact between individuals (environment).

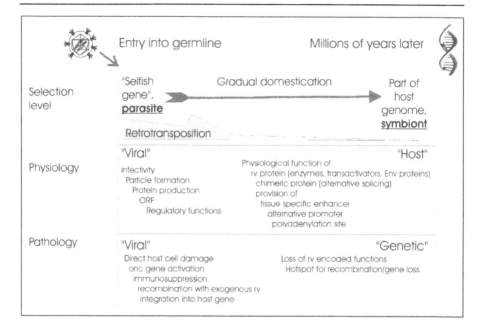

Figure 1. Events following endogenization of a retrovirus.

## Events following Endogenization of a Retrovirus

As seen in Figure 1, ERVs seem to undergo a more or less regular sequence of events. Initially, the provirus is a "selfish" gene,[16] i.e., it is optimising towards its own replication and replicates like an exogenous retrovirus in somatic cells, occasionally infecting other individuals, being more or less damaging to the host. In order to become an ERV, a retrovirus must infect a germ line cell. That cell must survive. It must give rise to offspring which is unhindered to reproduce. Already at this stage, harmless ERVs are selected. For example, endogenous replication competent HIV would be incompatible with life.[17] After a subsequent long period, the ERV may or may not be fixed in the species. This filters away most integrations, especially those with pathogenic properties. However, a population will never reach complete genetic equilibrium. Due to the evolution of both organism and its environment, the genome is in continuous transition, demanding genetic flexibility. Retroelements are prime providers of genetic flexibility, which gives a certain positive selection value to them. Therefore, rare cases of ERV pathogenicity will not be eliminated.

After an initial integration there may be intense reintegration into somatic cells, occasionally spilling over into the germ line. Within a short time, many follow-up integrations can accumulate in the germ line. Each integration may create havoc through the long range effects of retroviral enhancers.[18] This may reduce host fitness, encouraging crippling of the provirus(es), especially their LTRs. The initial pathology is centered around retroviral effects at the cellular or organismic levels, like those of an exogenous retrovirus.

Directly after integration and until severely mutated, ERVs may also serve as hotspots for gene loss through illegitimate homologous recombination, causing genetic disease (see below).

Initial beneficial effects of ERVs may be protection from XRVs by receptor interference (FV4),[6] or interference with formation of the preintegration complex (FV1).[19] Later, ERVs may become physiological servants to the organism. These events are briefly mentioned in Figures 1 and 2. Potentially useful functional modules offered by retroviruses are tissue specific

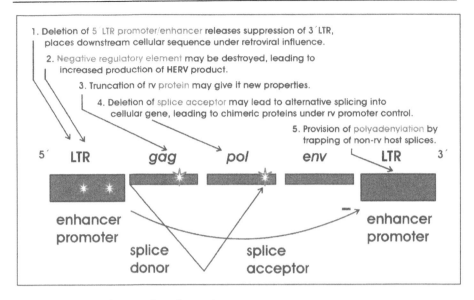

Figure 2. Activation of an ERV through mutation.

enhancers,[20,21] polymerase II promoters,[22] splice donors,[23] splice acceptors,[24] a myristoylated cytoplasmic membrane protein,[25] structural and nucleic acid binding proteins capable of packaging RNA, a dUTPase,[26] a protease,[27,28] an RNA dependent DNA polymerase with RNAse H[29,30] (reverse transcriptase may also be involved in the pathogenesis of nonretroviral RNA viruses due to their persistence as DNA copies[31]), an integrase,[32] an envelope protein which binds to a host surface protein, a spring-loaded transmembrane protein ready to fuse membranes and possibly cause immunosuppression, and a polyadenylation signal and site.[33-35] Retroviral single LTRs are very abundant[36] and probably have a great influence on the expression of many nonretroviral human genes. Several cloned HERV enzymes: protease,[27,28] dUTPase,[26] reverse transcriptase[29] and integrase[32] have shown function. So far, they all derive from the HERV-K(HML-2) group. The most likely scenario for such a function added relatively late during evolution is modulation of existing cellular mechanisms.

Even old retroviral integrations retain a part of their retroviral properties, which makes them "genetic jokers", a little less predictable than other portions of the genome. The silencing of ERVs may therefore be more eventful than the silencing of other genes. The sophisticated regulatory machinery of a retrovirus involves a careful balance of positive and negative regulatory functions, and interaction between structural proteins (Fig. 2). During the gradual mutational crippling of ERVs opportunities for deregulation occur, i.e., a deleted splice acceptor leads to splicing into downstream host genes.[37] HERV proviruses can also downregulate translation of a nearby host gene, presumably through strong secondary structure.[38,39] Thus, a price to be paid for the genetic flexibility provided by ERVs is a small pathogenic potential remaining even after many millions of years.

Factors which may prolong the activity of ERVs despite stop or frame shift mutations are i. copackaging of defective RNAs into particles encoded by other ERVs, ii. complementation of proteins from related elements which still have an active protein, iii. recreation of functional hybrid genomes by gene conversion[40,41] (see below) or recombination, iv. evasion from the nonsense mediated decay system[42,43] (Fig. 3), v. stop codon suppression like selenocysteine read-through of TGA ("Opal") stops. [44] Although the latter mechanism has

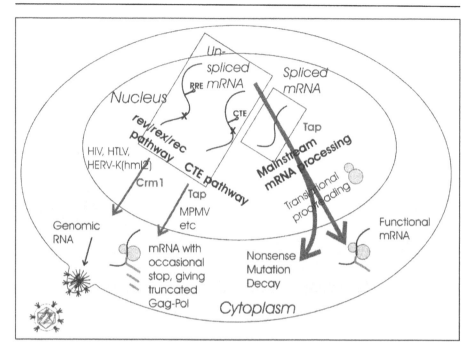

Figure 3. Removal of unspliced genomic retroviral mRNA from the mainstream of mRNA processing. The bulk of all cellular mRNAs are processed after translational proofreading in the nucleus. mRNA containing premature stop codons is then degraded in the cytoplasm by the nonsense mutation system. This probably terminates many defective ERV transcripts. However, the viral proteins *rev*, *rex* and *rec*, for HIV, HTLV and HERV-K(HML-2), respectively, bind to their cognate response elements, shown as hairpins, and are removed with unspliced RNA from the nucleolus and nucleus by the protein CRM1.[46] Other retroviruses, like the *betaretrovirus* MPMV use the binding of the cellular transport protein TAP to a hairpin named CTE to transport their unspliced RNA to the cytoplasm via a separate path.[47] A similar sequence, IAP expression element, IAPE, was recently described in the murine endogenous retroviral intracisternal type A particle genome.[48] Related mechanisms probably exist in other retroviruses, because all share the need to remove genomic RNA before splicing. A likely outcome is that genomic RNA, to be packaged into progeny particles, is not degraded even if nonsense mutations occur. Genomic RNA also encodes the Gag-Pol polyprotein. Therefore it is reasonable to assume that *gag-pol* transcripts evade NMD and can exist in truncated forms.[19] On the other hand, spliced messages, like those from *env*, should undergo nonsense mutation decay, preventing the production of truncated Env proteins. The scheme is an extrapolation from existing knowledge. Its consequences are testable by site directed mutagenesis.

not been demonstrated for HERVs, it potentially could prolong their activity. The validity of point i. was recently emphasized by a demonstration of pseudotyping of HIV capsids into HERV-W Env ex vivo, allowing HIV access to a new set of target cells.[45]

Many transcribed genes are reintegrated as processed pseudogenes through the LINE1 reverse transcription/integration mechanism. Evidently, some of the HERV integrations arose in this way.[49] It provides a pathway for intragenomic spread of defective elements.

Examples of events which resulted in a physiological interplay between HERVs and other genes have been found in genes involved in energy metabolism: transaldolase (which contains the repetitive element HRES-1 which has a weak similarity with retroviral sequences),[50] apoC-I lipoprotein,[22] leptin receptor,[51] leptin itself,[52 52] alcohol dehydrogenase,[53] and amylase.[20] However, no metabolic diseases due to lack of HERV function have yet been described.

A good illustration of how an ERV can modify function in a complex regulatory situation is the beta globin locus of primates.[54,55] An ERV-9 LTR integrated in the locus control region and assumed transcriptional control, i.e., beta globin transcripts initiate from the ERV-9 LTR promoter. It thereby shunted out several other promoters within the locus control region. These promoters remain cryptic, and can regain control if the ERV-9 LTR is experimentally damaged. It is surprising that the acquisition of a new retroviral promoter, with all its retrovirus-specific transcription factor binding sites, was compatible with function in the beta globin locus.

## Active HERVs, Expressed As Particles or As Proteins

HERVs which retain the ability to produce particles and proteins can be referred to as "active". Like the pathogenic animal ERVs they often are relatively recently endogenized. In principle, particle production may or may not be accompanied by transposition. Judging from the experience from animal ERVs, active HERVs are more likely to cause disease than more defective HERVs.

### *HERV Polymorphism*

In animals, a major cause of retrovirus-induced disease is reintegration (transposition). Unlike the LINE elements, transpositional activity of HERVs has not been directly demonstrated. However, if HERVs are integrationally active, integrational polymorphism should exist in the population. Several human-specific integrations of HERV-K(HML-2) was described,[56-58] highlighting them as recently acquired and as candidates for integrational polymorphism. It was therefore logical when the group of Jack Lenz recently described two cases of HERV-K(HML-2) integrational polymorphism.[59] The proviruses (HERV-K(113,HML-2) and HERV-K(115,HML-2) ) seemingly are intact, for the first time providing HERV candidates with potential infectivity. Using differential amplification techniques, several new unique HML-2 cluster 9 integrations[60] were detected. The presence of integrational polymorphism raises the possibility of active replication of HERV-K(cluster 9, HML-2), in somatic cells. If so, overt pathogenicity due to viral growth and even insertional oncogene activation may occur. The disease association of cluster 9 should be further investigated.

The occurrence of human-specific integrations in other HERV groups has not been systematically investigated. There is however one in the HML-4 group (LTR13).[61,62]

Another kind of HERV integrational polymorphism is seen in the MHC region.[63-66] It is likely that the HERV diversity in this region mostly is produced by HERVs being coselected as parts of cassettes, in linkage with actively selected MHC genes.[67] HERVs, like other repetitive elements, can give rise to illegitimate homologous recombination (see below). This has been observed in the MHC region.[68]

Single nucleotide polymorphism (SNP) also occurs in HERVs.[69-71] This is to be expected from genes who mostly mutate at a "pseudogene" rate.[72]

### *Retroviruslike Particles*

Retroviruslike particles are normally formed in placenta of many animals.[73-79] This is true also for humans[80] and also occurs in human teratocarcinoma cell lines[81] and in a breast cancer cell line (see below).[82,83] HERV-K(HML-2) genomes are major components of those particles. For the purpose of this review we will use the following distinctions: A "particle" is an entity which can be spun down at 100 000 x g. If the particle is filterable through 0.2 μ filters, bands at a density of 1.16 g/cm$^3$ in a sucrose density gradient, or at a corresponding density in another medium, and contains retroviral RNA and/or reverse transcriptase, it is a "retroviral particle". Other important but not obligate criteria are electron microscopy, which reveals the morphology, and antigenicity, which may reveal the source of the capsid or envelope proteins.

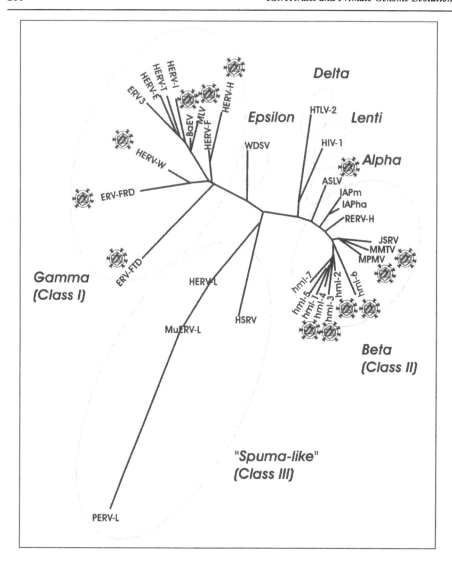

Figure 4. ERV sequences found as RNA in particles. Human ERVs are denoted HERV-X, hml-x, ERV3, ERV-FRD or ERV-FTD. Literature references are given in the text. The neighbour-joining tree was constructed from an alignment of 900 base pairs from the amino terminus of the pol gene. Retroviral sequences are either referred to as "HERV-" plus the tRNA specificity of the primer binding site,[12] or (for human betaretroviral sequences) as HML-X, where HML means "Human MMTV Like", and x stands for the HML group number (1-10).[11] Other abbreviations: ASLV; Avian Sarcoma-Leukosis Virus, BaEV; Baboon Endogenous Virus, HIV; Human Immunodeficiency Virus, HSRV; Human SpumaRetroVirus, HTLV; Human T Lymphotropic Virus, IAPm; Intracisternal type A Particle from mouse, IAPha; Intracisternal type A Particle from hamster, JSRV; Jaagsiekte Retrovirus, MLV; Murine Leukemia Virus, MPMV; Mason-Pfizer Monkey Virus, MMTV; Mouse mammary Tumour Virus, RERV-H; rabbit endogenous retrovirus – H (identical to the previous term HRV-5, human retrovirus 5),[91] WDSV; Walleye Dermal Sarcoma Virus,. The encircled branches correspond to the following retroviral genera: *Alpha-*, *Beta-*, *Gamma-*, *Delta-*, *Epsilon-*, *Lenti-* and *Spumavirus*. Class I and II HERVs correspond to *Beta-* and *Gammaretrovirus* genera. Class III HERVs are most similar to genus *Spumavirus* and are therefore referred to as "Spuma-like" in the figure. A thorough and still readable overview of HERVs was presented by Wilkinson et al.[92]

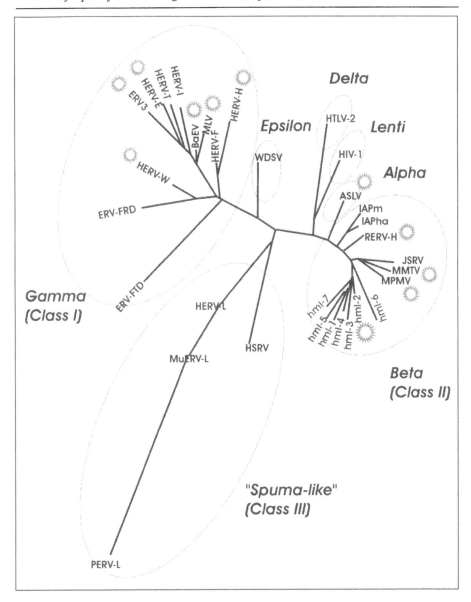

Figure 5. ERVs with ORFs. ERV sequences known to have at least one ORF in *gag-pro-pol* or *env* genes are symbolised with a sun. References: HERV-H *env* ORF (Lindeskog et al, 1999; de Parseval et al, 2001; Jern et al, 2002),[71,93,94] HML-2 *gag, pol* and *env* ORFs (Mayer et al, 1997; Mayer et al, 1999).[95,96] The tree is otherwise the same as in Figure 4.

The term "retroviruslike particle" is justified if most but not all of these properties are demonstrated.

Besides the retroviral particles mentioned above, retroviral particles are produced by the human breast cancer cell line T47D.[83] These particles contain RNA from HERV-K(T47D,HML-4), HERV-P(T47D) and HERV-K(HML-6). Retroviruslike particles have been found in human

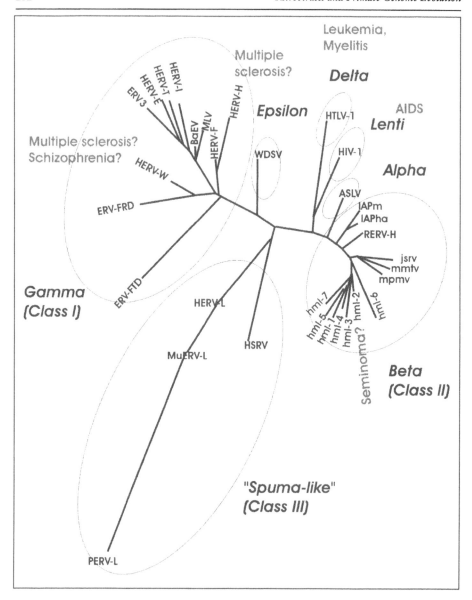

Figure 6. Connection of retroviruses to human disease. In cases where causality is not established the disease name is shown with a question mark. Details are given in the text.

serum and cerebrospinal fluid[84-88] (see also below, Figs. 4, 6 and 8). An important aspect of retroviral particle detection by reverse transcription PCR (RT-PCR) is sensitivity and specificity. Although excellent as discovery tools in an initial phase, broadly amplifying systems like the PAN primers,[89] or MOP primers[90] tend to have a good specificity but a relatively low sensitivity. They are also very sensitive to mutations in the primer target sequences, which can lead to an uneven representation of intended targets among the amplimers. In the context of this review, it is especially interesting that the PAN primer system can probably not amplify many

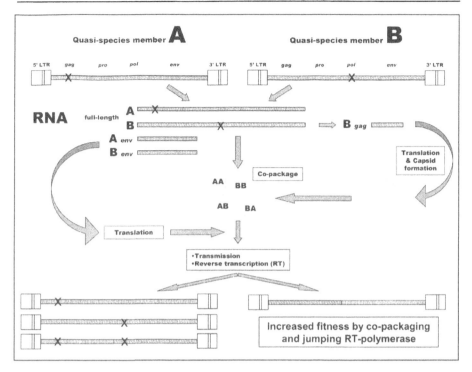

Figure 7. Generation of "Breakout" exogenous retrovirus from a proviral quasispecies of endogenous retroviral sequences. Defects in the original proviruses are depicted with an X. This is a hypothetical scheme which, however, has support from animal models (see the text). It is here suggested as an explanation for the occurrence of retroviral particles in some human diseases.

HERV-H sequences due to variation in the LPQG and YXDD motifs (Blomberg, unpublished). Thus, such studies should be followed up with studies using other RT-PCR systems, hybridisation microarrays, or cDNA libraries.

Particles from the latter sources (serum and cerebrospinal fluid), and from B cells of MS patients, have been reported to contain nucleic acid from several HERVs, notably HERV-H and HERV-W.[84,86] Although it has not been systematically investigated, many of the RNAs encapsidated in the particles from these sources have been defective. They may have been cross-packaged into particles encoded by one or several HERVs which have non-defective capsid or envelope genes.

*"Infectious" HERVs*, meaning passage of HERV production in cell culture, have been reported by three groups.[86,88,97,98] These "isolations" of infectious HERVs were done with samples containing particles with HERV-H or HERV-W sequences, respectively. Both of these class I HERVs have members with ORFs in *env* and relatively intact *gag* and *pol* frames. Because human target cells usually contain the same DNA sequence as the RNA in the particles, a passage into cell lines without corresponding HERVs is required to definitely ascertain the isolation. Such experiments have been initiated[99] (see below). Another requirement is that the produced particle-associated sequence should be virtually identical to the sequence in the input particles. These basic criteria have not been met. Much work is required before the infectiosity of the particles can be considered established. However, if the reports are right, a paradigm shift may be on its way. The possibility that such viruses are "breakout viruses" (see below and Fig. 7) should be addressed.

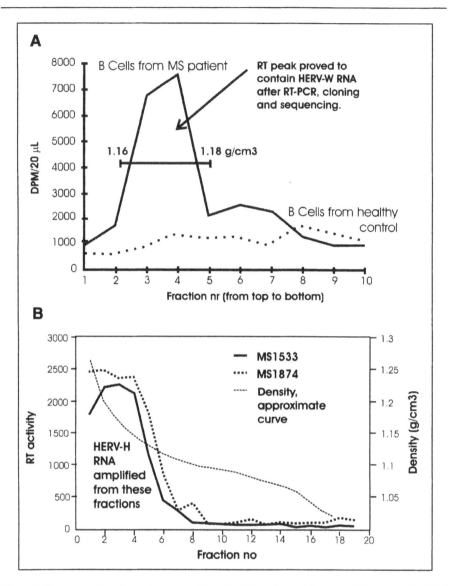

Figure 8. Characterisation of retrovirus secreted from B cell lines isolated from two MS patients. Modified from Moller-Larsen et al.[120]

## Intraindividual ERV Evolution, "ERV Breakout"

The thousands of ERV integrations can be viewed as a "proviral quasispecies" analogous to the quasispecies of XRVs. This genomic quasispecies is not static. From time to time it is modified by gene conversion, which pastes one sequence over a highly similar one.[40,41,94] Therefore the stops and frameshifts will vary from cell to cell and from individual to individual. In this ERV lottery, ORFs can both be obliterated and recreated. The extent of this type of somatic variation in humans is not known, but several observations indicate that it is not uncommon.[40,94] Although many retrotransposons become epigenetically silenced, often by methylation,[8] their degree of methylation varies, and is diminished during aging.[100,101] Indeed, the expression of endogenous retroviruses in mice is age-dependent.[102-106] Besides variation in methylation, there is evidence for a second mechanism of ERV activation. We would like to call it "ERV breakout" (Fig. 7). Starting from humble beginnings, maybe after a temporary demethylation,[101] one quasispecies member (B) may provide functional capsids and partially functional enzymes. Another quasispecies member (A) may provide the envelope proteins. It is conceivable that such particles can copackage RNAs from each other and other more or less defective members of the quasispecies (Fig. 7), which during reverse transcription recombine, infect with a low efficiency and reintegrate. Once the process starts, it is self optimizing, and may generate a more or less infectious exogenous retrovirus, in some cases progressing to a replicatory avalanche. This is reminiscent of the self optimizing nature of tumour cells, once they start to escape from extraneous control of cell division. The process will be more rapid the more replication competent the starting provirus is. A "breakout" process involving both mutation and recombination occurs regularly in AKR mice during leukemogenesis.[6,106] A related phenomenon is the so-called "patch repair" of MLV.[107] ERV breakout to XRV could be a common phenomenon in many species, restarting early, maybe after the conception of an individual, becoming more or less developed during its lifetime. A number of factors will be influential: initial presence of intact or nearly intact ERVs, the often restricted cross-packaging of defective ERV RNA,[108,109] the extent of mRNA degradation due to the nonsense mutation decay mechanism,[42,43] and the degree of ERV RNA and protein expression. A breakout XRV then could be pathogenic or not depending on its emergent properties. We suggest that this mechanism should be further studied in animal models and be addressed in studies on ERV pathogenesis in humans. It can be falsified by studies on substitutions and recombinations in viral particles, comparing the sequences of particle-associated RNA with that of related loci in genomic DNA.

## Diseases Where a Connection with HERVs Has Been Implicated

### *Neurological Diseases: MS and Schizophrenia*
The rapid and recent development of higher mental activity is a unique human event which probably required genetic flexibility and preparedness. Mobile genetic elements could therefore be suspected to have assisted in brain development.[110,111] A connection between axon growth and retrotransposon transcription in the developing Drosophila brain has been described.[112] Theoretically, remaining transposon activity could then be associated with disease.[113] The upcoming comparison of chimpanzee and human genomes will shed light on this issue.[114] Whether related to this or not, the neurological diseases MS and schizophrenia belong to the diseases where an increased production of retroviruslike particles and RNA have been reported (Figs. 3, 6 and 8).

In MS patients, particles containing reverse transcriptase were discovered at the end of the 1980s by a French and a Danish group in supernatants from cultured leptomeningeal and B cells from MS patients.[84-87,115,116] Two main groups of particle-bound sequences have been detected. One belongs to the group HERV-W (see above, Figs. 4, 6 and 8), and was initially called "Multiple Sclerosis Retrovirus" (MSRV).[84] The other belongs to the group HERV-H (see above, and Figs. 4, 6 and 8).[86]

Studying supernatants from two choroid plexus-derived cells and one B cell line, all three from MS patients, the French group found particles banding at 1.14-1.20 g/cm$^3$ in sucrose density gradients which had reverse transcriptase activity and had an RNA from which a retroviral *pol* stretch was amplifiable with a broad RT-PCR (Fig. 8A). The sequence was similar to the earlier published ERV-9 sequence,[117] and was fully sequenced by a primer walking procedure. It was initially called MSRV, later HERV-W. Using a HERV-W specific PCR, HERV-W RNA was detected in cell pellets of 5 of 10 cerebrospinal fluid (CSF) samples, while none of 10 CSF cell pellets from patients with other neurological diseases were positive. All of the five HERV-W RNA positive patients were not treated with immunosuppression. In a further study, using a nested RT-PCR approach targeted to the MSRV *pol* sequence,[84] 9 of 17 MS patients were positive for particle-bound HERV-W RNA in serum. Three of 44 controls were MSRV positive. A tendency towards a lower percentage of positivity in azathioprin/corticosteroid-treated persons was noted. A small Swedish study showed one MS patient with a variant of HERV-W RNA in serum,[118] out of nine analysed, while none of 21 blood donor controls were positive. The broadly amplifying PAN primer system was used.[89]

The Danish group found RT-associated particles in supernatants of B cell lines from 5 of 21 MS patients, and from 1 of 13 healthy controls.[119] Cultures which produced retrovirus-like particles also produced Epstein-Barr virus. The retrovirus-like particles had a type-C morphology, and banded at a density of 1.25 g/cm$^3$. HTLV-1 banded in a similar fashion (Fig. 8B).[120] Primers based on HERV-H clone RGH2 amplified HERV-H sequences from retroviral particles of supernatants from B cells of MS patients. They also amplified such sequences from particles in plasma of MS patients. Twenty four of 33 cell free plasma samples from MS patients were positive in the HERV-H PCR, whereas none of 29 plasmas from patients with other autoimmune disease and none of 20 plasmas of healthy controls were positive. Cloned amplimers from particles of four patients (two from B-cell supernatants and two from plasma samples) gave different sequences, but were still highly related to RGH-2. Thus, several HERV-H integrations may be involved. The nested HERV-W specific primer sets ST-1/2 did not amplify from the same particle preparations.

The occurrence of retroviral particles either containing HERV-H or HERV-W sequences is perplexing. However, the French group also found ERV-9 sequences in a minority of their particles. Both groups report heterogeneous sequences, i.e., more than one HERV-H, HERV-W and ERV-9 integration seem to contribute. A simple explanation is that several gammaretroviral HERV loci are activated in MS. Then the amplification specificity of the PCRs involved could favour either HERV-H or HERV-W/ERV-9 detection. This should be studied systematically with quantitative PCRs and sequencing of particle-bound RNA (Table 1).

The packaging of these RNAs is also astonishing. Although no candidate *gag-pol* HERV-H/W ORF is known (J Blomberg, unpublished observation) it is possible that such a protein can be created via somatic mutation (see "breakout RVs", above), and provide the capsids. The envelope protein may be supplied via one or several of the known *env* ORF genes.[71,93,121]

Both the Danish and the French groups maintain that they can grow the respective virus in human cells[97,98,116] (see the discussion on "infectious" HERVs). Considering the great number of microbes which have been associated with MS during the previous century these reports must be interpreted cautiously. However, the lead seems strong enough to motivate a thorough investigation.

A selective increase in HERV-W RNA expression was discovered by representational difference analysis (RDA) of monozygotic twin pairs discordant for schizophrenia in schizophrenic patients compared to those of mentally healthy controls.[122] Particles containing RNA from HERV-W-like retroviruses have also been detected in cerebrospinal fluid of newly debuted patients with schizophrenia.[88,99] However, also particles with RNA from related HERVs (i.e., ERV-9 and ERV-FRD, also known as HERV-P(T47D)) were detected. Ten of 35 newly debuted schizophrenia patients had HERV Class I RNA containing particles in CSF. Seven of the 10 had HERV-W, two had ERV-9 and one had ERV-FRD RNA. CSF from one of 20 patients with chronic schizophrenia, none of 22 patients suffering from non-schizophrenic neurological disease, and none of the healthy controls had particles containing HERV RNA. The PAN primer system[89] was used. Some of the patients also had retroviral RNA in serum. The full spectrum of gammaretroviral RNA sequence containing particles in CSF and serum from MS and schizophrenic patients as well as in controls has not been determined. It would be interest-

**Table 1. Models for a possible involvement of ERVs in MS and schizophrenia**

| Model | Mechanism | Evidence | Falsification |
|---|---|---|---|
| 1. Reintegration after recombination/ mutation to replication competent virus ("breakout" virus). | Integrational (in)activation of cellular gene | Particle formation, possibly infectious. | No proto-oncogene activation (malignancy) observed. No ORF in particle RNA. No recombination in particle RNA detected (not much studied, though). Proof of infectiosity weak. No reintegration described. |
| 2. Upregulation of ERV loci. No replication competent virus. | Production of ERV proteins, of which some immunise against brain by mimicry (MS). | Particle formation. RNA upregulated (schizophrenia). | No ERV-brain mimicry demonstrated. Only a few ERV encoded proteins described. Upregulated loci not characterised. Absence of inflammation in schizophrenia. |
| 3. Absence of protective ERV product + pathogenic XRV ("Lake Casitas model"). | XRV, or breakout ERV, allowed to infect in brain due to absence of receptor block. | Particle formation, possibly infectious. | Particle RNA very similar to ERV, no evidence for XRV. Multiple variants of ERV-like RNAs in particles. Proof of infectiosity weak. Absence of inflammation in schizophrenia. |
| 4. ERV polymorphism confers a disease predisposition. | ERV protein interaction with a host encoded molecule is present or absent. | Particle formation? | Absence of disease-related polymorphism. |
| 5. No pathogenic role, neither in MS nor in schizophrenia | The unspecific activation of certain ERV loci is an epiphenomenon, secondary to disturbances in brain metabolism/ regulation. | Particle formation. HERV RNA upregulation. | Only certain loci may be upregulated, a sign of specificity. ERV activation in cell culture due to cellular stress needs more study. |

All build on the observation of ERV RNA containing particles in CSF, plasma and in supernatant of cultivated B cells and cells from CNS, and upregulation of HERV trancripts in new-onset schizophrenia. As a first requisite, these observations should be confirmed and extended. Other requisites are listed under "falsification". Most falsification tests involve a more complete sequencing of the RNA of these particles, and a critical assessment of their infectiosity. The fifth hypothesis implies no pathogenic role at all for ERVs, as suggested or discussed.[131,132,140-142]

ing to know which loci and alleles that transcribed the RNA that got packaged into particles in MS and schizophrenia. It is appropriate to mention that HERV-W uses the glutamate transporter protein as a receptor.[123] Glutamate is an important brain signal molecule of relevance for schizophrenia.[124]

Also in schizophrenia there are reports of virus passage.[88,99] A particular strength of this claim is that owl monkey kidney cells were used. The owl monkey is a New World monkey, who does not have many of the more recent HERVs.

It is too early to judge whether the increased amount of HERV particles and HERV RNA in multiple sclerosis and schizophrenia is an epiphenomenon or a sign of a pathogenic role of certain gammaretroviral (Class I) HERVs. There is evidence (sometimes conflicting) that antigenic mimicry between brain proteins and antigens from some viruses elicits anti-brain autoimmunity in postinfectious encephalitides.[125-128] Theoretically, a similar mechanism could operate in MS (see Table 1). However, mouse and rat ERVs can be activated nonspecifically, for example by anoxia.[129,130] Recent publications[131,132] have claimed that HERV-H and HERV-W transcription is upregulated in macrophages at the site of inflammation in MS brains. Thus, the occurrence of these RNAs and particles containing them in MS would just be a sign of immune activation. However, this does not explain the activation of HERV-W in schizophrenia, a disease without CNS inflammation.

### How Could ERVs Cause MS and Schizophrenia? Theoretical Possibilities

Although MS and schizophrenia[133] are two distinct diseases both tend to debut in 18 – 28 year-old persons. In both there is support both for genetic[134] and for environmental, unevenly distributed, possibly infectious, etiological factors.[135,136] Schizophrenia debuts often are seasonal.[137] This is not an argument for a pathogenic role of HERVs which are present since conception in nearly all persons. The reasons for seasonality in viral infections are not entirely known. This epidemiological pattern is primarily connected with fecal-oral or airborne transmission. Such types of transmission are remote from currently known HERV biology. If HERVs, via a polymorphism or a reintegration event, would be the major etiological factors of MS and schizophrenia there should be a pronounced hereditary component and a minor environmental one. However, if HERVs modulate the disease by interaction with environmental factors, there could be equally important hereditary and environmental components of the etiology. The latter pattern is found in the neurological disease of Lake Casitas mice,[15] where a defective ERV prevents infection from a pathogenic XRV by receptor interference. XRVs tend to behave in three fashions; *i.* as sexually transmitted infections, ii. as blood borne diseases or iii. as nonseasonal childhood infections. The debut age and seasonality pattern therefore do not fit entirely with the Lake Casitas model either. However, a common infection, which mainly infects during childhood, when it is rather asymptomatic, but is symptomatic (mononucleosis; "kissing disease") when infecting teenagers, is Epstein-Barr virus. A combination of a host factor (age) and salivary transmission of an infectious agent (EBV) is necessary to cause disease. The debut ages of both schizophrenia and multiple sclerosis agree approximately with the mononucleosis pattern. A similar reasoning was published early by Danish authors, who indeed have implied a role for EBV in MS.[115,138] Another reason for a retarded appearance of disease caused by retroviruses is a persistent virus production with random integration, which eventually may cause neoplasia due to activation of a proto-oncogene.[6] In this type of pathogenesis, various neoplasms, primarily leukemia/lymphoma, are the main outcomes. Neither MS nor schizophrenia are associated with such neoplasms. A third retroviral pathogenic mechanism is recombination between exogenous and endogenous retroviruses.[6] Such recombinant viruses have new properties, some being pathogenic due to a large virus production and/or access to a new target cell. The immunosuppression which occurs in FeLV viremic cats may be of this kind.[139] A common retrovirus (a "breakout" ERV or XRV) with limited replicatory potential theoretically could undergo recombination with a low frequency, or gradually mutate back to greater replicatory potential, causing disease in a few despite of ubiquity. These hypotheses are to some degree testable (Table 1).

As seen from this speculative reasoning, the explanation will not be straightforward. As a starting point, the frequency and amount of retroviral particles in body fluids of these patients and controls must be unequivocally established before a role for HERVs in these diseases can be inferred.

### The HERV-Cancer Connection

There are many ways in which HERVs may be involved in normal and malignant growth. A striking dependence of a choriocarcinoma cell line on HERV-E expression for growth was elegantly shown[143] (Fig. 9). Most probably, this occurs through HERV-E.PTN driven expression of the growth stimulatory factor pleiotrophin. If human choriocarcinomas in general have the same dependence, preventing HERV-E.PTN expression could be a useful target for gene therapy.

In animals, retroviruses and ERVs are known to be involved in carcinogenesis. The most well-known examples are leukemia caused by the ecotropic gammaretrovirus murine leukemia

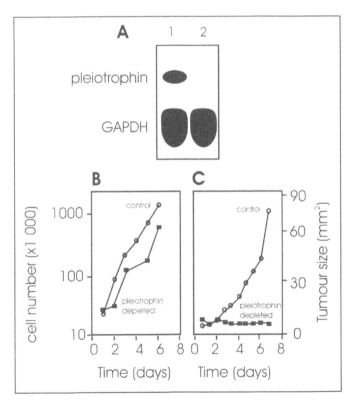

Figure 9. Effect of the depletion of HERV-E.PTN mRNA from JEG-3 choriocarcinoma cells. A) Northern blot analysis of total RNA from control JEG-3 cells (lane 1) and JEG-3 cells expressing a ribozyme targeted to pleiotrophin mRNA (lane 2) is shown. B) Proliferation in vitro of control and pleiotrophin mRNA-depleted JEG-3 cells. C) Growth curves of subcutaneous tumours in athymic nude mice. Two million control or PTN mRNA-depleted JEG-3 cells were inoculated. The data represent mean tumour size. Modified from Schulte et al.[143]

virus in AKR mice and the mammary tumours caused by the betaretrovirus endogenous mouse mammary tumour virus in GR mice, which primarily act by insertional enhancement of protooncogene promoters. These ERVs are intact and can produce infectious virus. Although a few HERVs seem to be intact, none of them are yet known to produce infectious virus. The class II HERV sequences are relatively closely related to MMTV and it has therefore been natural to look for expression of such viral sequences in human breast cancer. There are, however, no clear signs of a causal relationship between HERV class II and breast cancer,[144,145] but there are a few reports on expression of HERV-K(HML-2) like RNA in leukemic leukocytes.[146-148] Those reports need to be confirmed and extended. In the following we discuss some specific examples of possible involvement of HERVs in carcinogenesis.

## Germ Cell (GC) Tumours, in Particular Seminomas

Even if HERVs are a type of transposons, stimulation via HERV integration close to a proto-oncogene has yet to be observed. This is in contrast to other retrotransposons, primarily the LINE1 elements, where several such cases have been discovered.[5] However, there is a correlation between occurrence of antibodies to HERV in seminomas, a type of testicular carcinoma. During active disease, high titers of antibodies binding to several HERV-K(HML-2) proteins (Env and Gag) occur (Fig. 10 A,B). These antibodies disappear after treatment.[149,150] Immuno EM confirms that anti-HERV particle (from teratocarcinoma cells) antibodies occur in GC tumours of males.[151]

HERV-K(HML-2) virus is expressed in teratocarcinoma cell lines.[81,153] HERV-K(HML-2) transcripts and proteins are also abundant in testicular and ovarian GC tumours except for teratomas and spermatogenic seminomas.[154] HERV-K(HML-2) Gag proteins are also expressed in testicular and ovarian GC tumours.[155,156] Human teratocarcinoma derived virus, HTDV, is a term coined after the demonstration of particles containing HERV-K10-like RNA,[81] and Gag proteins antigenically related to HERV-K10 Gag in supernatant from a teratocarcinoma cell line.[153] Although HTDV is not derived as a specific clone, this virus will here be referred to as HERV-K(HTDV,HML-2) ). Thus, the presence of both virus expression in GC cells and antiretroviral antibodies in male patients with GC tumours indicate a connection, etiological or not, with HERV-K(HTDV, HML-2). The most likely loci for production of HERV-K(HTDV, HML-2) particles are either HERV-K(113, HML-2) on chromosome 19p13 or HERV-K(HML-2.HOM), (here called HERV-K(HOM,HML-2)) on chromosome 7.

Although an oncogenic mechanism is not known, the research groups in Homburg and Langen focus on two novel HML-2 proteins, which arise by alternative splicing, (i) *rec*, earlier called c-orf[81,157,158] and, (ii) np9[159] (Fig. 11). HERV-K(HML-2) *rec* transforms cultured cells to grow as tumours in nude mice. It binds to a Zn finger gene, *PLZF*.[160] *PLZF* is associated with retinoic-acid resistant forms of promyelocytic leukemia (PML). Hypothetically, PLZF acts via cyclin A2, a regulator of meiosis, which precedes formation of spermatozoa. Increased expression of *rec* indirectly could influence the maturation of GC to spermatozoa, and increase the likelihood of malignant transformation.[160] The np9 protein is a newly discovered alternative splice product of HERV-K(HML-2)[159] (Fig. 11). It is a nuclear protein which is expressed in GC tumours. Its oncogenic role in GC tumours is not defined.

Among the candidates for a HERV-cancer connection, seminomas are the strongest. Even in this case, much work remains before an etiological connection can be considered as established.

## A Malignancy-Associated Genotype of Some Betaretroviruses

The most common function of the *env* SU is binding of a receptor.[161] Even though SU is a highly variable sequence, a conserved region in the N-terminal has been shown[144] (Fig. 12). According to one research group this region encodes a putative Vβ7 T cell receptor specific superantigen, pSAG, as a product of HERV-K18, here called HERV-K (IDDM, HML-2).[162]

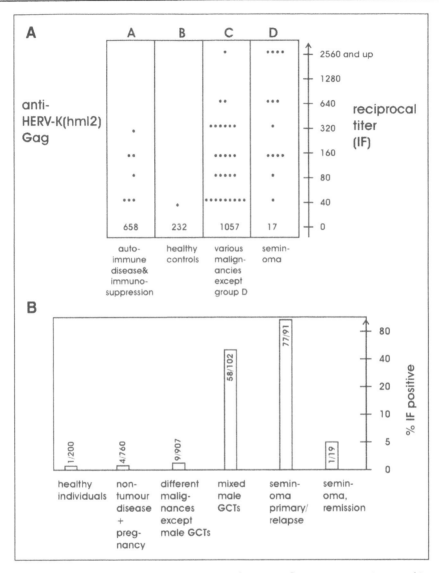

Figure 10. A) Distribution of IgG antibody reactivity in the immunofluorescence test, using recombinant HERV-K(HML-2) Gag protein in a baculovirus system.[152] B) Prevalence of IgG antibody reactivity in the immunofluorescence (IF) test, using recombinant HERV-K(HML-2) Env protein in a baculovirus system.[149] Although pregnancy is not a disease, it is shown together with non-tumour diseases. Male GCT: Germ Cell Tumours of male. This includes testicular tumours except for seminoma. Primary seminoma cases, and cases with seminoma relapsing after treatment are shown together. An IF titer of 40 or higher was considered positive.

The conserved motif is also present in HERV-K(HOM,HML-2), discussed above, and in the Jaagsiekte sheep retrovirus (JSRV). JSRV is a contagious, possibly airborne, retrovirus which causes lung cancer in sheep.[163,164] The tumorigenesis may be mediated through (a) cell surface interaction with the tumor suppressor HYAL2, which is a receptor for JSRV in the bronchi,[165] and (b) participation in an intracellular tyrosine kinase signal cascade leading to growth

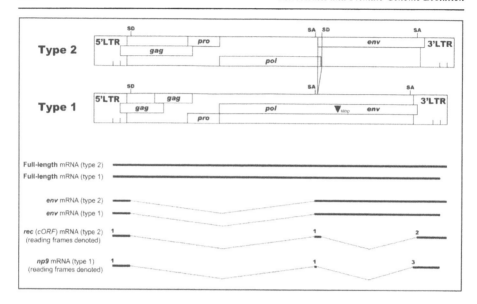

Figure 11. Two types of HERV-K(HML-2) and their spliced mRNA transcripts. In addition to full-length mRNA and *env* mRNA, HERV-K(HML-2) type 2 produce *rec* (*c-ORF*) as a product of splicing and frameshift from ORF-1 to ORF-2 (numbers over mRNA). HERV-K(HML-2) type 1, with a 292-bp deletion, produce np9 instead after splicing and frameshift from ORF-1 to ORF-3.

stimulation.[166] Endogenous JSRV variants do not have the tyrosine crucial for this pathway. JSRV is rather closely related to class II (genus *betaretrovirus*) HERVs (Fig. 4). The striking conservation in the amino-terminal portion of Env of class II HERVs and JSRV indicates that they may have the same receptor, i.e., HYAL2 (Fig. 12). Thus, there is reason to investigate if Env proteins of betaretroviral HERVs have any of these properties. Whether the growth modulating properties of this group of betaretroviral envelope proteins are related to its reported superantigen properties is obscure (see below). It is reasonable to assume that the original viruses which became HERV-K(HML-X) had some properties in common with JSRV.

### Evasion of Antitumor Cytotoxicity Mediated via Gammaretroviral Transmembrane Protein

p15E is the transmembrane protein of feline leukemia virus (FeLV). The "p15E story" dates back to early work on FeLV pathogenesis and functions of retroviral proteins.[167-170] The following effects were reported: i. an immunosuppressive effect of the transmembrane proteins of many (but not all) gammaretroviral transmembrane proteins and peptides derived (e.g CKS-17, consisting of 17 conserved amino acids) from the so-called immunosuppressive domain (ISU)[170] (Fig. 13), ii. an increased expression of p15E-like epitopes in animal and human tumours as defined by monoclonal antibodies to p15E,[171-173] which may be a sign of retrovirus-assisted evasion of anti-tumour cytotoxicity. Since then, "p15E" became a sometimes uncritically used catchword. Direct evidence for such a mechanism has however been published by the group of Heidmann.[174,175] The p15E expression in some mouse and rat tumours has been reported by several authors.[173,176] Whether a p15E mediated escape from tumour immunity is common in mice is less certain.[176] Although a detailed appraisal of the p15E effect[177,178] is outside the scope of this review, ISU peptides from p15E may inhibit protein kinase C,[179,180] and binding of TGFβ, to its receptor.[181]

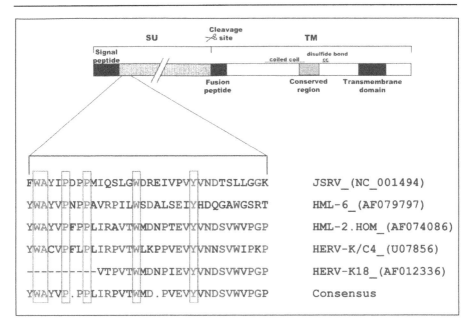

Figure 12. Schematic structure of a subset of *betaretrovirus* Env proteins with a conserved motif in the N-terminus of SU.

The first molecular evidence for the presence of p15E-like sequences in the human genome was published by Lindeskog et al[182] The human genome contains a few HERV *env* sequences with open reading frame which are candidates for formation of a functional p15E-like human protein.[71,93,183] The data on expression of p15E-like epitopes in human tumours are not as extensive as in animals. They are mostly based on immunofluorescence (IF) or RNA data. Foulds et al found p15E expression in 24 of 30 colorectal carcinomas and 4 of 4 gastric carcinomas.[184] However, the molecular nature of the p15E like RNA detected in this work is obscure.

Sera from patients with head and neck carcinomas have been shown to contain p15E-like low molecular weight factors that inhibit the response of monocytes and macrophages to chemotaxis and that the bioactivity of these factors decreased to normal values after removal of the tumour.[171] IFNγ is generally known to activate NK cells, but the subtype IFNγJ blocks this activation and thereby the NK-cells from boosting other IFN· subtypes.[185] Anti-retroviral p15E antibodies decrease these immunosuppressive activities of IFN· suggesting a relationship between p15E-like factors and human IFN·.[172] There is also a weak structural similarity between IFN· and p15E.[172]

In conclusion, there are interesting lead observations on a possible p15E mediated tumour evasion mechanism in humans, but much work on the signaling pathway of immunosuppression via TM-ISU and the frequency of TM-ISU mediated evasion from antitumour cytotoxicity, in humans and other animals, remains. More precise measurements regarding TM-ISU expression at the RNA and protein levels are also needed.

## HERV-Encoded Tumor Antigens

In mice, it is well known that murine ERVs (MLV and IAP) often encode tumour antigens recognized by cytotoxic T cells.[186,187] A corresponding mechanism was recently found in humans.[188] Cytotoxic T lymphocytes specific for a short Env-derived peptide (MLAVISCAV)

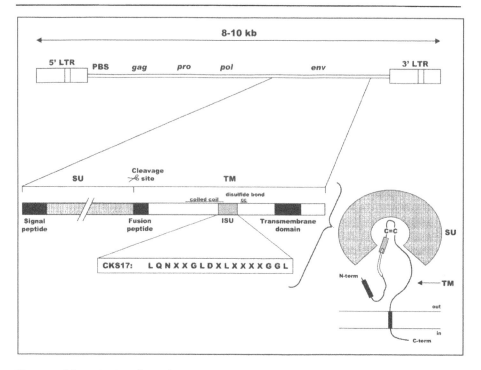

Figure 13. Schematic view of an endogenous *gammaretrovirus* with its envelope (*env*) gene. The immunosuppressive unit (ISU) in *env*TM containing the 17 amino acids long conserved CKS17 motif, is highlighted.

was detected in two patients with malignant melanoma. It is encoded from a 2578 bp retroviral RNA stemming from chromosome 16. The provirus belongs to the HERV-K(HML-6) group. Its name is HERV-K-MEL, or HERV-K(MEL,HML-6) according to the nomenclature of Andersson et al [11] The finding shows that even highly mutated ERVs can give rise to protein products with biological activity. HERV-K(MEL,HML-6) RNA was especially abundant in metastatic melanomas and in testis tissue.[188] Moreover, the MLAVISCAV sequence is situated in the amino terminus of the HML-6 envelope protein (see Fig. 12), which is similar to the oncogenic envelope protein of JSRV, and to the disputed HML-2 superantigen of Conrad et al[162] The tumor antigenicity, the similarity to known or alleged growth stimulatory betaretroviral envelope proteins, and the high expression of HERV-K(HML-6) located on chromosome 6 in a breast cancer,[189] may mean that HERV-K(HML-6) encodes a surface oncoprotein, which is upregulated in a variety of tumours. It is too early to speculate on a functional role for HERV-K(HML-6) in human cancer. It does, however, merit further investigation. An important aspect is that HERV-K(HML-6) peptides could be used to elicit tumor immunity.

### Herpesvirus—ERV Interaction: Does EBV Use a HERV for Stimulation of B Cell Growth?

The Marek disease herpesvirus of turkeys (MDV) can transduce avian leukosis virus (ALV).[190] Such recombinant MDV is more lymphomagenic than MDV which does not contain ALV. This is an example of herpesvirus pathology enhanced by a retrovirus. Herpesviruses are highly prevalent in many species. In humans, human herpesvirus 7 (HHV7), Epstein-Barr Virus (EBV), human herpesvirus 6 (HHV6), Varicella-Zoster Virus (VZV) and Herpes Simplex type 1 (HSV1) are the most prevalent. They infect over 60% of the population and can then remain latent for a lifetime (for an overview, see Richman et al ref. 191). In this respect the herpesviruses are almost as "endogenous" as HERVs which often are present in 99-100% of the population. Thus, it is not unlikely that these viruses may utilize each other. The love and hate relationship between the human organism and a HERV then becomes a triangle drama. The MDV – ALV interaction may not be the only herpes-retrovirus interaction. It was recently found that the immediate early protein ICP0 of herpes simplex can transactivate the LTR of HERV-K(HML-2)[192] (a similar finding was however reported much earlier[193]). In an intriguing series of experiments, Thorley-Lawson's group got evidence for a HERV-K(HML-2) *env* superantigen mediated stimulation of T cells after EBV infection.[194] The growth stimulation was abrogated in the presence of anti-HERV-K(HML-2) antibodies. The suggested chain of effects is reminiscent of that of MMTV, which also activates a T cell to secrete B cell growth factors via a viral superantigen in order to stimulate growth of its host cell, the B cell. In this case, the benefactor allegedly is EBV. These findings need confirmation, but they emphasize the potential of interactions between highly prevalent selfish genes/retroviruses and ubiquitous exogenous viruses like the herpesviruses.

EBV is considered a major cause of lymphomas. The putative involvement of the HERV protein in EBV-related growth stimulation of B cells raises the possibility that immunization against the HERV protein, or prevention of its expression via transduction of anti-sense constructs, might inhibit EBV lymphomagenesis. This would be a logical consequence of the results of Sutkowski et al[194] There are many reasons to clarify this issue (see below).

## Autoimmune Disease

Chronic virus infections, including retroviral ones, may induce autoimmune phenomena.[195] It is conceivable that a variable virus occasionally will present self-like epitopes, which may lead to a breach of tolerance. HERV encoded proteins basically should be perceived as self. Despite this, IgM anti-HERV antibodies occur in healthy persons (Lawoko et al,[196] who used synthetic peptides), and in disease (IgG; Sauter et al,[149] who used recombinant proteins). The occurrence of anti-HERV IgG during HIV infection may be due to presentation of variable epitopes more or less similar to HERV epitopes.[196]

### The HERV-Diabetes Connection

An increased virus particle production in explanted pancreatic islets from juvenile diabetes, containing genomes of an hml2 variant here referred to as HERV-K(IDDM,HML-2), earlier called IDDMK1,2-22 or HERV-K18, was reported.[162] This provirus is located on chromosome 1q21.2 – q22.[197] A superantigen in the aminoterminal portion of the predicted Env protein of this virus was also reported. Several negative reports have accumulated since then.[198,199] There was no difference in frequency of HERV-K(IDDM,HML-2) RNA in plasma between IDDM and controls. The superantigen effect could not be reproduced. The frequency of HERV-K(HML-2) antibodies in plasma was the same in patients and controls.[200,201] Two recent publications did however succeed in demonstrating a superantigen effect.[194,202] The same authors also connected the HERV-K(HML-2) superantigen with the B cell growth stimulatory effect of EBV (see above).

Thus, the possibility of a superantigen property in the aminoterminal portion of HERV K(HML-2) Env remains open. Whether it corresponds to a conserved amino acid motif in the amino terminal portion of some human endogenous betaretroviral Env sequences[189] (Fig. 12; see above) is uncertain. Clarification of this issue is important.

Presence of two other partial Class II sequences, HERV-K(HML-6)[11,13,145,189,203]and HERV-K(HML-4), also referred to as HERV-K(T47D)[83], among the histocompatibility genes on chromosome 6[65] confers an increased risk of juvenile diabetes. The correlation may be caused by histocompatibility antigen alleles, MHC-associated HERVs, or a combination of them.

## Systemic Lupus Erythematosus (SLE)

A HERV may have a physiological function within the *complement system,* which is also situated in the MHC region. Some alleles of complement factor 4 contain HERV-K(C4,HML-10). Absence of HERV-K(C4,HML-10) is related to systemic lupus erythematosus (SLE), a serious autoimmune disease.[204] It is however not known if the HERV-K(C4, HML-10) gene is directly involved in the pathogenesis of this type of SLE. It could simply be a marker of other genetic events at the C4 locus.

Another aspect of SLE is DNA methylation. Certain drugs used in humans, i.e., prokainamide, are known to be inhibitors of DNA methylation. They can induce SLE.[205] An increased RNA expression of HERV-E(4-1) was found in SLE patients.[206] Treatment of PBMC with a demethylating agent gave a pronounced increase of HERV-E(4-1) expression in healthy controls, but no further increase in SLE patients.[207] How this relates to autoimmunity is not known, but it does emphasize a possible involvement of demethylation of normally methylated sequences (which often are retrotransposons) in SLE. Examples which advocate methylation as a central mechanism for control of retrotransposons are i. a dramatic increase in retrotransposon activity, probably due to undermethylation, which was described in an hybrid of two wallaby species,[208] and ii. undermethylation of retrotransposons in some human tumours.[209]

A third aspect is immune dysregulation. SLE patients have increased amounts of IgG antibodies reactive with certain HERV peptides.[210,211] However, a similar increase was also found in HIV infected patients.[196] In contrast, IgM antibodies reactive with the same peptides occurred in most healthy controls. IgG reactive with the same peptides did not occur in them. It was therefore suggested that HERVs may induce broadly reactive "natural" IgM against retroviruses. However, HIV infected persons frequently had IgG directed against the same set of peptides, suggesting an Ig chain switching event due to antigenic stimulation. Natural immunity in the form of IgM is an important protective factor against several microbes and it is conceivable that it can influence the course of retrovirus-induced disease. We currently believe that the increased levels of certain anti-HERV IgGs are symptoms of immune dysregulation rather than causes of SLE. The connection between HERVs and SLE has been reviewed,[212] as well as between HERVs and autoimmune disease.[195]

## Other Autoimmune Diseases

HERV RNA (ERV-9, HERV-K, HERV-L) expression in synovial fluid cells has been reported.[213] The connection to rheumatoid arthritis (RA) seems tentative. The HLA-DQB1 locus has alleles associated with RA. They contain an HERV-K(HML-6) LTR.[214]

## *Placentation Defects*

The evidence for a retroviral contribution to human physiology is relatively strong in some cases, more tentative in others. Theoretically, any of these functions can have its malfunction, i.e., disease. The case for a functional contribution is rather strong for placenta. Pleiotrophin,[143] endothelin B receptor[22] and syncytin[121] are three important proteins which are highly expressed in placenta under the control of retroviral promoters. Moreover, many HERVs are highly expressed in placenta both as RNA[144,215] and protein.[216,217] Loss of retroviral mechanisms may therefore be serious for placental function. Indeed, several recent reports indicate that abnormal placentation and function (i.e., preeclampsia) are associated with abnormal expression of the HERV-W envelope protein, Syncytin.[218]

These data indicate a role for ERVs in placenta function. However, many questions remain. The placenta organ evolved gradually during the last 130 million years.[219,220] The primate variant (a placenta with three tissue layers between fetal and maternal circulation, a hemochorial placenta) formed 40-50 million years ago. This structure is different from that of rodents who have a hemendothelial (1 layer-) placenta. Dogs and cats have an endotheliochorial (4 layer-) placenta.

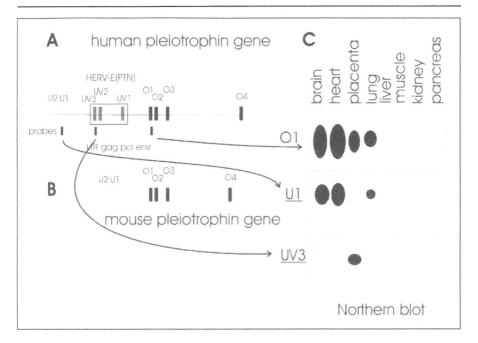

Figure 14. The HERV-E.PTN LTR confers strict placenta specificity of pleiotrophin (PTN) expression. HERV-E.PTN integrated into the human, but not the mouse, PTN gene. Homologous exons corresponding to the ORF (O1 to O4) or 5-UTR (U2, U1) of the human (A) and the murine (B) gene, as well as the HERV-derived exons (UV3, UV2, and UV1) and the position of exon-specific probes, are shown. C) Northern blot analysis of mRNA from various human tissues using the exon-specific probes. Modified from Schulte et al.[143]

Ruminants have a syndesmochorial (5 layer-) placenta. Pig, horse and donkey have an epitheliochorial (6 layer-) placenta structure.[221] All known HERVs implicated in the growth and differentiation of the human placenta (HERV-E.PTN, HERV-E (Endothelin B receptor) and HERV-W) integrated after the formation of the present type of placenta in human ancestors. HERV-E.PTN integrated 25 Mya i.e., long after this, and even after the split of Old and New World monkeys. The evolutionary and functional importance of HERV-E and HERV-W integrations should be studied by a comparison of placentas of Old and New World monkeys, because the latter ones do not have HERV-E and HERV-W. Thus it is likely that these relatively recent events only modulated preexisting mechanisms. Despite this, HERV-E.PTN LTR driven transcription is necessary for growth of trophoblast and choriocarcinoma cells in cell culture[143] (Fig. 9). The HERV-E.PTN LTR also confers strict placenta specificity of expression (Fig. 14).

## Genetic Disease (Gene Loss) Due to Illegitimate Recombination between HERV Sequences

Illegitimate recombination occurs between adjacent and highly similar repetitive elements. Several recent reports indicate that HERV participation in this elimination process is not uncommon.[222]

The Sertoli-only syndrome with azoospermia due to loss of 792 kb which include the Azoospermia factor a (AZFa, also named DYS11) is caused by illegitimate recombination between HERV15yq1 and HERV15yq2, both situated at Yq11.[223,224] This is a common cause of male infertility (Fig. 15).

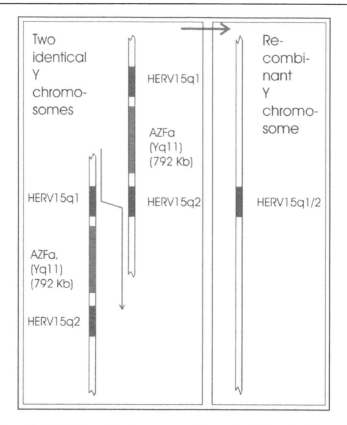

Figure 15. Loss of the DYS11 (containing the azoospermia factor, the AZFa gene) region on chromosome Yq11 through illegitimate intrachromosomal recombination between two HERV-I, also known as HERV15, genes. This leads to total loss of sperm production (azoospermia).

## Other Diseases

It was recently reported that HERV-E env was transcribed in alveolar macrophages obtained by bronchoalveolar lavage from 3 cases of idiopathic pulmonary fibrosis and 3 cases of sarcoidosis, but not in PBL. Antibodies to HERV-E was also detected in the same fluid.[225]

## What Can Be Done to Demonstrate a HERV-Disease Connection?

Most HERVs obviously are harmless because all human cells and all individuals have essentially the same set of HERVs. Because of their ubiquity it is especially difficult to establish a relationship between HERVs and disease. The correlations may be both positive and negative. Both heredity and environment can be influential. The example with the neurological disease in the mice at Lake Casitas shows how an ERV can be tied to a disease by detection of its genotypic and phenotypic properties. Potential genotypic markers of a HERV are new integrations and point mutations. Only a few of the former have been found.[59] Integrational polymorphism of ERVs is common in mice[226] and fowl.[227] Point mutations (SNPs) in HERVs seem to be relatively common.[70,71,228,229] An average HERV sequence gets 0.2% point mutations per million years. [189] HERVs are often 10-40 million years of age. Thus, they often contain several per cent mutations of which some differ between individuals. On top of this, there may be a somatic HERV variation (see above). Phenotypic changes can be detected at the

RNA or protein levels. There are several reports on a correlation between HERV expression and disease[84-87,97,99,115,118,122,152,162,230] (see also in other portions of this article).

Proof of a HERV etiological involvement beyond statistical correlations may be hard to achieve. The Koch postulates or similar rules adapted to present-day techniques mainly pertain to exogenous agents.[231] HERV gene knockouts other than naturally occurring polymorphisms are impossible in humans, but may be done in cell culture. The strategy therefore often must be one of guilt by association. The tools will be sequencing of genomic DNA and cDNA, RNA quantification, promoter/enhancer studies and protein detection. There is precedence in animals for disease caused by XRV infection just after birth, or early activation of ERVs, followed by a chronic viremia which later develops into disease. Such temporally separated events should therefore be looked for.

## Conclusions

Because endogenous retroviruses are a type of transposon, "jumping gene", one expects them to integrate into new loci in the human genome. Such "jumps" have hitherto not been observed in vivo. Individual-specific integrations observable at the germ line level indicate that more frequent transpositions also occur at the somatic level. Most somatic integrations will not be seen in the germ line. The selection pressure at the somatic level is more varied. The likelihood that transposition at this level will be compatible with cellular survival therefore is greater. The finding of Jack Lenz'group[59] has increased the likelihood that humans have intact and active HERVs which theoretically are infectious and therefore could have a similar pathogenic status as MMTV in mice.

Humans, like all other eukaryotes, have a number of more or less functional endogenous retroviruses or retrovirus-like transposons. The plenitude of HERVs (Fig. 2) shows that there have been many more exogenous retroviruses in the forefathers of humans than the four human exogenous retroviruses known today (HTLV-1 and –2, HIV-1 and –2). Many of them may have died out as exogenous retroviruses, but some may survive as exogenous, infectious, viruses in an animal more or less related to humans, or even in humans themselves.

The new information leaves many questions: Are there really infectious HERV particles? Which are the physiological roles played by HERVs? In this review we have analysed the pathogenic potential of HERVs from an evolutionary perspective. Although the great majority of HERVs probably will prove not be pathogenic,[140] there is now enough information to justify a search for correlation to human disease. There are leads for further research on seminoma, MS, schizophrenia, placental defects, gene loss and SLE.

### *Acknowledgements*

We thank Tove Airola, Dixie Mager and Robert Yolken for critical reading of the manuscript and valuable suggestions.

### References

1. Tristem M, Herniou E, Summers K et al. Three retroviral sequences in amphibians are distinct from those in mammals and birds. J Virol 1996; 70:4864-4870.
2. Herniou E, Martin J, Miller K et al. Retroviral diversity and distribution in vertebrates. J Virol 1998; 72:5955-5966.
3. Smit AF. Interspersed repeats and other mementos of transposable elements in mammalian genomes. Curr Opin Genet Dev 1999; 9:657-663.
4. Whitelaw E and Martin DI. Retrotransposons as epigenetic mediators of phenotypic variation in mammals. Nature Genet 2001; 27:361-365.
5. Kazazian HH Jr, Moran JV. The impact of L1 retrotransposons on the human genome. Nature Genet 1998; 19:19-24.

6. Boeke J, Stoye JP. Retrotransposons, endogenous retroviruses, and the evolution of retroelements, In: Coffin JM, Hughes SH, Varmus HE, eds. Retroviruses. New York: Cold Spring Harbor Laboratory Press, 1997: 345-435

7. Lander ES, Linton LM, Birren B et al. Initial sequencing and analysis of the human genome. Nature 2001; 409:860-921.

8. Yoder JA, Walsh CP, Bestor TH. Cytosine methylation and the ecology of intragenomic parasites. Trends Genet 1997; 13:335-340.

9. Jensen S, Gassama MP, Heidmann T. Taming of transposable elements by homology-dependent gene silencing. Nature Genet 1999; 21:209-212.

10. Best S, Le Tissier PR, Stoye JP. Endogenous retroviruses and the evolution of resistance to retroviral infection. Trends Microbiol 1997; 5:313-318.

11. Andersson ML, Lindeskog M, Medstrand P et al. Diversity of human endogenous retrovirus class II-like sequences. J Gen Virol 1999; 80(Pt 1):255-260.

12. Cohen M, Larsson E. Human endogenous retroviruses. Bioessays 1988; 9:191-196.

13. Medstrand P, Blomberg J. Characterization of novel reverse transcriptase encoding human endogenous retroviral sequences similar to type A and type B retroviruses: differential transcription in normal human tissues. J Virol 1993; 67:6778-6787.

14. Bittner JJ. Some possible effects of nursing on the mammary gland tumour incidence in mice. Science 1936; 84:162.

15. Gardner MB, Kozak CA, O'Brien SJ. The Lake Casitas wild mouse: evolving genetic resistance to retroviral disease. Trends Genet 1991; 7:22-27.

16. Dawkins R. The Selfish Gene. Oxford: Oxford Univerity Press, 1976.

17. Abramczuk JW, Pezen DS, Leonard JM et al. Transgenic mice carrying intact HIV provirus: biological effects and organization of a transgene. J Acquir Immune Defic Syndr 1992; 5:196-203.

18. Bartholomew C, Ihle JN. Retroviral insertions 90 kilobases proximal to the Evi-1 myeloid transforming gene activate transcription from the normal promoter. Mol Cell Biol 1991; 11:1820-1828.

19. Towers G, Bock M, Martin S et al. A conserved mechanism of retrovirus restriction in mammals. Proc Natl Acad Sci USA 2000; 97:12295-12299.

20. Ting CN, Rosenberg MP, Snow CM et al. Endogenous retroviral sequences are required for tissue-specific expression of a human salivary amylase gene. Genes Dev 1992; 6:1457-1465.

21. Ling J, Pi W, Bollag R et al. The solitary long terminal repeats of ERV-9 endogenous retrovirus are conserved during primate evolution and possess enhancer activities in embryonic and hematopoietic cells. J Virol 2002; 76:2410-2423.

22. Medstrand P, Landry JR, Mager DL. Long terminal repeats are used as alternative promoters for the endothelin B receptor and apolipoprotein C-I genes in humans. J Biol Chem 2001; 276:1896-1903.

23. Kato N, Pfeifer-Ohlsson S, Kato M et al. Tissue-specific expression of human provirus ERV3 mRNA in human placenta: two of the three ERV3 mRNAs contain human cellular sequences. J Virol 1987; 61:2182-2191.

24. Feuchter-Murthy AE, Freeman JD, Mager DL. Splicing of a human endogenous retrovirus to a novel phospholipase A2 related gene. Nucleic Acids Res 1993; 21:135-143.

25. Juang SH, Huang J, Li Y et al. Molecular cloning and sequencing of a 58-kDa membrane- and microfilament-associated protein from ascites tumor cell microvilli with sequence similarities to retroviral Gag proteins. J Biol Chem 1994; 269:15067-15075.

26. Harris JM, McIntosh EM, Muscat GE. Structure/function analysis of a dUTPase: catalytic mechanism of a potential chemotherapeutic target. J Mol Biol 1999; 288:275-287.

27. Mueller-Lantzsch N, Sauter M, Weiskircher A et al. Human endogenous retroviral element K10 (HERV-K10) encodes a full-length gag homologous 73-kDa protein and a functional protease. AIDS Res Hum Retroviruses 1993; 9:343-350.

28. Towler EM, Gulnik SV, Bhat TN et al. Functional characterization of the protease of human endogenous retrovirus, K10: can it complement HIV-1 protease? Biochemistry 1998; 37:17137-17144.

29. Berkhout B, Jebbink M, Zsiros J. Identification of an active reverse transcriptase enzyme encoded by a human endogenous HERV-K retrovirus. J Virol 1999; 73:2365-2375.

30. Awad R. Retroviral and related nucleic acid processing enzymes expression and detection; Applications on clinical studies of HIV & HERVs. [dissertation]. Uppsala. 2001

31. Klenerman P, Hengartner H, Zinkernagel RM. A non-retroviral RNA virus persists in DNA form. Nature 1997; 390:298-301.

32. Kitamura Y, Ayukawa T, Ishikawa T et al. Human endogenous retrovirus K10 encodes a functional integrase. J Virol 1996; 70:3302-3306.

33. Mager DL. Polyadenylation function and sequence variability of the long terminal repeats of the human endogenous retrovirus-like family RTVL-H. Virology 1989; 173:591-599.

34. Mager DL, Hunter DG, Schertzer M et al. Endogenous retroviruses provide the primary polyadenylation signal for two new human genes (HHLA2 and HHLA3). Genomics 1999; 59:255-263.

35. Baust C, Seifarth W, Germaier H et al. HERV-K-T47D-Related long terminal repeats mediate polyadenylation of cellular transcripts. Genomics 2000; 66:98-103.

36. Leib-Mosch C, Haltmeier M, Werner T et al. Genomic distribution and transcription of solitary HERV-K LTRs. Genomics 1993; 18:261-269.

37. Kowalski PE, Freeman JD, Mager DL. Intergenic splicing between a HERV-H endogenous retrovirus and two adjacent human genes. Genomics 1999; 57:371-379.

38. Kowalski PE and Mager DL. A human endogenous retrovirus suppresses translation of an associated fusion transcript, PLA2L. J Virol 1998; 72:6164-6168.

39. Landry JR, Medstrand P, Mager DL. Repetitive elements in the 5' untranslated region of a human zinc-finger gene modulate transcription and translation efficiency. Genomics 2001; 76:110-116.

40. Liao D, Pavelitz T, Kidd JR et al. Concerted evolution of the tandemly repeated genes encoding human U2 snRNA (the RNU2 locus) involves rapid intrachromosomal homogenization and rare interchromosomal gene conversion. EMBO J 1997; 16:588-598.

41. Johnson WE, Coffin JM. Constructing primate phylogenies from ancient retrovirus sequences. Proc Natl Acad Sci USA 1999; 96:10254-10260.

42. Byers PH. Killing the messenger: new insights into nonsense-mediated mRNA decay. J Clin Invest 2002; 109:3-6.

43. Wilkinson MF, Shyu AB. RNA surveillance by nuclear scanning? Nature Cell Biol 2002; 4:E144-147.

44. Moreno-Pelayo MA, Fernandez-Soria VM, Paz-Artal E et al. Complete cDNA sequences of the DRB6 gene from humans and chimpanzees: a possible model of a stop codon readingthrough mechanism in primates. Immunogenetics 1999; 49:843-850.

45. An DS, Xie Y, Chen IS. Envelope gene of the human endogenous retrovirus HERV-W encodes a functional retrovirus envelope. J Virol 2001; 75:3488-3489.

46. Yang J, Bogerd HP, Peng S et al. An ancient family of human endogenous retroviruses encodes a functional homolog of the HIV-1 Rev protein. Proc Natl Acad Sci USA 1999; 96:13404-13408.

47. Pasquinelli AE, Ernst RK, Lund E et al. The constitutive transport element (CTE) of Mason-Pfizer monkey virus (MPMV) accesses a cellular mRNA export pathway. EMBO J 1997; 16:7500-7510.

48. Wodrich H, Bohne J, Gumz E et al. A new RNA element located in the coding region of a murine endogenous retrovirus can functionally replace the Rev/Rev-responsive element system in human immunodeficiency virus type 1 Gag expression. J Virol 2001; 75:10670-10682.

49. Pavlicek A, Paces J, Elleder D et al. Processed pseudogenes of human endogenous retroviruses generated by LINEs: their integration, stability, and distribution. Genome Res 2002; 12:391-399.

50. Banki K, Halladay D, Perl A. Cloning and expression of the human gene for transaldolase. A novel highly repetitive element constitutes an integral part of the coding sequence. J Biol Chem 1994; 269:2847-51.

51. Kapitonov VV, Jurka J. The long terminal repeat of an endogenous retrovirus induces alternative splicing and encodes an additional carboxy-terminal sequence in the human leptin receptor. J Mol Evol 1999; 48:248-251.

52. Bi S, Gavrilova O, Gong DW et al. Identification of a placental enhancer for the human leptin gene. J Biol Chem 1997; 272:30583-30588.

53. Chen HJ, Carr K, Jerome RE et al. A retroviral repetitive element confers tissue-specificity to the Human Alcohol Dehydrogenase 1C (ADH1C) Gene. DNA Cell Biol 2002; 21:793-801.

54. Plant KE, Routledge SJ, Proudfoot NJ. Intergenic transcription in the human beta-globin gene cluster. Mol Cell Biol 2001; 21:6507-6514.

55. Routledge SJ, Proudfoot NJ. Definition of transcriptional promoters in the human beta globin locus control region. J Mol Biol 2002; 323:601-611.

56. Medstrand P, Mager DL. Human-specific integrations of the HERV-K endogenous retrovirus family. J Virol 1998; 72:9782-9787.

57. Zhu ZB, Jian B, Volanakis JE. Ancestry of SINE-R.C2 a human-specific retroposon. Hum Genet 1994; 93:545-551.

58. Barbulescu M, Turner G, Seaman MI et al. Many human endogenous retrovirus K (HERV-K) proviruses are unique to humans. Curr Biol 1999; 9:861-868.

59. Turner G, Barbulescu M, Su M et al. Insertional polymorphisms of full-length endogenous retroviruses in humans. Curr Biol 2001; 11:1531-1535.

60. Lebedev YB, Belonovitch OS, Zybrova NV et al. Differences in HERV-K LTR insertions in orthologous loci of humans and great apes. Gene 2000; 247:265-277.

61. Liao D, Pavelitz T, Weiner AM. Characterization of a novel class of interspersed LTR elements in primate genomes: structure, genomic distribution, and evolution. J Mol Evol 1998; 46:649-660.

62. Fanning T, Alves G. A family of repetitive DNA sequences in Old World primates. Gene 1997; 199:279-282.

63. Andersson G, Svensson AC, Setterblad N et al. Retroelements in the human MHC class II region. Trends Genet 1998; 14:109-114.

64. Horton R, Niblett D, Milne S et al. Large-scale sequence comparisons reveal unusually high levels of variation in the HLA-DQB1 locus in the class II region of the human MHC. J Mol Biol 1998; 282:71-97.

65. Donner H, Tonjes RR, Bontrop RE et al. Intronic sequence motifs of HLA-DQB1 are shared between humans, apes and Old World monkeys, but a retroviral LTR element (DQLTR3) is human specific. Tissue Antigens 1999; 53:551-558.

66. Kulski JK, Gaudieri S, Inoko H et al. Comparison between two human endogenous retrovirus (HERV)-rich regions within the major histocompatibility complex. J Mol Evol 1999; 48:675-683.

67. Kupfermann H, Satta Y, Takahata N et al. Evolution of Mhc-DRB introns: implications for the origin of primates. J Mol Evol 1999; 48:663-674.

68 Gaudieri S, Kulski JK, Dawkins RL et al. Different evolutionary histories in two subgenomic regions of the major histocompatibility complex. Genome Res 1999; 9:541-549.

69. Goodchild NL, Freeman JD, Mager DL. Spliced HERV-H endogenous retroviral sequences in human genomic DNA: evidence for amplification via retrotransposition. Virology 1995; 206:164-173.

70. Deb P, Klempan TA, O'Reilly RL et al. A single-primer PCR-based retroviral-related DNA polymorphism shared by two distinct human populations. Genome 1998; 41:662-668.

71. Jern P, Lindeskog M, Karlsson D et al. Full-Length HERV-H Elements with env SU Open Reading Frames in the Human Genome. AIDS Res Hum Retroviruses 2002; 18:671-676.

72. Futuyma DJ, ed(s). Evolutionary biology. 3rd ed. Sunderland, Massachussetts: Sinauer Associates Inc., 1998: 325

73. Schidlovsky G, Ahmed M. C-type virus particles in placentas and fetal tissues of Rhesus monkeys. J Natl Cancer Inst 1973; 51:225-233.

74. Vernon ML, McMahon JM, Hackett JJ. Additional evidence of type-C particles in human placentas. J Natl Cancer Inst 1974; 52:987-989.

75. de Harven E. Remarks on the ultrastructure of type A, B, and C virus particles. Adv Virus Res 1974; 19:221-264.

76. Kalter SS, Heberling RL, Helmke RJ et al. A comparative study on the presence of C-type viral particles in placentas from primates and other animals. Bibl Haematol 1975; 391-401.

77. Kalter SS, Heberling RL, Smith GC et al. C-type viruses in chimpanzee (Pan sp.) placentas. J Natl Cancer Inst 1975; 55:735-736.

78. Gross L, Schidlovsky G, Feldman D et al. C-type virus particles in placenta of normal healthy Sprague-Dawley rats. Proc Natl Acad Sci USA 1975; 72:3240-3244.

79. Imamura M, Phillips PE, Mellors RC. The occurrence and frequency of type C virus-like particles in placentas from patients with systemic lupus erythematosus and from normal subjects. Am J Pathol 1976; 83:383-394.

80. Simpson GR, Patience C, Lower R et al. Endogenous D-type (HERV-K) related sequences are packaged into retroviral particles in the placenta and possess open reading frames for reverse transcriptase. Virology 1996; 222:451-456.

81. Lower R, Boller K, Hasenmaier B et al. Identification of human endogenous retroviruses with complex mRNA expression and particle formation. Proc Natl Acad Sci USA 1993; 90:4480-4484.

82. Keydar I, Ohno T, Nayak R et al. Properties of retrovirus-like particles produced by a human breast carcinoma cell line: immunological relationship with mouse mammary tumor virus proteins. Proc Natl Acad Sci USA 1984; 81:4188-4192.

83. Seifarth W, Skladny H, Krieg-Schneider F et al. Retrovirus-like particles released from the human breast cancer cell line T47-D display type B- and C-related endogenous retroviral sequences. J Virol 1995; 69:6408-6416.

84. Perron H, Garson JA, Bedin F et al. Molecular identification of a novel retrovirus repeatedly isolated from patients with multiple sclerosis. The Collaborative Research Group on Multiple Sclerosis. Proc Natl Acad Sci USA 1997; 94:7583-7588.

85. Komurian-Pradel F, Paranhos-Baccala G, Bedin F et al. Molecular cloning and characterization of MSRV-related sequences associated with retrovirus-like particles. Virology 1999; 260:1-9.

86. Christensen T, Dissing Sorensen P, Riemann H et al. Expression of sequence variants of endogenous retrovirus RGH in particle form in multiple sclerosis. Lancet 1998; 352:1033.

87. Christensen T, Tonjes RR, zur Megede J et al. Reverse transcriptase activity and particle production in B lymphoblastoid cell lines established from lymphocytes of patients with multiple sclerosis. AIDS Res Hum Retroviruses 1999; 15:285-291.

88. Karlsson H, Bachmann S, Schroder J et al. Retroviral RNA identified in the cerebrospinal fluids and brains of individuals with schizophrenia. Proc Natl Acad Sci USA 2001; 98:4634-4639.

89. Tuke PW, Perron H, Bedin F et al. Development of a pan-retrovirus detection system for multiple sclerosis studies. Acta Neurol Scand Suppl 1997; 169:16-21.

90. Seifarth W, Krause U, Hohenadl C et al. Rapid identification of all known retroviral reverse transcriptase sequences with a novel versatile detection assay. AIDS Res Hum Retroviruses 2000; 16:721-729.

91. Griffiths DJ, Venables PJ, Weiss RA et al. A novel exogenous retrovirus sequence identified in humans. J Virol 1997; 71:2866-2872.

92. Wilkinson DA, Mager DL, Leong JC. Endogenous human retroviruses, In: Levy J, ed. The Retroviridae. New York: Plenum Press, 1994: 465-535

93. Lindeskog M, Mager DL, Blomberg J. Isolation of a human endogenous retroviral HERV-H element with an open env reading frame. Virology 1999; 258:441-450.

94. de Parseval N, Casella J, Gressin L et al. Characterization of the three HERV-H proviruses with an open envelope reading frame encompassing the immunosuppressive domain and evolutionary history in primates. Virology 2001; 279:558-569.

95. Mayer J, Meese E, Mueller-Lantzsch N. Chromosomal assignment of human endogenous retrovirus K (HERV-K) env open reading frames. Cytogenet Cell Genet 1997; 79:157-161.

96. Mayer J, Sauter M, Racz A et al. An almost-intact human endogenous retrovirus K on human chromosome 7. Nature Genet 1999; 21:257-258.

97. Perron H, Firouzi R, Tuke P et al. Cell cultures and associated retroviruses in multiple sclerosis. Collaborative Research Group on MS. Acta Neurol Scand Suppl 1997; 169:22-31.

98. Christensen T, Pedersen L, Sorensen PD et al. A transmissible human endogenous retrovirus. AIDS Res Hum Retroviruses 2002; 18:861-866.

99. Yolken RH, Karlsson H, Yee F et al. Endogenous retroviruses and schizophrenia. Brain Res Brain Res Rev 2000; 31:193-199.

100. Ono T, Shinya K, Uehara Y et al. Endogenous virus genomes become hypomethylated tissue—specifically during aging process of C57BL mice. Mech Ageing Dev 1989; 50:27-36.

101. Barbot W, Dupressoir A, Lazar V et al. Epigenetic regulation of an IAP retrotransposon in the aging mouse: progressive demethylation and de-silencing of the element by its repetitive induction. Nucleic Acids Res 2002; 30:2365-2373.

102. Huebner RJ. Identification of leukemogenic viruses: specifications for vertically transmitted, mostly 'switched off' RNA tumor viruses as determinants of the generality of cancer. Bibl Haematol 1970; 22-44.

103. Gardner MB, Henderson BE, Rongey RW et al. Spontaneous tumors of aging wild house mice. Incidence, pathology, and C-type virus expression. J Natl Cancer Inst 1973; 50:719-734.

104. Barker AD, Dennis AJ, Moore VS et al. Effects of thymosin on cytotoxicity and virus production in AKR mice. Ann N Y Acad Sci 1979; 332:70-80.

105. Wada Y, Tsukada M, Kamiyama S et al. Retroviral gene expression as a possible biomarker of aging. Int Arch Occup Environ Health 1993; 65:S235-239.

106. Bartman T, Murasko DM, Sieck TG et al. A murine leukemia virus expressed in aged DBA/2 mice is derived by recombination of the Emv-3 locus and another endogenous gag sequence. Virology 1994; 203:1-7.

107. Mikkelsen JG, Pedersen FS. Genetic reassortment and patch repair by recombination in retroviruses. J Biomed Sci 2000; 7:77-99.

108. Patience C, Takeuchi Y, Cosset FL et al. Packaging of endogenous retroviral sequences in retroviral vectors produced by murine and human packaging cells. J Virol 1998; 72:2671-2676.

109. Chong H, Starkey W, Vile RG. A replication-competent retrovirus arising from a split-function packaging cell line was generated by recombination events between the vector, one of the packaging constructs, and endogenous retroviral sequences. J Virol 1998; 72:2663-2670.

110. Crow TJ. Left brain, retrotransposons, and schizophrenia. Br Med J (Clin Res Ed) 1986; 293:3-4.

111. Crow TJ. Schizophrenia as the price that homo sapiens pays for language: a resolution of the central paradox in the origin of the species. Brain Res Rev 2000; 31:118-129.

112. Mozer BA, Benzer S. Ingrowth by photoreceptor axons induces transcription of a retrotransposon in the developing Drosophila brain. Development 1994; 120:1049-1058.

113. Crow TJ. A re-evaluation of the viral hypothesis: is psychosis the result of retroviral integration at a site close to the cerebral dominance gene? Br J Psychiatry 1984; 145:243-253.

114. Gagneux P, Varki A. Genetic differences between humans and great apes. Mol Phylogenet Evol 2001; 18:2-13.

115. Haahr S, Sommerlund M, Christensen T et al. A putative new retrovirus associated with multiple sclerosis and the possible involvement of Epstein-Barr virus in this disease. Ann N Y Acad Sci 1994; 724:148-156.

116. Christensen T, Dissing Sorensen P, Riemann H et al. Molecular characterization of HERV-H variants associated with multiple sclerosis. Acta Neurol Scand 2000; 101:229-238.

117. La Mantia G, Maglione D, Pengue G et al. Identification and characterization of novel human endogenous retroviral sequences prefentially expressed in undifferentiated embryonal carcinoma cells. Nucleic Acids Res 1991; 19:1513-1520.

118. Olsson P, Ryberg B, Awad R et al. Retroviral RNA related to ERV9/MSRV in a human serum: a new sequence variant. AIDS Res Hum Retroviruses 1999; 15:591-593.

119. Munch M, Moller-Larsen A, Christensen T et al. B-lymphoblastoid cell lines from multiple sclerosis patients and a healthy control producing a putative new human retrovirus and Epstein-Barr virus. Mult Scler 1995; 1:78-81.

120. Moller-Larsen A, Christensen T. Isolation of a retrovirus from multiple sclerosis patients in self-generated Iodixanol gradients. J Virol Methods 1998; 73:151-161.

121. Mi S, Lee X, Li X et al. Syncytin is a captive retroviral envelope protein involved in human placental morphogenesis. Nature 2000; 403:785-789.

122. Deb-Rinker P, Klempan TA, O'Reilly RL et al. Molecular characterization of a MSRV-like sequence identified by RDA from monozygotic twin pairs discordant for schizophrenia. Genomics 1999; 61:133-144.

123. Lavillette D, Marin M, Ruggieri A et al. The envelope glycoprotein of human endogenous retrovirus type W uses a divergent family of amino acid transporters/cell surface receptors. J Virol 2002; 76:6442-6452.

124. Goff DC, Coyle JT. The emerging role of glutamate in the pathophysiology and treatment of schizophrenia. Am J Psychiatry 2001; 158:1367-1377.

125. Hahn AF. Guillain-Barre syndrome. Lancet 1998; 352:635-641.

126. Wucherpfennig KW, Strominger JL. Molecular mimicry in T cell-mediated autoimmunity: viral peptides activate human T cell clones specific for myelin basic protein. Cell 1995; 80:695-705.

127. Levin MC, Lee SM, Kalume F et al. Autoimmunity due to molecular mimicry as a cause of neurological disease. Nature Med 2002; 8:509-513.

128. Benoist C, Mathis D. Autoimmunity provoked by infection: how good is the case for T cell epitope mimicry? Nature Immunol 2001; 2:797-801.

129. Rasheed S. Retroviruses and oncogenes in rats, In: Gallo RC, Varnier OE, Stehelin D, eds. Retroviruses and Human Pathology. Clifton, New Jersey: Humana Press, 1986: 153-175

130. Kamme F. Changes in gene expression during dalayed neuronal death after cerebral ischemia in the rat. [dissertation]. Lund 1998

131. Johnston JB, Silva C, Holden J et al. Monocyte activation and differentiation augment human endogenous retrovirus expression: implications for inflammatory brain diseases. Ann Neurol 2001; 50:434-442.

132. Rieger F, Pierig R, Cifuentes-Diaz C et al. New perspectives in multiple sclerosis: retroviral involvement and glial cell death. Pathol Biol (Paris) 2000; 48:15-24.

133. Crow TJ, Done DJ. Age of onset of schizophrenia in siblings: a test of the contagion hypothesis. Psychiatry Res 1986; 18:107-117.

134. Shastry BS. Schizophrenia: a genetic perspective (review). Int J Mol Med 2002; 9:207-212.

135. Hunter SF, Hafler DA. Ubiquitous pathogens: links between infection and autoimmunity in MS? Neurology 2000; 55:164-165.

136. Steiner I, Nisipianu P, Wirguin I. Infection and the etiology and pathogenesis of multiple sclerosis. Curr Neurol Neurosci Rep 2001; 1:271-276.

137. Yolken RH, Torrey EF. Viruses, schizophrenia, and bipolar disorder. Clin Microbiol Rev 1995; 8:131-145.

138. Munch M, Hvas J, Christensen T et al. A single subtype of Epstein-Barr virus in members of multiple sclerosis clusters. Acta Neurol Scand 1998; 98:395-399.

139. Hoover EA, Mullins JI. Feline leukemia virus infection and diseases. J Am Vet Med Assoc 1991; 199:1287-1297.

140. Stoye JP. The pathogenic potential of endogenous retroviruses: a sceptical view. Trends Microbiol 1999; 7:430; discussion 431-432.

141. Mager DL. Human endogenous retroviruses and pathogenicity: genomic considerations. Trends Microbiol 1999; 7:431; discussion 431-432.

142. Lower R. The pathogenic potential of endogenous retroviruses: facts and fantasies. Trends Microbiol 1999; 7:350-356.

143. Schulte AM, Lai S, Kurtz A et al. Human trophoblast and choriocarcinoma expression of the growth factor pleiotrophin attributable to germ-line insertion of an endogenous retrovirus. Proc Natl Acad Sci USA 1996; 93:14759-14764.

144. Yin H, Medstrand P, Andersson ML et al. Transcription of human endogenous retroviral sequences related to mouse mammary tumor virus in human breast and placenta: similar pattern in most malignant and nonmalignant breast tissues. AIDS Res Hum Retroviruses 1997; 13:507-516.

145. Yin H. Human mouse mammary tumour virus like elements and their relation to breast cancer. [dissertation]. Uppsala 1999

146. Brodsky I, Foley B, Gillespie D. Expression of human endogenous retrovirus (HERV-K) in chronic myeloid leukemia. Leuk Lymphoma 1993; 11(Suppl 1):119-123.

147. Brodsky I, Foley B, Haines D et al. Expression of HERV-K proviruses in human leukocytes. Blood 1993; 81:2369-2374.

148. Depil S, Roche C, Dussart P et al. Expression of a human endogenous retrovirus, HERV-K, in the blood cells of leukemia patients. Leukemia 2002; 16:254-259.

149. Sauter M, Roemer K, Best B et al. Specificity of antibodies directed against Env protein of human endogenous retroviruses in patients with germ cell tumors. Cancer Res 1996; 56:4362-4365.

150. Goedert JJ, Sauter ME, Jacobson LP et al. High prevalence of antibodies against HERV-K10 in patients with testicular cancer but not with AIDS. Cancer Epidemiol Biomarkers Prev 1999; 8:293-296.

151. Boller K, Janssen O, Schuldes H et al. Characterization of the antibody response specific for the human endogenous retrovirus HTDV/HERV-K. J Virol 1997; 71:4581-4588.

152. Sauter M, Schommer S, Kremmer E et al. Human endogenous retrovirus K10: expression of Gag protein and detection of antibodies in patients with seminomas. J Virol 1995; 69:414-421.

153. Boller K, Konig H, Sauter M et al. Evidence that HERV-K is the endogenous retrovirus sequence that codes for the human teratocarcinoma-derived retrovirus HTDV. Virology 1993; 196:349-353.

154. Herbst H, Sauter M, Kuhler-Obbarius C et al. Human endogenous retrovirus (HERV)-K transcripts in germ cell and trophoblastic tumours. Apmis 1998; 106:216-220.

155. Herbst H, Sauter M, Mueller-Lantzsch N. Expression of human endogenous retrovirus K elements in germ cell and trophoblastic tumors. Am J Pathol 1996; 149:1727-1735.

156. Gotzinger N, Sauter M, Roemer K et al. Regulation of human endogenous retrovirus-K Gag expression in teratocarcinoma cell lines and human tumours. J Gen Virol 1996; 77(Pt 12):2983-2990.

157. Lower R, Tonjes RR, Korbmacher C et al. Identification of a Rev-related protein by analysis of spliced transcripts of the human endogenous retroviruses HTDV/HERV-K. J Virol 1995; 69:141-149.

158. Magin C, Lower R, Lower J. cORF and RcRE, the Rev/Rex and RRE/RxRE homologues of the human endogenous retrovirus family HTDV/HERV-K. J Virol 1999; 73:9496-9507.

159. Armbruester V, Sauter M, Krautkraemer E et al. A novel gene from the human endogenous retrovirus k expressed in transformed cells. Clin Cancer Res 2002; 8:1800-1807.

160. Boese A, Sauter M, Galli U et al. Human endogenous retrovirus protein cORF supports cell transformation and associates with the promyelocytic leukemia zinc finger protein. Oncogene 2000; 19:4328-4336.

161. Coffin JM. The viruses and their replication, In: Fields BN, Knipe DM, Howley PM, eds. Fields Virology. 3rd ed. Philadelphia: Lippincott-Raven, 1996: 1767-1847

162. Conrad B, Weissmahr RN, Boni J et al. A human endogenous retroviral superantigen as candidate autoimmune gene in type I diabetes. Cell 1997; 90:303-313.

163. Palmarini M, Sharp JM, de las Heras M et al. Jaagsiekte sheep retrovirus is necessary and sufficient to induce a contagious lung cancer in sheep. J Virol 1999; 73:6964-6972.

164. Maeda N, Palmarini M, Murgia C et al. Direct transformation of rodent fibroblasts by jaagsiekte sheep retrovirus DNA. Proc Natl Acad Sci USA 2001; 98:4449-4454.

165. Alberti A, Murgia C, Liu SL et al. Envelope-induced cell transformation by ovine betaretroviruses. J Virol 2002; 76:5387-5394.

166. Rai SK, Duh FM, Vigdorovich V et al. Candidate tumor suppressor HYAL2 is a glycosylphosphatidylinositol (GPI)-anchored cell-surface receptor for jaagsiekte sheep retrovirus, the envelope protein of which mediates oncogenic transformation. Proc Natl Acad Sci USA 2001; 98:4443-4448.

167. Perryman LE, Hoover EA, Yohn DS. Immunologic reactivity of the cat: immunosuppression in experimental feline leukemia. J Natl Cancer Inst 1972; 49:1357-1365.

168. Mathes LE, Olsen RG, Hebebrand LC et al. Abrogation of lymphocyte blastogenesis by a feline leukaemia virus protein. Nature 1978; 274:687-689.

169. Oroszlan S, Nowinski RC. Lysis of retroviruses with monoclonal antibodies against viral envelope proteins. Virology 1980; 101:296-299.

170. Cianciolo GJ, Copeland TD, Oroszlan S et al. Inhibition of lymphocyte proliferation by a synthetic peptide homologous to retroviral envelope proteins. Science 1985; 230:453-455.

171. Tas MP, Laarman D, Haan-Meulman MD et al. Retroviral p15E-related serum factors and recurrence of head and neck cancer. Clin Otolaryngol 1993; 18:324-328.

172. Simons PJ, Oostendorp RA, Tas MP et al. Comparison of retroviral p15E-related factors and interferon alpha in head and neck cancer. Cancer Immunol Immunother 1994; 38:178-184.

173. Lindvall M, Sjogren HO. Inhibition of rat yolk sac tumour growth in vivo by a monoclonal antibody to the retroviral molecule P15E. Cancer Immunol Immunother 1991; 33:21-27.

174. Mangeney M, Heidmann T. Tumor cells expressing a retroviral envelope escape immune rejection in vivo. Proc Natl Acad Sci USA 1998; 95:14920-14925.

175. Mangeney M, de Parseval N, Thomas G et al. The full-length envelope of an HERV-H human endogenous retrovirus has immunosuppressive properties. J Gen Virol 2001; 82:2515-2518.

176. Huang AY, Gulden PH, Woods AS et al. The immunodominant major histocompatibility complex class I-restricted antigen of a murine colon tumor derives from an endogenous retroviral gene product. Proc Natl Acad Sci USA 1996; 93:9730-9735.

177. Oostendorp RA, Meijer CJ, Scheper RJ. Immunosuppression by retroviral-envelope-related proteins, and their role in non-retroviral human disease. Crit Rev Oncol Hematol 1993; 14:189-206.

178. Haraguchi S, Good RA, Day NK. Immunosuppressive retroviral peptides: cAMP and cytokine patterns. Immunol Today 1995; 16:595-603.

179. Ruegg CL, Clements JE, Strand M. Inhibition of lymphoproliferation and protein kinase C by synthetic peptides with sequence identity to the transmembrane and Q proteins of visna virus. J Virol 1990; 64:2175-2180.

180. Takahashi A, Day NK, Luangwedchakarn V et al. A retroviral-derived immunosuppressive peptide activates mitogen-activated protein kinases. J Immunol 2001; 166:6771-6775.

181. Huang SS, Huang JS. A pentacosapeptide (CKS-25) homologous to retroviral envelope proteins possesses a transforming growth factor-beta activity. J Biol Chem 1998; 273:4815-4818.

182. Lindeskog M, Medstrand P, Blomberg J. Sequence variation of human endogenous retrovirus ERV9-related elements in an env region corresponding to an immunosuppressive peptide: transcription in normal and neoplastic cells. J Virol 1993; 67:1122-1126.

183. Benit L, Dessen P, Heidmann T. Identification, phylogeny, and evolution of retroviral elements based on their envelope genes. J Virol 2001; 75:11709-11719.

184. Foulds S, Wakefield CH, Giles M et al. Expression of a suppressive p15E-related epitope in colorectal and gastric cancer. Br J Cancer 1993; 68:610-616.

185. Langer JA, Ortaldo JR, Pestka S. Binding of human alpha-interferons to natural killer cells. J Interferon Res 1986; 6:97-105.

186. Hayashi H, Matsubara H, Yokota T et al. Molecular cloning and characterization of the gene encoding mouse melanoma antigen by cDNA library transfection. J Immunol 1992; 149:1223-1229.

187. de Bergeyck V, De Plaen E, Chomez P et al. An intracisternal A-particle sequence codes for an antigen recognized by syngeneic cytolytic T lymphocytes on a mouse spontaneous leukemia. Eur J Immunol 1994; 24:2203-2212.

188. Schiavetti F, Thonnard J, Colau D et al. A human endogenous retroviral sequence encoding an antigen recognized on melanoma by cytolytic T lymphocytes. Cancer Res 2002; 62:5510-5516.

189. Yin H, Medstrand P, Kristofferson A et al. Characterization of human MMTV-like (HML) elements similar to a sequence that was highly expressed in a human breast cancer: further definition of the HML-6 group. Virology 1999; 256:22-35.

190. Brunovskis P, Kung HJ. Retrotransposition and herpesvirus evolution. Virus Genes 1995; 11:259-270.

191. Richman DD, Whitley RJ, Hayden FG, eds. Clinical Virology. Churchill Livingstone Inc, 1997:375-524

192. Kwun HJ, Han HJ, Lee WJ et al. Transactivation of the human endogenous retrovirus K long terminal repeat by herpes simplex virus type 1 immediate early protein 0. Virus Res 2002; 86:93-100.

193. Perron H, Suh M, Lalande B et al. Herpes simplex virus ICP0 and ICP4 immediate early proteins strongly enhance expression of a retrovirus harboured by a leptomeningeal cell line from a patient with multiple sclerosis. J Gen Virol 1993; 74(Pt 1):65-72.

194. Sutkowski N, Conrad B, Thorley-Lawson DA et al. Epstein-Barr virus transactivates the human endogenous retrovirus HERV-K18 that encodes a superantigen. Immunity 2001; 15:579-589.

195. Sekigawa I, Ogasawara H, Kaneko H et al. Retroviruses and autoimmunity. Intern Med 2001; 40:80-86.

196. Lawoko A, Johansson B, Rabinayaran D et al. Increased immunoglobulin G, but not M, binding to endogenous retroviral antigens in HIV-1 infected persons. J Med Virol 2000; 62:435-444.

197. Hasuike S, Miura K, Miyoshi O et al. Isolation and localization of an IDDMK1,2-22-related human endogenous retroviral gene, and identification of a CA repeat marker at its locus. J Hum Genet 1999; 44:343-347.

198. Lower R, Tonjes RR, Boller K et al. Development of insulin-dependent diabetes mellitus does not depend on specific expression of the human endogenous retrovirus HERV-K. Cell 1998; 95:11-4; discussion 16.

199. Lapatschek M, Durr S, Lower R et al. Functional analysis of the env open reading frame in human endogenous retrovirus IDDMK(1,2)22 encoding superantigen activity. J Virol 2000; 74:6386-6393.

200. Badenhoop K, Donner H, Neumann J et al. IDDM patients neither show humoral reactivities against endogenous retroviral envelope protein nor do they differ in retroviral mRNA expression from healthy relatives or normal individuals. Diabetes 1999; 48:215-218.

201. Murphy VJ, Harrison LC, Rudert WA et al. Retroviral superantigens and type 1 diabetes mellitus. Cell 1998; 95:9-11; discussion 16.

202. Stauffer Y, Marguerat S, Meylan F et al. Interferon-alpha-induced endogenous superantigen. a model linking environment and autoimmunity. Immunity 2001; 15:591-601.

203. Donner H, Tonjes RR, Van der Auwera B et al. The presence or absence of a retroviral long terminal repeat influences the genetic risk for type 1 diabetes conferred by human leukocyte antigen DQ haplotypes. Belgian Diabetes Registry. J Clin Endocrinol Metab 1999; 84:1404-1408.

204. Blanchong CA, Zhou B, Rupert KL et al. Deficiencies of human complement component C4A and C4B and heterozygosity in length variants of RP-C4-CYP21-TNX (RCCX) modules in caucasians. The load of RCCX genetic diversity on major histocompatibility complex-associated disease. J Exp Med 2000; 191:2183-2196.

205. Le Goff P and Saraux A. Drug-induced lupus. Rev Rhum Engl Ed 1999; 66:40-45.

206. Ogasawara H, Okada M, Kaneko H et al. Quantitative comparison of human endogenous retrovirus mRNA between SLE and rheumatoid arthritis. Lupus 2001; 10:517-518.

207. Okada M, Ogasawara H, Kaneko H et al. Role of DNA methylation in transcription of human endogenous retrovirus in the pathogenesis of systemic lupus erythematosus. J Rheumatol 2002; 29:1678-1682.

208. O'Neill RJ, O'Neill MJ, Graves JA. Undermethylation associated with retroelement activation and chromosome remodelling in an interspecific mammalian hybrid. Nature 1998; 393:68-72.

209. Florl AR, Lower R, Schmitz-Drager BJ et al. DNA methylation and expression of LINE-1 and HERV-K provirus sequences in urothelial and renal cell carcinomas. Br J Cancer 1999; 80:1312-1321.

210. Blomberg J, Nived O, Pipkorn R et al. Increased antiretroviral antibody reactivity in sera from a defined population of patients with systemic lupus erythematosus. Correlation with autoantibodies and clinical manifestations. Arthritis Rheum 1994; 37:57-66.

211. Bengtsson A, Blomberg J, Nived O et al. Selective antibody reactivity with peptides from human endogenous retroviruses and nonviral poly(amino acids) in patients with systemic lupus erythematosus. Arthritis Rheum 1996; 39:1654-1663.

212. Adelman MK, Marchalonis JJ. Endogenous retroviruses in systemic lupus erythematosus: candidate lupus viruses. Clin Immunol 2002; 102:107-116.

213. Nakagawa K, Brusic V, McColl G et al. Direct evidence for the expression of multiple endogenous retroviruses in the synovial compartment in rheumatoid arthritis. Arthritis Rheum 1997; 40:627-638.

214. Seidl C, Donner H, Petershofen E et al. An endogenous retroviral long terminal repeat at the HLA-DQB1 gene locus confers susceptibility to rheumatoid arthritis. Hum Immunol 1999; 60:63-68.

215. Willer A, Saussele S, Gimbel W et al. Two groups of endogenous MMTV related retroviral env transcripts expressed in human tissues. Virus Genes 1997; 15:123-133.

216. Venables PJ, Brookes SM, Griffiths D et al. Abundance of an endogenous retroviral envelope protein in placental trophoblasts suggests a biological function. Virology 1995; 211:589-592.

217. Kitamura M, Maruyama N, Shirasawa T et al. Expression of an endogenous retroviral gene product in human placenta. Int J Cancer 1994; 58:836-840.

218. Lee X, Keith JC Jr, Stumm N et al. Downregulation of placental syncytin expression and abnormal protein localization in pre-eclampsia. Placenta 2001; 22:808-812.

219. Murphy WJ, Eizirik E, O'Brien SJ et al. Resolution of the early placental mammal radiation using Bayesian phylogenetics. Science 2001; 294:2348-2351.

220. Harris JR. The evolution of placental mammals. FEBS Lett 1991; 295:3-4.

221. Tizard IR, ed. Veterinary Immunology: An Introduction. Philadelphia, Pennsylvania: Saunders publisher, 2000: Chapter 19.

222. Hughes JF, Coffin JM. Evidence for genomic rearrangements mediated by human endogenous retroviruses during primate evolution. Nature Genet 2001; 29:487-489.

223. Kamp C, Hirschmann P, Voss H et al. Two long homologous retroviral sequence blocks in proximal Yq11 cause AZFa microdeletions as a result of intrachromosomal recombination events. Hum Mol Genet 2000; 9:2563-2572.

224. Kamp C, Huellen K, Fernandes S et al. High deletion frequency of the complete AZFa sequence in men with Sertoli-cell-only syndrome. Mol Hum Reprod 2001; 7:987-994.

225. Tamura N, Iwase A, Suzuki K et al. Alveolar macrophages produce the Env protein of a human endogenous retrovirus, HERV-E 4-1, in a subgroup of interstitial lung diseases. Am J Respir Cell Mol Biol 1997; 16:429-437.

226. Sawby R, Wichman HA. Analysis of orthologous retrovirus-like elements in the white-footed mouse, Peromyscus leucopus. J Mol Evol 1997; 44:74-80.

227. Dimcheff DE, Drovetski SV, Krishnan M et al. Cospeciation and horizontal transmission of avian sarcoma and leukosis virus gag genes in galliform birds. J Virol 2000; 74:3984-3995.

228. Rasmussen HB, Clausen J. Large number of polymorphic nucleotides and a termination codon in the env gene of the endogenous human retrovirus ERV3. Dis Markers 1998; 14:127-133.

229. de Parseval N, Heidmann T. Physiological knockout of the envelope gene of the single-copy ERV-3 human endogenous retrovirus in a fraction of the Caucasian population. J Virol 1998; 72:3442-3445.

230. Christensen T, Jensen AW, Munch M et al. Characterization of retroviruses from patients with multiple sclerosis. Acta Neurol Scand Suppl 1997; 169:49-58.

231. Fredericks DN, Relman DA. Sequence-based identification of microbial pathogens: a reconsideration of Koch's postulates. Clin Microbiol Rev 1996; 9:18-33.

# Index

# M

MAD family 174

MADS-box gene 179, 181

Major histocompatibility complex (MHC) 56, 110, 134, 209, 225, 226

Major homology region (MHR) 189, 194-196, 200

Mammaliam genome 12

Mammalian LTR retroelement (MaLR) 21, 22, 105, 106, 113, 116

Mammalian somatic cell genome 101

Mammalian type B, C, or D retroviruses 186

Marek disease (MDV) 224, 225

Mariner elements 108

Mason-pfizer monkey virus (MPMV) 98, 208, 210

Master gene 146, 150, 151, 153

Matrix protein (MA) 95, 96

Maximum parsimony 188-190, 193, 196-198

Maximum parsimony tree 188, 190, 196-198

Metazoa 16, 18

Methylation 6, 21, 22, 35, 36, 40, 45, 101, 106, 129, 133, 214, 215, 226

Methylation pattern 6

MHC-DRB 56

Microchip 17, 27, 83, 86, 164, 165, 167

Microchip hybridization 167

Microdissected tumor 164

Microsatellite 14, 36, 37, 73, 79

Microsatellite variation 73

Microtubule-associated protein 134

MIR 13, 22, 105, 145

Mitochondria 78, 79

Mitochondrial DNA (mtDNA) 40, 53-58, 62, 63, 65, 73, 77, 78

Mitochondrial genome 40, 59, 78, 79

Mitochondrial oxidative phosphorylation complex 137

Mitochondrial protein 137, 167

Mitochondrial sequence polymorphism 79

Mobile element 13, 93, 106, 107, 109, 110, 118, 145, 146

Modular cell biology 4

Modularity 15, 16, 18

Modulation of cellular gene expression 130

Module 1, 8-10, 15-18, 20, 22-25, 27, 148, 158, 206, see also Domain

Molecular clock 12, 74, 76, 77, 189

Monophyletic group 13, 21, 53, 54, 187, 197, 198

Monophyletic HERV families 187

Monozygotic twin (MZ) 36, 40, 216

Morphological changes 2, 148

Most recent common ancestor (MRCA) 54, 62, 82

Mouse coat color *agouti* gene 101

Mouse genome 9-13, 21, 22, 99

Mouse mammary tumour virus (MMTV) 98, 107, 123, 128, 189, 196, 197, 205, 210, 220, 225, 229

Mouse otoconin-90 (*OC90*) gene 135

mRNA 5, 23, 32, 71, 83, 94, 95, 101, 102, 123, 128, 134, 138, 151, 157, 165, 166, 169, 170, 173-175, 178, 179, 181, 208, 215, 219, 222, 227

mtDNA *see* Mitochondrial DNA

Multiple sclerosis (MS) 127, 130, 165, 204, 213-218, 229

Multiple sclerosis-associated retrovirus (MSRV) 127, 165, 215, 216

Multiregional evolution model 74

Murine leukemia virus (MLV) 98, 107, 123, 205, 210, 215, 219, 223

*Mus musculus* 5

Mutation 3, 7, 10-12, 14, 15, 18, 20, 22, 27, 32, 40, 41, 44, 59, 63, 70, 72, 73, 75-83, 98, 100, 104, 107, 118, 123, 127, 129, 132, 145, 153, 156, 189, 194, 195, 198, 199, 204, 207, 208, 212, 215,-217, 228

Mutation bias 15

Mutation rate 12, 72, 73, 7-77, 82, 189, 198